INTRODUCTION TO
STRUCTURAL MECHANICS

FOR BUILDING AND ARCHITECTURAL
STUDENTS

by the same authors

Structural Steelwork
for
Building and Architectural
Students

INTRODUCTION TO
STRUCTURAL MECHANICS
FOR BUILDING AND ARCHITECTURAL STUDENTS

TREFOR J. REYNOLDS
B.Sc., A.M.I.C.E., M.I.Struct.E., M.Sth.W.Inst.E.
Formerly Lecturer-in-Charge of Structural Engineering,
L.C.C. School of Building, Brixton

and

LEWIS E. KENT
O.B.E., B.Sc., F.I.C.E., F.I.Struct.E.
Consulting Structural Engineer,
Formerly Lecturer in Structural Engineering,
L.C.C. School of Building, Brixton

with

DAVID W. LAZENBY
D.I.C., M.I.C.E., M.I.Struct.E.
Consulting Engineer

THE ENGLISH UNIVERSITIES PRESS LTD

ISBN 0 340 11540 8 (Boards Edition)
ISBN 0 340 11541 6 (Paperback Edition)

First printed 1944, Reprinted 1945, 1946, 1947, 1949,
Second Edition 1951, Reprinted 1953, Third Edition 1955,
Reprinted 1956, 1958, Fourth Edition 1961, Reprinted 1963,
Fifth Edition 1965, Reprinted 1966, Sixth Edition 1973, Reprinted 1974

The English Universities Press Ltd
St Paul's House Warwick Lane
London EC4P 4AH

PRINTED AND BOUND IN ENGLAND
FOR THE ENGLISH UNIVERSITIES PRESS LTD
BY HAZELL WATSON AND VINEY LTD, AYLESBURY

PREFACE TO SIXTH EDITION

Lewis E. Kent, who alone had been responsible for the 4th and subsequent editions of this book, died in February 1972. By that time, the work of revision for this edition had almost been completed.

The change to a metric system brings with it the necessity for text books to be be brought into line with that system. 'Going metric' is not merely a case of simple conversion from imperial to metric measure, but involves the acceptance of some entirely different units for the evaluation of qualities and quantities. Moreover, simple conversion would lead to some arithmetic values which could be difficult to handle, and a certain amount of 'rounding off' is necessary.

The 'SI metric' system is being adopted in practice, and the intention of this edition is to make a direct conversion to SI metric units; the previous editions using imperial units are thereby superseded.

British Standard 449:1959 was revised in 1969 and 1970, to become a two-part document, BS 449:1970, Part 1 being the imperial unit version of the new Part 2; BS 449:1969, Part 2 is the metric version, which has in some degree involved 'rounding off,' whilst not being a technical revision.

A table of conversion factors has been added immediately before the main text.

In revising the book to accord with this major change, the opportunity has been taken of making minor alterations to the text, and of replacing many of the illustrations; acknowledgements for the photographs are noted with their titles. The standard tables have again been included, as currently issued by the B.C.S.A.

L. E. K.
D. W. L.

PREFACE TO FIRST EDITION

THE subject of 'Structural Mechanics' in the more advanced stages is adequately covered by the many excellent books dealing with the 'theory of structures' and the 'strength of materials.' The authors trust that the present volume, introductory and fundamental in character, will be of assistance to building and architectural students in the early stages of their professional training.

The aim of the book is to present, in simple language and in a logical sequence, the basic principles of building mechanics. The temptation to enlarge unduly on certain topics has been resisted. Too much detail in the early stages of a student's reading is apt to make him lose sight of the natural development of the subject, stage by stage.

An endeavour has been made to reduce the amount of mere arithmetical working in problems by selection of suitable data. It is difficult for an elementary student to keep unobscured the underlying principles involved in an exercise, when he has to deal with a mass of awkward figures.

The text is thoroughly interspersed with numerical examples and diagrams, the only effective method of sustaining interest and enthusiasm.

In addition to the exercises at chapter ends, a chapter is devoted to revision examples. To these examples abridged solutions are provided.

The last chapter contains test papers. Completely worked solutions to the numerical portions of these papers are supplied. Page references are given for points of theory raised in the papers. Readers are urged to attempt these question papers in the spirit of examination tests. Reference to the solutions should not be made until the paper attempted has been fully worked.

Students preparing for the Intermediate examination of the Royal Institute of British Architects, the Graduateship examination of the Institution of Structural Engineers, the examinations of the Chartered Surveyors' Institution, and similar examinations, should find the book helpful.

Teachers proposing to use the book as a class book in preliminary courses approved for National Certificates and Diplomas in Building

vii

will have no difficulty in sealing-up the answers and worked solutions, if this were considered to be desirable.

The book has not been written to comply with any one set of building regulations. It has been necessary, however, to refer to typical regulations occasionally in order to demonstrate the relationship between theory and practice. Having grasped the basic principles of structural mechanics the reader will be in a position to apply his knowledge to the interpretation of any new regulations which may be issued from time to time by the British Standards Institution, etc.

T. J. R.
L. E. K.

CONTENTS

TABLE OF CONVERSION FACTORS

A **newton** is the force required to accelerate a mass of **1 kilogram** by **1 metre per second per second.**

1 N	$= 0.224\ 81$ lbf
1 kN	$= 224.81$ lbf
1 lbf	$= 4.448\ 22$ N
1 tonf	$= 9964$ N
1 inch	$= 25.4$ mm $= 0.0254$ m
1 foot	$= 304.8$ mm $= 0.3048$ m
1 m	$= 3.280\ 84$ ft $= 39.37$ in
1 lbf/ft	$= 14.593\ 86$ N/m
1 tonf/ft	$= 32\ 690.2464$ N/m $= 32.69$ kN/m
1 lbf/in^2	$= 0.006\ 894\ 76$ N/mm^2 $= 6894.76$ N/m^2
1 tonf/in^2	$= 15.44$ N/mm^2
1 lbf/ft^2	$= 47.8803$ N/m^2
1 tonf/ft^2	$= 107\ 251.872$ N/m^2 $= 107.252$ kN/m^2
1 lbf/ft^3	$= 157.088$ N/m^3

CHAPTER I

COMPOSITION OF FORCES. RESULTANTS

Introduction.—'Structural mechanics' is mainly concerned with forces, how they combine together, how they keep a body at rest and, in general, with the effect they have on the stability of the parts of the structure to which they are applied. In this subject we will have to consider also the effect a force has on the size and shape of the actual material upon which it acts.

The object of studying the subject of structural mechanics is to learn how to build structures with a view to both *economy* and *strength*.

Force

It is usual to define *force* as that which tends to alter the state of rest of a body or its uniform motion in a straight line. As building students we are chiefly concerned with bodies at rest. It must be realised that if a single force acts on a body, it will produce motion in that body. Also, if a number of forces act on a body in such a manner as to be equivalent to a '*net resultant force*,' the body cannot possibly remain at rest. From our point of view, the important fact to remember is that if any unit in a structure, e.g. a girder, is to remain in *equilibrium* (i.e. at rest), there must be no resultant force acting on it. It is our duty to provide for such a member a set of forces that shall satisfy the '*laws of equilibrium.*'

Types of Forces met with in Structural Calculations

(*a*) In all practical problems there is one force which should always be carefully noted. This force is the '*self-weight*' of the member involved. For example, in beam problems the weight of the beam itself should always be considered in estimating the total load the beam has to carry. The load due to the weight of a member is usually termed a '*dead load.*'

(*b*) Structural members have to support external loads which are known as '*superimposed loads.*' In floor-design calculations, superimposed loads are frequently described as '*live loads*.' The following represents a typical load table for a reinforced concrete floor:

$$
\begin{aligned}
\text{Dead load} &= 3\,500 \text{ N/m}^2 \\
\text{Floor finish} &= 500 \text{ N/m}^2 \\
\text{Live load} &= 11\,000 \text{ N/m}^2 \\
\hline
\text{Total} &= \overline{15\,000 \text{ N/m}^2}
\end{aligned}
$$

In this case we have a floor slab 150 mm thick with an additional top surface designed to resist wear. The term 'live' in this example merely indicates that the load of '11 000 N/m²' *may* be applied all over the floor.

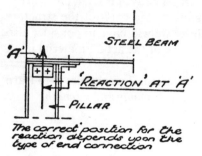

Fig 1.—Beam Support Reaction

Strictly speaking, a '*live load*' is of a dynamic character, such as the load a pile receives when it is hit by the falling 'monkey.' A 'live load' of this type may be considerably more damaging to a structure than a 'dead load' of the same magnitude.

(*c*) When one structural member rests upon another, the force exerted by the supporting member is termed a '*reaction*.' Fig. 1 shows a steel beam simply supported at its end 'A'. The force exerted in this case by the pillar on the beam at 'A,' indicated in the diagram by an arrow, would be termed the '*reaction at A.*'

(*d*) In framed structures, some of the members will be pulling at their end connections and others will be exerting a thrust (Fig. 2). Those members which pull are termed '*ties*,' and those which push are known as '*struts*.' The arrow heads

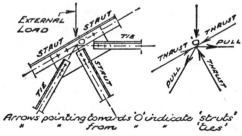

Fig. 2.—Struts and Ties.

in Fig. 2 indicate the nature of the forces which the members exert at the joint. The reader should carefully note the relationship between the way in which the respective arrows are pointing and he type of the member concerned.

(*e*) Walls sometimes have to be built to retain liquids, or granular materials like earth. Such structures are known as '*retaining walls*.' The forces involved in retaining-wall calculations are dealt with in Chapter XV.

(*f*) When a member is subjected to load, the fibres of the material transmit the load from section to section throughout the member. Such a system of internal force transmission is termed a '*stress*.' Chapter IX is concerned with the subject of 'stress.'

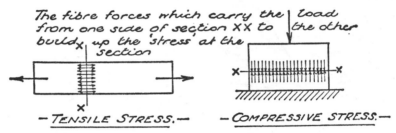

FIG. 3.—THE NATURE OF STRESS.

Units for Expressing Force Values

Two systems of units are employed.

(i) *Gravitational or Engineers' system.*

(ii) *Absolute system.*

The latter system is based upon Newton's laws of motion, and in this system a force is measured by the acceleration it produces in a given mass. This method of giving force values is not commonly used in building calculations and will not be further referred to.

Gravitational System.—Every body is attracted towards the earth's centre by a force which depends upon (i) the amount of matter in the body, i.e. its '*mass*,' and (ii) the distance the body is from the centre.

We may ignore the slight difference in the '*force of gravity*' which a given mass would experience at different positions on the earth's surface and which is occasioned by the fact of the earth not being a perfect sphere. For all practical purposes we may regard the mass of a body as the only factor which influences the value of the earth's pull on it.

In London there is stored a piece of platinum the mass of which is defined to be '*one pound*.' If we were to measure the pull of gravity on this standard unit of mass we would get the force value known as '*one pound weight*.' From this primary unit the system of force units commonly used, heretofore, in structural calculations was built up.

The unit of force in the SI metric system is the *newton*, symbolised by the letter N. The use of the letter N to denote force will obviate the need, in previous systems, to denote force by the use of the letter *f* after the unit, e.g. *lbf, kgf.*

Resultant of a Force System.—The 'resultant' of a system of forces (i.e. of a specified number of given forces) is the single force which we could replace for the given system without altering the net effect the system has on the state of rest (*'equilibrium'*) of the body upon which it acts (see Fig. 4 (*a*)).

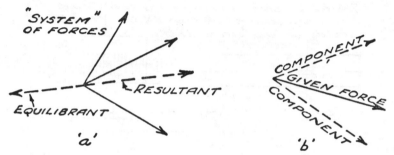

FIG. 4.—RESULTANT, EQUILIBRANT AND COMPONENT.

It is often important to be able to substitute one hypothetical force for an existing group of forces. In all problems of the *'equilibrium'* type, this is a permissible substitution.

Equilibrant.—The *'equilibrant'* of a system of forces is the force which would have to be introduced into the system in order to bring it into the state of equilibrium. It is clear that the 'equilibrant' of a system must balance the *resultant* of that system. The relationship between these two important forces is, therefore, that they are *'equal in magnitude'* and act in *'exactly opposite directions.'* Both resultant and equilibrant will be of zero value for a system in equilibrium.

Components.—The components of a force are the separate forces which acting all together will have the given force as their resultant.

The operation of resolving a force into its components, termed *'resolution,'* is frequently employed in structural calculations. Fig. 4 (*b*) illustrates a force resolved into two components.

Determination of Resultants

It is intended in this book to consider only the type of force system in which all the forces act in the same plane. Such a system is known as a 'co-planar system of forces.' It is the type which commonly occurs in building calculations.

Force systems may be divided into two classes:

(i) 'Concurrent systems,' in which all the 'lines of action' of the various forces pass through one common point.

(ii) 'Non-concurrent systems,' in which the lines of action have no common point of concurrence.

Fig. 5 illustrates these two main types and also the special important case of the 'parallel system.'

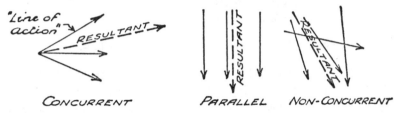

FIG. 5.—FORCE SYSTEMS WITH TYPICAL RESULTANTS.

Problems in structural work are solved either by methods of calculation or by 'graphical methods.' Graphical methods are widely used and will be considered first.

Graphical Representation of a Force

A force has 'magnitude,' 'direction,' and 'position.' For example, we might have a force of 1 000 N (magnitude) acting vertically downwards (direction) at the apex of a roof truss (position). It is possible to represent these three properties by a straight line drawn to a convenient force scale (see Fig. 6).

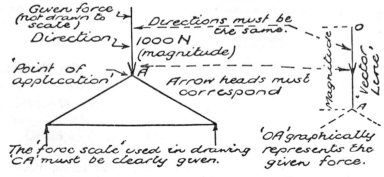

FIG. 6.—GRAPHICAL REPRESENTATION OF A FORCE.

EXAMPLE.—*Represent graphically a force of* 400 *N acting, at a given point, horizontally towards the right.*

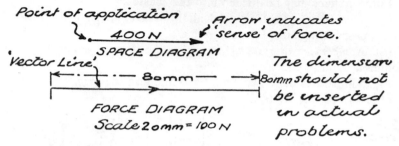

FIG. 7.—A FORCE REPRESENTED BY A STRAIGHT LINE.

In Fig. 7, the original scale chosen was 20 mm = 100 N. The length of the '*vector line*' representing the force was therefore '80 mm.' The reader should not calculate the length of line required but use the side of his scale rule marked off in millimetre graduations and plot '80' of these graduations in this case. Lines drawn to scale to represent force values will be termed '*vector lines*.' A '*vector quantity*' is one (like a force) which has '*direction*' as well as '*magnitude*.'

It is absolutely essential in graphical work that **a vector line shall be drawn parallel to the line of action of the force it represents.** In drawing a vector line, as far as is possible, the pencil point should travel in the direction of the '*sense*' of the given force, as indicated by the arrow head in the space diagram.

Always indicate clearly, near the diagram concerned, the scale used in its construction.

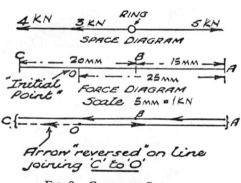

FIG. 8.—GRAPHICAL SOLUTION.

Resultant of a Co-linear Force System

EXAMPLE.—*Fig. 8 shows three horizontal forces pulling on a ring. Find the resultant pull on the ring* (a) *by calculation,* (b) *by 'graphics.'*

In graphical work a *space diagram* should always be drawn, showing the correct directions of the forces acting, with indications of the force

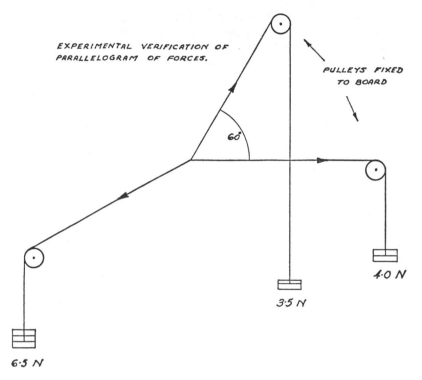

EXPERIMENTAL VERIFICATION OF
PARALLELOGRAM OF FORCES.

PULLEYS FIXED
TO BOARD

60°

4·0 N

3·5 N

6·5 N

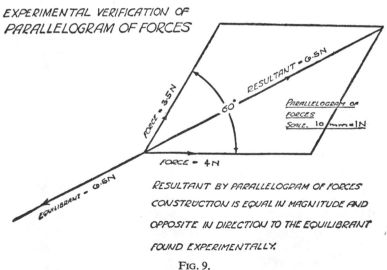

EXPERIMENTAL VERIFICATION OF
PARALLELOGRAM OF FORCES

RESULTANT = 6·5N

FORCE = 3·5N

60°

PARALLELOGRAM OF
FORCES
SCALE. 10 /mm = 1N

FORCE = 4N

EQUILIBRANT = 6·5N

RESULTANT BY PARALLELOGRAM OF FORCES
CONSTRUCTION IS EQUAL IN MAGNITUDE AND
OPPOSITE IN DIRECTION TO THE EQUILIBRANT
FOUND EXPERIMENTALLY.

FIG. 9.

values where known. The forces are not represented to scale in a space diagram, but the diagram may require a linear scale to set out correctly the various force directions (see page 35).

(a) Total force to right = 5 kN.
   Total force to left = (3 + 4) kN.
   Net force to left = (3 + 4 − 5) kN = 2 kN.

∴ Resultant = 2 kN, acting horizontally towards the left.

(b) From a convenient initial point 'O' draw a vector line 'OA,' horizontally towards the right, to represent the '5 kN' force to a suitable scale. To the same scale draw 'AB' (to left) to represent '3 kN' and 'BC' (to left) to represent '4 kN.'

Join the final point of the diagram (C) to the initial point (O) and reverse the arrow on this line.

'OC' will represent the resultant of the three given forces. The reader will note that this agrees with the calculated result.

This example has been explained in detail because the rule given in heavy type above will be shown later to be applicable to any system of forces acting at a point.

Resultant of Two Intersecting Forces

The solution of this case is effected by means of a theorem known as the 'parallelogram of forces.' The experimental verification of the theorem may be conducted by apparatus such as that illustrated in Fig. 9.

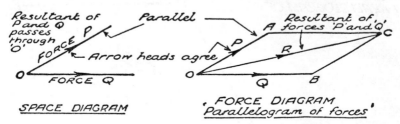

FIG. 10.—RESULTANT OF TWO INTERSECTING FORCES.

Rule.—To find the resultant of two forces 'P' and 'Q' (Fig. 10) proceed as follows. From a convenient point 'O' draw 'OA' (to suitable scale) to represent force 'P' and also draw 'OB' (to same scale) to represent force 'Q.' Complete the parallelogram 'OACB' and draw-in the diagonal 'OC.' 'OC' will then represent (to the chosen force scale) the resultant of the forces 'P' and 'Q.'

EXAMPLE.—*A peg, fixed in the ground, has two ropes attached to it, both ropes being parallel to the ground. The angle between the ropes is 60°. In one rope there is a tension (i.e. pull) of 70 N and in the other a tension of 80 N. Find the resultant pull on the peg.*

By means of the parallelogram of forces shown in Fig. 11, the resultant is found to be a force of 130 N, acting in the direction indicated.

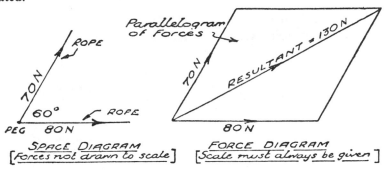

FIG. 11.—PARALLELOGRAM OF FORCES.

In order to remove all pull off the peg, a force of 130 N acting *in the opposite direction* to that of the resultant would have to be applied. This force is, of course, the *'equilibrant'* of the system.

Special Cases.—Fig. 12 shows a *'thrust'* and a *'pull'* acting at a point. To use the *'parallelogram of forces'* we must reduce the system either to two 'pulls' or to two 'thrusts.' Such cases may be solved more conveniently by the method given in the next paragraph.

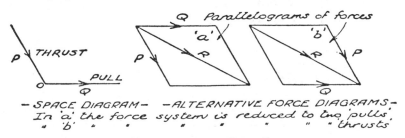

FIG. 12.—REDUCTION OF FORCE SYSTEM.

Second Method for Two Intersecting Forces

The vector line 'OC,' which represents the resultant in Fig. 10, may be obtained as follows: draw 'OA' (Fig. 13) to represent force 'P' and at

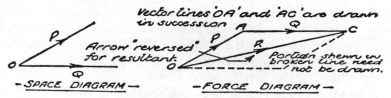

Fig. 13.—Alternative Method for two Intersecting Forces.

'A' draw 'AC' to represent force 'Q.' **Join the final point of the diagram (C) to the initial point (O) and reverse the arrow on this line.** This rule agrees with that given for co-linear forces on page 8.

Example.—*In Fig. 14 the rafter exerts a thrust of* 2 000 *N and the horizontal tie pulls with* 1 730 *N force. Find the resultant force the truss transmits to its end support.*

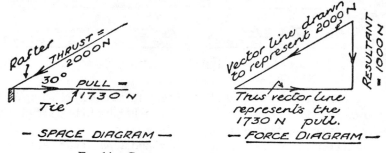

Fig. 14.—Compounding a Thrust and a Pull.

The force diagram in Fig. 14 indicates that the support will have to carry a vertical force of 1 000 N.

Note.—Fig. 15 shows a body pulled by two strings. The strings do not

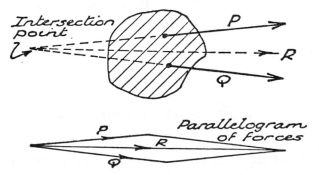

Fig. 15.—Production of Lines of Action.

actually intersect. In such cases the lines of action must be produced until they do intersect. The parallelogram of forces (or alternative method) may then be applied in the usual manner. The resultant will pass through the intersection point of the two lines of action. In all problems of this type, concerning equilibrium, a force may be assumed to be acting at any point in its own line of action.

Resultant of any Number of Concurrent Forces

Fig. 16 illustrates how the previous method of successive vector line construction can be applied to any number of forces. 'OB' represents

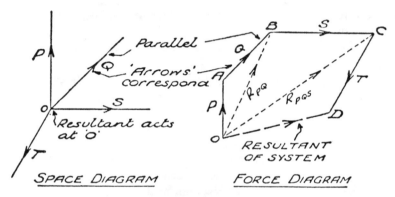

FIG. 16.—RESULTANT OF A CONCURRENT SYSTEM.

R_{PQ}, the resultant of forces 'P' and 'Q.' This resultant force is then combined with force 'S,' giving 'OC' as the vector line representing the resultant of forces 'P,' 'Q' and 'S.' This procedure is continued until all the forces are taken. 'OD' represents the final resultant in the example taken. In practical examples there will be no need to draw in the intermediate resultant vector lines 'OB,' 'OC,' etc. It will be necessary simply to construct the polygonal outline 'OABCD.'

The rule is therefore the same as for the previous force systems considered, viz: **draw vector lines in succession to represent the forces, join the final point of diagram to the initial point and reverse the arrow on this vector line.**

EXAMPLE.—*Find the resultant of the concurrent force system given in Fig. 17.*

The resultant is a force of 38·26 kN acting in the direction indicated

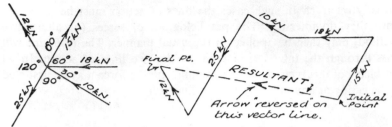

The forces may be represented in the force diagram in any desired sequence provided the arrows follow round the force diagram 'in order'

FIG. 17.—CONCURRENT SYSTEM.

in the force diagram. Its 'point of application' is the point of concurrence of the forces forming the system.

Recapitulation.—The reader should note carefully the following important points:

(i) **It is essential to construct the vector lines in a force diagram parallel to the respective forces they represent in the space diagram.**

(ii) **The arrow head on a vector line must agree 'in sense' with that indicated in the space diagram for the particular force represented, i.e. the two arrows must point in the same direction.**

(iii) **The arrow heads must follow round the force diagram 'in order' (i.e. all point ahead as you proceed round the diagram), except on the vector line which represents the resultant of the system. In this case it is directly 'reversed' to all the others.**

(iv) **The forces given in the space diagram may be represented in any sequence when constructing the force diagram. The vector lines of the force diagram may cross one another.**

Calculation Methods

As an alternative to graphical methods, methods involving the employment of simple trigonometrical functions and formulæ may be used for the solution of problems requiring the compounding of forces.

Readers unacquainted with trigonometry should omit the following paragraphs and continue with the study of the graphical methods given in the next and subsequent chapters.

If we can express by a formula the length of the diagonal of a parallelogram, or the closing line of a polygonal diagram, there will be no need to draw such diagrams to scale.

Two Intersecting Forces

In Fig. 18, if 'θ' be the angle 'AOB,' the length 'OC' of the parallelogram is given by the formula:

$$OC^2 = OA^2 + OB^2 + 2.OA.OB.\cos\theta.$$

Expressed in terms of force values this formula becomes:

$$R^2 = P^2 + Q^2 + 2.P.Q.\cos\theta.$$

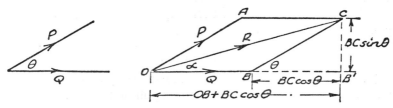

FIG. 18.—RESULTANT BY CALCULATION METHOD.

Applying this method to the example given in Fig. 11 we have:

$$
\begin{aligned}
R^2 &= 70^2 + 80^2 + 2 \times 70 \times 80 \times \cos 60° \\
&= 4\,900 + 6\,400 + 2 \times 70 \times 80 \times 0{\cdot}5 \\
&= 16\,900 \\
R &= 130 \text{ N.}
\end{aligned}
$$

If forces 'P' and 'Q' act at right angles to one another we get the simple formula $R^2 = P^2 + Q^2$.

Referring to Fig. 18, let 'α' be the angle the diagonal 'OC' makes with the side 'OB.'

$$\tan\alpha = \frac{CB'}{OB'} = \frac{BC\sin\theta}{OB + BC\cos\theta}.$$

Expressing in force values:

$$\tan\alpha = \frac{P\sin\theta}{Q + P\cos\theta}.$$

Applying this formula to the case given in Fig. 11:

$$\tan\alpha = \frac{70\sin 60°}{80 + 70\cos 60°} = \frac{70 \times 0{\cdot}866}{80 + 70 \times 0{\cdot}5} = \frac{60{\cdot}62}{115} = 0{\cdot}5271.$$

$$\therefore \quad \alpha = 27° 48'.$$

The resultant makes an angle of 27° 48' with the 80 N force.

When Chapter II, dealing with the '*resolution of forces*,' has been studied, the formula given above for 'tan α' will be more readily appreciated. We will then also be in a position to consider the derivation of the resultant of a system of several concurrent forces by the calculation method.

The examples given in Exercises 1 are intended to be solved by graphical methods. Calculation methods may be employed, if desired, as a check on the graphical work. The answers given at the end of the book have been *calculated*. Results obtained by graphical methods are sufficiently accurate for all practical purposes, provided the various diagrams are carefully drawn.

Exercises I

(1) Fig. 19 shows a bolt to which are attached two chains. Find the magnitude and direction of the resultant pull on the bolt.

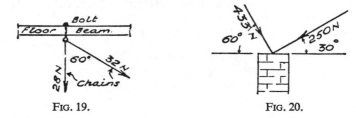

FIG. 19.   FIG. 20.

(2) A vertical post has fixed to it, at the top, two horizontal wires in which the pulls are respectively 80 N and 60 N. The angle between the wires is 90°. Find the magnitude of the resultant pull on the post. Obtain the angle between the line of action of the resultant force and that of the 60 N force.

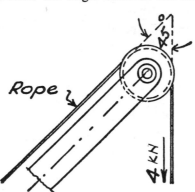

FIG. 21.

(3) A wall supports two thrusts in the manner indicated in Fig. 20. Show that the wall will have no tendency to move horizontally. Find the vertical force the wall has to support.

(4) A rope passes over a pulley (which may be assumed to have frictionless bearings) placed at the end of the jib of a crane (Fig. 21). Find the resultant force on the jib end due to the rope.

(*Note.*—The tension (i.e. the pull) in the rope may be taken to be constant so that at the end of the jib there are two forces, each of value 4 kN, acting in the respective rope directions. Produce these rope directions to intersect.)

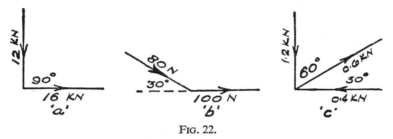

FIG. 22.

(5) Find the resultant force in each of the cases given in Fig. 22.

(6) In Fig. 23 is shown a vertical dead load of 1 200 N and a wind load

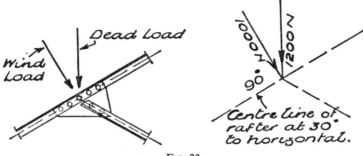

FIG. 23.

of 1 000 N acting at a rafter joint in a roof truss. Find the resultant load at the joint.

(7) A particle is acted upon by three forces as follows: (i) 10 N vertically downwards, (ii) 15 N horizontally towards the right, (iii) 20 N acting towards the right and making an angle of 30° with, and above, the horizontal. Prove (i) that the particle has no tendency to move vertically, (ii) that it will begin to move horizontally towards the right as if a single force of 32·32 N were acting on it.

(8) Find the equilibrant of the concurrent force system given in Fig. 24.

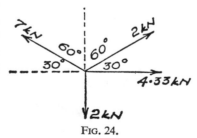

FIG. 24.

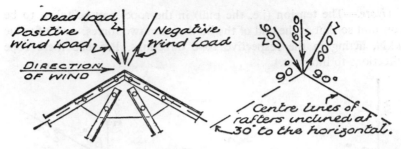

FIG. 25.

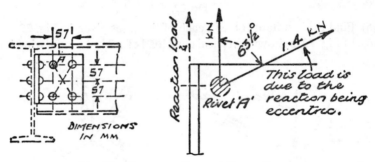

FIG. 26.—ECCENTRIC LOAD ON RIVET GROUP.

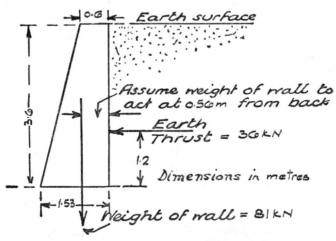

FIG. 27.—RETAINING WALL.

(9) At the apex joint of a truss (Fig. 25) three forces are acting: (i) a dead load of 1 600 N, (ii) a positive wind load of 900 N, and (iii) a negative (or suction) wind load of 600 N. Find the magnitude of the resultant load at the joint. (Wind loads act at right angles to the roof slope.)

(10) Fig. 26 illustrates a riveted connection in which the rivets, owing to the nature of the applied reaction load, are subjected to two simultaneous loads. Taking rivet 'A', which carries loads of 1 kN and 1·4 kN in the respective directions shown, determine the resultant load it must be capable of supporting.

(11) The diagram given in Fig. 27 represents the section of a retaining wall. The earth thrust and the self-weight of the wall are computed on the basis of one-metre length of wall. Obtain the resultant force on the wall per metre of length. Determine the distance, from the vertical back of the wall, at which the resultant cuts the wall base.

(Produce the line of action of the earth thrust to cut the vertical line of action of the weight of the wall (see Fig. 15).)

CHAPTER II

RESOLUTION OF FORCES. RECTANGULAR COMPONENTS

In order to solve structural problems, not only must we be able to *compound* several forces into one equivalent resultant force, but we should be familiar with the methods whereby a single force may be *resolved* into two or more components.

A given force system has only one resultant force, but a single force may be resolved into component forces in an indefinite number of ways. It is necessary, therefore, to specify some particulars of the required components.

The usual problem is to have to resolve a given force into *two* components whose directions are specified. The most important case of all is when these two components are themselves at *right angles* to one another.

EXAMPLE.—*A force of 6 kN acts, at a certain point, vertically downwards. Resolve this force into two components, one acting at 30° to the left and the other at 45° to the right of the given force direction (Fig. 28).*

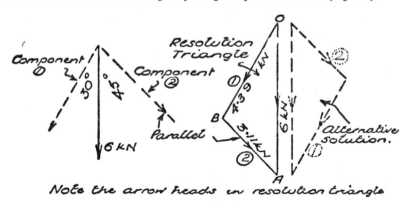

FIG. 28.—COMPONENTS OF A FORCE.

First of all, draw a space diagram showing the lines of action of the forces concerned. An arrow head should be placed on the given force. The correct placing of the arrow heads on the component lines is explained below. Now draw a vector line 'OA,' to a suitable force scale, to represent the '6 kN force.'

On this line it is essential to place an arrow head (i.e. pointing downwards) to indicate the *'sense'* of the force. Construct a triangle having the drawn vector line 'OA' for one side and having the other two sides respectively parallel to the required directions of the components. Place the arrow heads on these latter two sides, *'in order'* with one another but *'reversed'* in respect to the arrow head on the '6 kN force' vector line. The magnitude of each component may now be scaled off the 'resolution triangle,' being represented by 'OB' and 'BA' respectively. From Fig. 28 we see that the components are respectively 4·39 kN and 3·11 kN, acting in the directions indicated. It is clear, by the rule riven on page 10, that these two components have the given force of '6 kN' as their resultant.

Rectangular Components

When a force is resolved into two components, and these two components are at right angles to one another, they are termed *'rectangular components.'* The term *'resolved parts'* is also sometimes used in this case. It is important to remember, when considering one rectangular component of a force, that with this component there is always associated another, acting in a direction at right angles to it (Fig. 29).

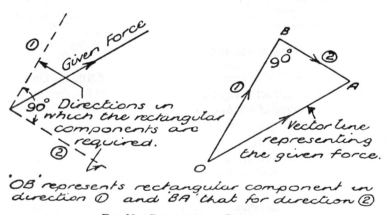

FIG. 29.—RECTANGULAR COMPONENTS.

Important Property possessed by Rectangular Components

In Fig. 30, the force 'F' exerts its full forward effect on the slider block only when it is applied horizontally. As the angle made by force 'F' with the horizontal increases, the effective forward pull on the block decreases. When force 'F' acts vertically it has no tendency to move the block forward horizontally.

A force, therefore, has no effect in a direction at right angles to its own line of action.

If we resolve a force into two components at right angles to one another, each component has no effect in the direction of the other. In such a case each component represents the *total effect* the original given force has in the particular component direction. Rectangular components are not merely components—each rectangular component represents a *net effective value* of the given force.

EXAMPLE (i).—*Taking the case referred to above, in which a slider block is constrained to move by guides in a horizontal direction, find the effective forward pull on the block under the conditions given in Fig. 30.*

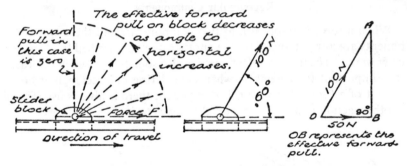

FIG. 30.—EFFECTIVE VALUE OF A FORCE.

Method.—Draw 'OA' to represent the given force of '100 N.' On 'OA,' as hypotenuse, construct a right-angled triangle 'OAB,' *with one side 'OB' in the direction in which the effective force is required*, i.e. horizontal in this case. 'OB' will represent the required forward pull on the block. The value of the pull will be found to be 50 N. If the resistance to forward motion, due to friction, etc., should equal 50 N, the block would not move forward in spite of the fact that the total force acting on it is 100 N.

EXAMPLE (ii).—*What tendency has the force given in Fig.* 31 (a) *to lift the body at right angles to the inclined plane,* (b) *to pull the body up the plane?*

Note the construction of the right-angled triangle in Fig. 31, with the sides containing the right angle drawn in the directions referred to in the question. Vector line 'OB' scales '2 kN' and represents the lifting tend-

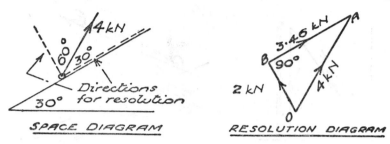

FIG. 31.—EFFECTIVE FORCE VALUES.

ency at right angles to the plane. 'BA,' which is parallel to the plane, represents the up-plane tendency and scales 3·46 kN.

EXAMPLE (iii).—*Fig. 32 shows a block of stone weighing 200 N resting on an inclined banker. Find the effective down-plane weight of the block.* In this case we must construct a right-angled triangle with one of the sides (containing the right angle) in the down-plane direction. As indicated in Fig. 32, the effective force of gravity down the plane is 68·4 N.

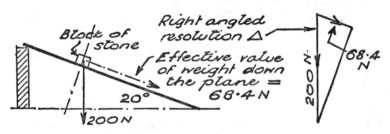

FIG. 32.—RESOLUTION OF SELF-WEIGHT.

Horizontal and Vertical Components

Very often in structural problems the rectangular components are required to be in the horizontal and vertical directions respectively.

It is not always convenient to find these components graphically, by the construction of resolution triangles. A simple formula involving the trigonometrical *cosine* is often employed (see Fig. 33).

Rule.—**To find the effective value of a force in a direction making an angle 'θ' with its own line of action, multiply the force by the cosine of 'θ.'**

In Fig. 33 (left)—where 'θ' is the angle made by the force 'F' with the horizontal—the horizontal component is 'F cos θ.' Similarly, the vertical component is 'F cos α' where 'α' is the angle made by force 'F' with the vertical. As $\alpha = 90° - \theta$, we may express the vertical component, if

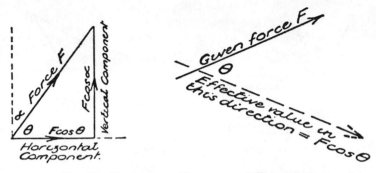

FIG. 33.—FORMULA FOR RECTANGULAR COMPONENTS.

desired, as 'F sin θ.' The reader is advised, at first, to use only the *cosine rule* in finding rectangular components and to ascertain in each problem the angle between the given force direction and that in which the effective component value is required.

EXAMPLE.—*A force of* 400 *N acts at an angle of* 60° *to the horizontal* (*Fig.* 34). *Find its horizontal and vertical components.*

Angle between the given force direction and the horizontal = 60°. Cos 60° = 0·5. Using the cosine rule, the horizontal component = F cos θ = (400 × 0·5) N = 200 N.

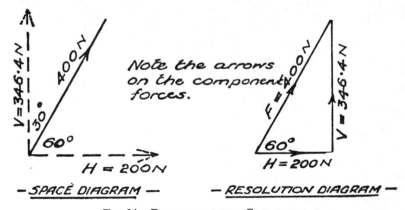

FIG. 34.—DETERMINATION OF COMPONENTS.

Angle between 'F' and the vertical = 30°. Cos 30° = 0·866, therefore the vertical component = F cos θ = F cos 30° = (400 × 0·866) N = 346·4 N.

There should be no difficulty in determining, in any given resolution example, whether the horizontal component acts towards the left or the right, or whether the vertical component acts upwards or downwards. Apart from the rules already given for arrow heads, imagine a particle to be acted upon by the given force. The general tendencies of motion of the particle, left or right, up or down, can be readily decided upon.

EXAMPLE.—*Fig. 35 shows a buttress subjected to an arch thrust. Find the tendency the thrust has* (i) *to move the abutment horizontally,* (ii) *to increase the load on the buttress foundation.*

The problem is solved graphically in Fig. 35.

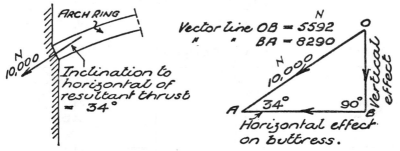

FIG. 35.—THRUST OF AN ARCH.

Using the cosine rule:
(i) horizontal component of force = F cos θ = (10 000 × cos 34°) N = (10 000 × 0·829) N = 8 290 N.
(ii) vertical component of force = F cos θ = (10 000 × cos 56°) N = (10 000 × 0·5592) N = 5 592 N.

Concurrent Force System reduced to Rectangular Components

The net or resultant horizontal effect of a system of forces will be the *algebraic* sum of the horizontal components of the various forces forming the system. Similarly the algebraic addition of the vertical components will give the real vertical effect of the system.

'*Algebraic addition*' simply means: *adding together all the components in one direction (say, to the right) and then adding together all the components in the opposite direction (i.e. the left), and finally subtracting the smaller total from the larger.*

EXAMPLE.—*Find graphically, and by the cosine rule, the resultant horizontal and vertical components, respectively, of the force system given in Fig. 36.*

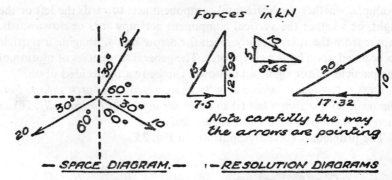

— SPACE DIAGRAM. —   — RESOLUTION DIAGRAMS

FIG. 36.—'H' AND 'V' FOR A CONCURRENT SYSTEM.

A tabular method may be used with advantage in this case.

Force. kN	Horizontal Component.		Vertical Component.	
	To Right.	To Left.	Up.	Down.
	kN	kN	kN	kN
15	7·50	—	12·99	—
10	8·66	—	—	5·00
20	—	17·32	—	10·00
TOTALS	16·16	17·32	12·99	15·00

Net horizontal force = (17·32 − 16·16) kN to left
= 1·16 kN to left.
Net vertical force = (15·00 − 12·99) kN down
= 2·01 kN down.

It is sometimes convenient to adopt a convention of signs, plus and minus, to indicate force direction.

Fig. 37 shows the usual convention. Applying these signs, and

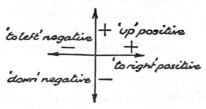

FIG. 37.—CONVENTION OF SIGNS.

employing the symbols 'H' and 'V' respectively for the total of horizontal and vertical components, we have:

$$H = (7{\cdot}5 + 8{\cdot}66 - 17{\cdot}32)\, kN = -1{\cdot}16\, kN,$$
i.e. $1{\cdot}16$ kN acting towards the left.
$$V = (12{\cdot}99 - 5 - 10)\, kN = -2{\cdot}01\, kN,$$
i.e. $2{\cdot}01$ kN acting downwards.

The figures in the table may be filled in by graphical or by calculation methods.

The usual practical method for dealing with such problems as these is to use the 'cosine rule' with the prefixing of signs.

Thus $H = 15 \cos 60° + 10 \cos 30° - 20 \cos 30°$
$$= (15 \times 0{\cdot}5) + (10 \times 0{\cdot}866) - (20 \times 0{\cdot}866)\, kN$$
$$= (7{\cdot}5 + 8{\cdot}66 - 17{\cdot}32)\, kN$$
$$= -1{\cdot}16\, kN, \text{ i.e. } 1{\cdot}16\, kN \text{ to left.}$$
$V = 15 \cos 30° - 10 \cos 60° - 20 \cos 60°$
$$= (15 \times 0{\cdot}866) - (10 \times 0{\cdot}5) - (20 \times 0{\cdot}5)\, kN$$
$$= (12{\cdot}99 - 5 - 10)\, kN$$
$$= -2{\cdot}01\, kN, \text{ i.e. } 2{\cdot}01\, kN \text{ downwards.}$$

EXAMPLE.—*Fig. 38 illustrates a portion of a retaining wall acted upon by three forces, viz. its own weight and two earth thrusts. Obtain (i) the resultant force tending to cause the wall to slide over its foundation, (ii) the total vertical thrust on the subsoil under the wall base.*

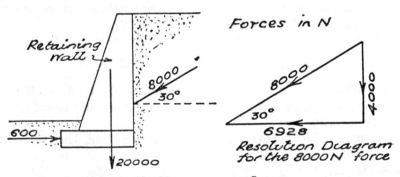

FIG. 38.—NON-CONCURRENT SYSTEM.

The forces in this example are not concurrent but the principles of horizontal and vertical resolution may be applied.

Solution by Graphical Method.—The only inclined force is the '8 000 N' earth thrust. By the resolution diagram of Fig. 38, the horizontal and vertical components of this force are '6 928 N' and '4 000 N' respectively. The other two forces require no resolution.

(i) Total horizontal component of system tending to cause sliding = (6 928 − 600) N = 6 328 N.

(ii) Total vertical thrust on subsoil = (20 000 + 4 000) N = 24 000 N.

Solution by Cosine Rule:

(i) Total horizontal component = (8 000 cos 30° − 600) N
                                = (8 000 × 0·866 − 600) N
                                = (6 928 − 600) N = 6 328 N.

(ii) Total vertical component   = (20 000 + 8 000 cos 60°) N
                                = (20 000 + 8 000 × 0·5) N
                                = (20 000 + 4 000) N = 24 000 N.

This type of problem is further considered in Chapter XV.

Resultant of a Concurrent System of Forces

The resultant of a concurrent system of forces may be conveniently obtained by means of rectangular components. It is usual to arrange for the axes of resolution to be in the horizontal and vertical directions respectively.

Rule.—Resolve all the forces into their horizontal and vertical components (see Fig. 39). Find 'H' and 'V,' the *algebraic sum* of the components in each case respectively. Using the parallelogram of forces law, compound 'H' and 'V' into their resultant 'R.'

Alternatively, we may obtain 'R' by the expression:

$$R^2 = H^2 + V^2 \quad \text{or} \quad R = \sqrt{H^2 + V^2}.$$

The graphical method will give the direction of the resultant. If the formula above be used, the direction of the resultant may be obtained by trigonometry.

Let 'θ' be inclination to the horizontal of the resultant. Then tan θ =

$$V \div H, \quad \text{i.e.} \quad \frac{V}{H}.$$

EXAMPLE.—*Find the resultant of the concurrent force system given in Fig. 39, by means of horizontal and vertical components.*

The total horizontal and vertical components, as obtained in Fig. 39, are 0·884 kN *to left* and 0·932 kN *upwards*, respectively. It is left as an

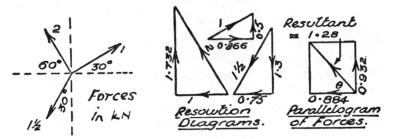

FIG. 39.—RESULTANT OF CONCURRENT SYSTEM.

exercise for the reader to check the separate components by the cosine rule.

The graphical solution gives 1·28 kN as the resultant, acting in the direction indicated in the parallelogram of forces. By calculation:

$$R^2 = H^2 + V^2, \quad \text{i.e.} \quad R^2 = 0·884^2 + 0·932^2$$
$$\therefore \quad R = \sqrt{0·884^2 + 0·932^2} \text{ kN}$$
$$= 1·28 \text{ kN}.$$

If the resultant acts at 'θ' to the horizontal,

$$\tan \theta = V/H = \frac{0·932}{0·884} = 1·054$$

$$\therefore \quad \theta \text{ (from trig. tables)} = 46\tfrac{1}{2}° \text{ nearly.}$$

It is advisable, in the calculation method, to draw a sketch diagram showing 'H' and 'V' in their correct directions, so as to avoid any risk of error in the direction of 'R.'

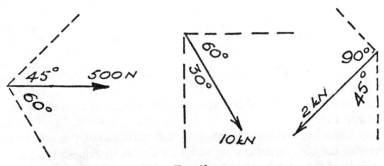

FIG. 40.

EXERCISES 2

(1) Resolve each of the forces given in Fig. 40 into components which act in the directions of the respective broken lines shown.

(Note: *The cosine rule must not be used unless the two required components are at right angles to one another.*)

(2) A skip is pulled along its rail track in the manner indicated in Fig. 41. Find the effective tractive force in the direction of travel.

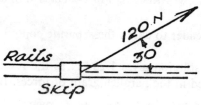

FIG. 41.

(3) In Fig. 42 the peg is inclined at 30° to the vertical and the rope makes an angle of 45° with the horizontal. If the friction of the earth round the peg can only exert a resistance to withdrawal (in the direction of the length of the peg) of 40 N, find the maximum safe pull in the rope.

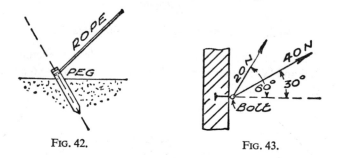

FIG. 42.   FIG. 43.

(4) What is the effective horizontal withdrawal force on the bolt given in Fig. 43?

(5) Find the horizontal and vertical components of each of the forces given in Fig. 44.

(6) Calculate the total horizontal and vertical components respectively of the concurrent system of forces shown in Fig. 45.

(7) Find the magnitude and direction of the resultant of the system of

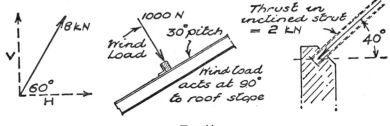

FIG. 44.

dead and wind loads shown in Fig. 46: (i) by graphical construction, (ii) by calculation of components.

(8) A retaining wall is subjected to a resultant thrust of 10 000 N for every metre length of wall. The resultant thrust makes an angle of 70° with the horizontal. Calculate the resistance to sliding which the foundation must be able to provide if it is to exceed the horizontal sliding tendency by 50 per cent.

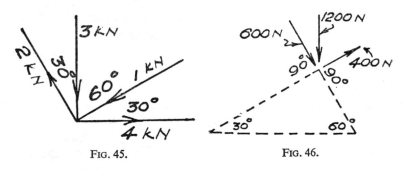

FIG. 45.   FIG. 46.

(9) A single timber shore is supported at the bottom by a horizontal timber balk. The thrust in the shore is 2 000 N. If the connection at the bottom of the shore has a safe resistance to horizontal movement of 1 000 N, find the minimum permissible angle the shore may make with the horizontal.

(10) The main rafters in a roof truss are inclined at 30° to the horizontal. The truss is subjected to a resultant wind load, on the left slope, of 6000 N. The left end of the truss is supported on a roller and provides no resistance to horizontal movement. Calculate the horizontal force which the right end reaction must provide to maintain equilibrium. (A wind load acts at right angles to the roof slope).

CHAPTER III

CONCURRENT FORCES. GRAPHICAL LAWS OF EQUILIBRIUM

Bow's Notation

IT is essential that the reader should become familiar with an extremely useful method of designating forces. We have hitherto described forces as 'force P' or 'force (1),' etc. An engineer named 'Bow' devised the following notation scheme, which is largely used in the solution of graphical problems.

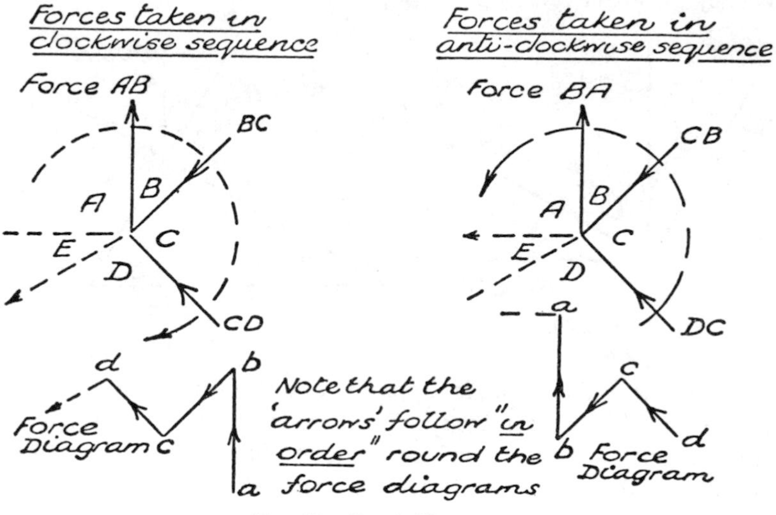

FIG. 47.—BOW'S NOTATION.

In each space, formed by the various lines of action of the forces in the 'space diagram,' a capital letter is placed. Any given force is then described by the pair of letters which lie on either side of its line of action. Fig. 47 illustrates the use of this method and gives the notation for each force, corresponding to the two possible sequences for taking the forces, viz. 'clockwise' and 'anticlockwise.'

It is extremely important to note the order of the precedence of the capital letters, e.g. it would be a serious error to confuse 'AB' with 'BA' in practical problems.

In the force diagram, corresponding small letters are used *at the ends of the vector lines*. Thus '*ab*' in the force diagram would be drawn parallel to, and represent to scale, the force 'AB' in the space diagram. Similarly, '*ba*' would represent 'BA,' in magnitude and direction.

Triangle of Forces

The two upper diagrams in Fig. 48 show the derivation of 'R', the resultant of forces 'P' and 'Q,' by means of the type of vector diagram explained in Chapter I. A force 'E' ('*equilibrant*'), equal in magnitude to 'R' and acting in the opposite direction, would balance 'R' and therefore balance 'P' and 'Q.' Forces 'P,' 'Q' and 'E' *acting together* would therefore constitute a system of three concurrent forces in equilibrium. It will be clear that if we draw vector lines in succession to represent forces 'P,' 'Q' and 'E,' a *triangle will be formed*. This triangle is known as the '**triangle of forces**' for the three given forces.

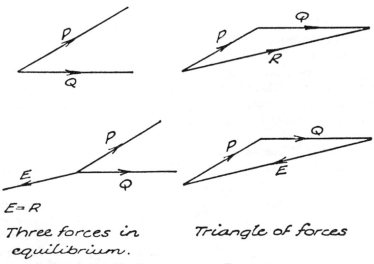

$E = R$

Three forces in equilibrium.   Triangle of forces

FIG. 48.—THREE FORCES IN EQUILIBRIUM.

Law.—**If three forces, acting at a point, be in equilibrium, they may be represented in magnitude and direction by the three sides of a triangle, taken in order.** The 'triangle of forces' law is best expressed for structural calculations in the terms given, and not in the more usual converse form, which gives the necessary qualification for three concurrent forces to be in equilibrium, viz. that a *'triangle should be formed.'*

Any triangle whose sides are parallel to the lines of action of the three given forces will have its sides respectively proportional to the magnitudes of the forces. If, therefore, we represent one of the forces to a certain scale, the remaining sides of the triangle will represent the two other respective forces to the same scale (Fig. 49).

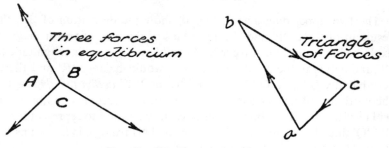

FIG. 49.—THE TRIANGLE OF FORCES.

The usual problem, involving the employment of the 'triangle of forces,' is to be given one of the three concurrent forces completely, but only the lines of action of the other two. The law enables us to find (i) the magnitudes of the two unknown forces, (ii) the *'sense'* in which the unknown forces act, i.e. whether they act respectively *'towards'* or *'away from'* the point of concurrence. The latter information decides whether the corresponding member is a 'strut' or a 'tie.'

Experimental Verification of Law

The triangle of forces law may be experimentally verified by means of the apparatus shown in Fig. 50.

The three concurrent forces are represented by (i) the suspended weight, 5 N, (ii) the pull in the left-hand string, 4·5 N, and (iii) the pull in the right-hand string, 4 N. The string pulls are obtained from the corresponding weights suspended over the side pulleys, on the assumption that the friction at the pulley bearings may be neglected.

As shown in Fig. 50, a space diagram is drawn for the three forces. The 5 N is represented by a vertical vector line drawn to a suitable scale. Vector lines are then drawn *in succession* to represent the string pulls of 4·5 N and 4 N respectively. It will be seen that the force diagram thus

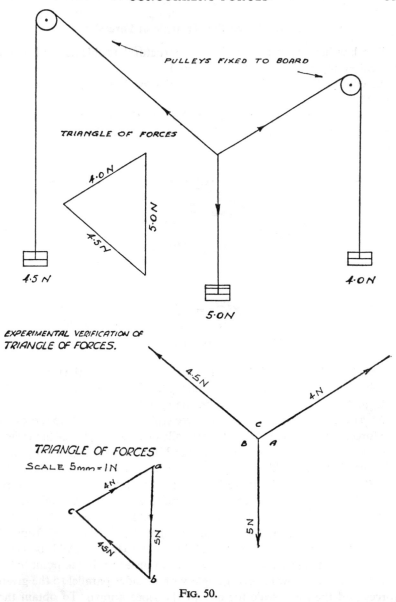

PULLEYS FIXED TO BOARD

TRIANGLE OF FORCES

4·0 N

5·0 N

4·5 N

4·5 N

4·0 N

5·0 N

EXPERIMENTAL VERIFICATION OF
TRIANGLE OF FORCES.

4·5 N

4 N

C

B   A

TRIANGLE OF FORCES

SCALE 5mm = 1N

a

4N

C

5 N

4·5N

b

5 N

FIG. 50.

constructed is a 'triangle.' The reader will note the employment of '*Bow's*
notation' in the drawing of the force diagrams in the experiment and in
Fig. 49 respectively.

Examples on the 'Triangle of Forces'

The law has many applications in structural problems, and a few typical examples will now be considered.

EXAMPLE (i).—*Find the tension in each rope in the example given in Fig. 51.*

Step (i).—Draw a space diagram showing the correct disposition in space of the forces concerned. In this case, the angles '30°' and '60°' determine the rope directions.

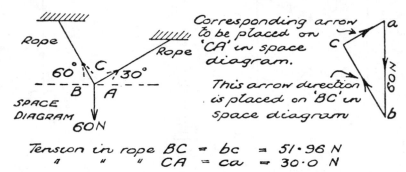

FIG. 51.—EXAMPLE ON TRIANGLE OF FORCES.

Step (ii).—Insert Bow's letters in the spaces formed by the three forces involved in the example. In this solution, 'AB' denotes the force '60 N,' 'BC' and 'CA' denote, respectively, the rope tensions.

Step (iii).—Choose a convenient force scale and draw '*ab*' to represent the force, '60 N.' Force 'AB' acts vertically downwards, hence the vector line '*ab*' should be drawn in a downward direction by the pencil point. The first letter in 'AB' is 'A,' hence the letter to be placed *where the pencil point starts* is '*a*.' If we had decided to take the forces in anti-clockwise sequence, the force would be represented by 'BA,' and the top of the vector line representing this force would have been '*b*.'

Step (iv).—Complete the triangle of forces by drawing '*bc*' from '*b*' parallel to force 'BC,' and '*ac*' from '*a*' parallel to force 'CA.' It is necessary to draw the last vector line *from* '*a*' in order to fix the point '*c*.'

Step (v).—We now have a triangle with its sides parallel to the given forces and the force scale for one of the sides known. To obtain the magnitudes of the unknown tensions we simply have to scale off (to force scale used for constructing '*ab*') the vector lines '*bc*' and '*ca*.'

EXAMPLE (ii).—*A weight of 2 kN, suspended from the end of a crane jib,*

is pulled by a horizontal force in the manner shown in Fig. 52. Determine
the value of the horizontal force and the tension in the upper portion of the
chain.

The problem is solved in Fig. 52. The reader should follow the
solution, step by step, as indicated in the previously worked example.

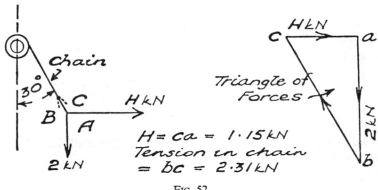

$H = ca = 1 \cdot 15 kN$
Tension in chain
$= bc = 2 \cdot 31 kN$

FIG. 52.

EXAMPLE (iii).—*A wall bracket is used to carry a weight of* 100 *N as*
indicated in Fig. 53. Determine the force in each projecting arm of the
bracket and state its nature.

In this example we require a linear scale for the space diagram in order
to construct correctly the directions of the three forces acting at the
extremity of the bracket. The arrow directions, derived from the triangle
of forces, are inserted in the space diagram on the appropriate members,

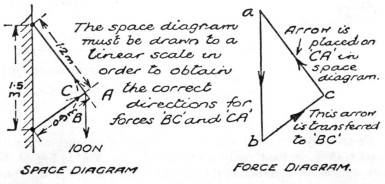

FIG. 53.

near the junction of the three force lines. Member 'BC' is a strut (thrust = 60 N) and member 'CA' is a tie (pull = 80 N).

EXAMPLE (iv).—*Determine the force in the principal rafter and the tension in the tie of the roof truss given in Fig. 54.*

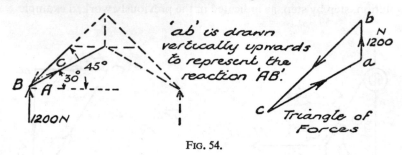

FIG. 54.

The principal rafter is a strut and exerts a thrust of 4 016 N. The tension in the inclined tie is 3 279 N.

Theorem relating to Three Forces in Equilibrium

The following theorem will be found useful in solving problems involving three forces in equilibrium.

If three forces acting on a body keep it in equilibrium, they must either be parallel forces or all their lines of action must pass through one common point.

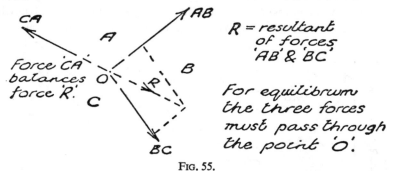

FIG. 55.

We are only concerned with the latter part of the theorem in this chapter. Consider the three forces 'AB,' 'BC,' and 'CA' in Fig. 55. We may replace 'AB' and 'BC' by their resultant 'R.' Force 'CA' will therefore balance 'R,' i.e. it will act in exactly the opposite direction. Clearly, then, it will pass through 'O,' the point in which forces 'AB' and 'BC' intersect.

EXAMPLE (i).—*In Fig. 56 it may be assumed that the reaction at the left end of the truss is vertical. Determine the value of each reaction for the conditions given.*

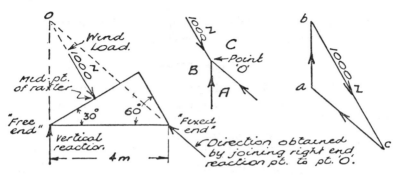

FIG. 56.

The direction of the reaction at the right end of the truss not being given, we make use of the foregoing theorem to determine it. The problem then reduces to a system of three forces (acting at point 'O') in equilibrium, and is solved as in Fig. 56 by the 'triangle of forces.'

The left-end reaction = 433 N. The right-end reaction acts in the direction indicated and its magnitude = 661 N.

EXAMPLE (ii).—*A ladder (Fig. 57) rests against a smooth wall. Taking the conditions given, determine the reactions at the wall and at the ground respectively.*

As the wall is smooth and cannot exert any frictional force on the ladder end, the reaction at the wall is horizontal. The lines of action of this

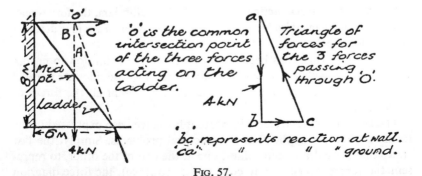

FIG. 57.

reaction and the weight respectively intersect at the point 'O'. Hence the ground reaction passes through 'O.' A triangle of forces is now drawn for the three forces, in equilibrium, passing through 'O.' The direction of the ground reaction is as given in the space diagram, its magnitude = ca = 4·27 kN. The wall reaction = bc = 1·5 kN.

A method of solution, by calculation, of problems involving three forces in equilibrium is given in Chapter V.

Polygon of Forces

The graphical method for finding the resultant of a concurrent force system is explained on page 11. Let us suppose that we applied this construction to a force system known to be in equilibrium—a system which has, of course, no resultant force. It is clear that there must be no distance between the final and initial points of the diagram, as the resultant equals zero. It appears, therefore, that for a system in equilibrium the force diagram must be a closed figure (Fig. 58). We have, thus, the following law of equilibrium, known as the '**polygon of forces**' law.

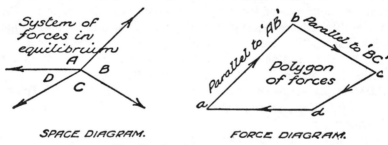

SPACE DIAGRAM.   FORCE DIAGRAM.

FIG. 58.—THE POLYGON OF FORCES.

'**If a number of forces acting at a point be in equilibrium, the forces may be represented in magnitude and direction by the sides, taken in order, of a certain closed polygon.**'

The word '*certain*' has been introduced into the definition to ensure that the reader will not assume that *any* polygon whose sides are parallel to the given force directions will necessarily represent all the forces to some given scale.

The terms of this extremely important law have been expressed in the form required for application to structural problems. Simply, the law states that if we draw vector lines, one at the end of the other, to represent the forces (taken in any convenient sequence), the force diagram

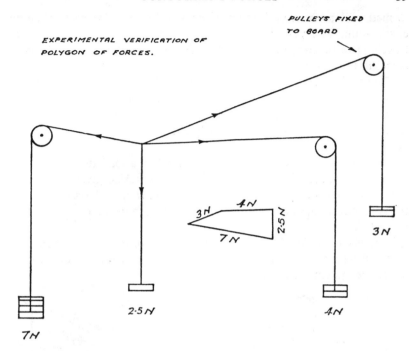

EXPERIMENTAL VERIFICATION OF
POLYGON OF FORCES.

PULLEYS FIXED
TO BOARD

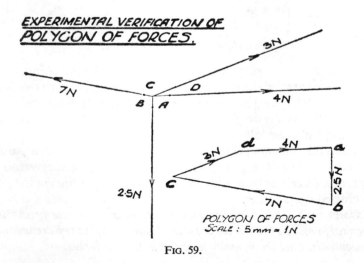

EXPERIMENTAL VERIFICATION OF
POLYGON OF FORCES.

POLYGON OF FORCES
SCALE : 5mm = 1N

FIG. 59.

formed will be a *closed polygon,* provided the force system be in equilibrium. The vector lines must be drawn with the pencil point travelling in the *sense* indicated by the arrows on the corresponding forces in the space diagram.

The usual problem is to have a number of forces, in equilibrium, acting at a point (such as a joint in a roof truss), the lines of action, but not the magnitudes, of all the forces being known. Not more than *two* forces must be unknown in magnitude. The 'polygon of forces' law is applied to determine the unknown magnitudes.

The polygon of forces law may be verified by apparatus such as that shown in Fig. 59. The special wall-board apparatus, given in Fig. 50, for the 'triangle of forces' law, is also suitable. In actual experiments of the type indicated, fine string should be used, as the weight of the string is neglected.

EXAMPLE (i).—*Test for equilibrium the system of forces given in Fig.* 60.

It will be seen, from the figure given, that the force diagram closes, hence the system is in equilibrium.

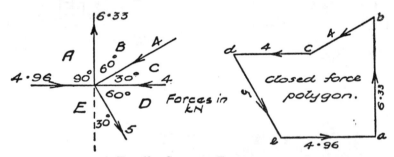

FIG. 60.—SYSTEM IN EQUILIBRIUM.

It is usual in employing Bow's notation to take the forces in a clockwise sequence around the common intersection point, though this is not essential. After drawing '*ab*' upwards to represent force 'AB,' i.e. '6·33 kN,' '*bc*' is drawn *from* '*b*' to represent 'BC,' i.e. '4 kN.' The vector line '*ea*' eventually closes the diagram.

Note (i) the correspondence of letters throughout, e.g. '*de*' is *parallel* to force 'DE'; (ii) the arrows go round the figure without reversal, i.e. they are '*in order*'; (iii) it would not matter if the vector lines in the polygon crossed one another.

EXAMPLE (ii).—*Fig.* 61 *represents four concurrent forces acting in a structural joint. The four forces are given in direction, but two are unknown in magnitude. Find the magnitude of the latter two forces.*

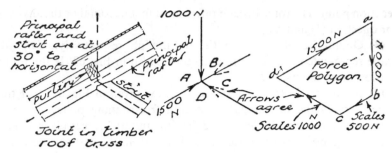

FIG. 61.—JOINT IN TIMBER TRUSS.

This problem, which is solved in Fig. 61, is the basic problem involved in *stress diagram* construction (which is considered in Chapter VIII).

Vector line '*da*' is drawn to represent force 'DA' (1 500 N) and '*ab*' is drawn from '*a*' to represent force 'AB' (1 000 N). We cannot draw '*bc*' directly because force 'BC' is unknown. If we draw from '*b*' a line parallel to 'BC' and from '*d*' a line parallel to 'CD,' the point '*c*' will be where these two lines intersect. Vector line '*bc*' (to scale of force diagram') scales '500 N' and '*cd*' scales '1 000 N.' The arrows are continued round the force diagram '*in order*' and transferred to the space diagram.

'BC' is a strut with a thrust of 500 N, and 'CD' is a strut with a thrust of 1 000 N.

EXERCISES 3

(1) Determine the force in each rafter of the truss shown in Fig. 62. Verify that the rafters act as struts.

(2) Find the force in the tie member of the truss (Fig. 62) by drawing a triangle of forces (i) for the left-end reaction point, (ii) for the right-end

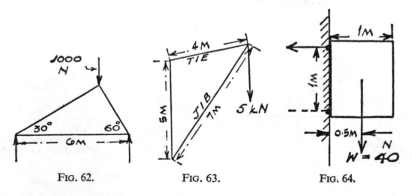

FIG. 62. FIG. 63. FIG. 64.

reaction point. The force in each rafter must first be found or the answers for question (1) assumed.

(3) Obtain the forces, in the 'tie' and the 'jib' respectively, of the jib crane represented by the outline diagram given in Fig. 63. Is the jib in tension or compression?

(4) Assuming, in the case of the jib crane given in Fig. 63, that the rope carrying the weight passed over a free-running pulley and ran parallel to the crane jib, determine the 'jib' and 'tie' forces respectively.

(5) A signboard is supported outside a wall in the manner indicated in Fig. 64. Assuming the upper hinge to be capable of exerting a horizontal reaction only, find (i) the direction of the reaction at the lower hinge, (ii) the magnitude of each reaction.

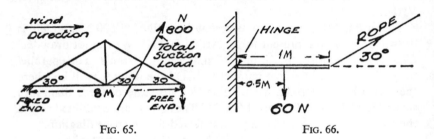

FIG. 65.   FIG. 66.

(6) Certain building regulations require the 'suction' effect on a roof, due to wind pressure, to be taken separately from the 'windward' loads. Taking the case given in Fig. 65, determine the support reactions for the truss. The reaction at the free end is vertical.

(7) A trap door is temporarily held in a horizontal position by a rope (Fig. 66). Assuming the particulars given, find the reaction at the hinge and the tension in the rope.

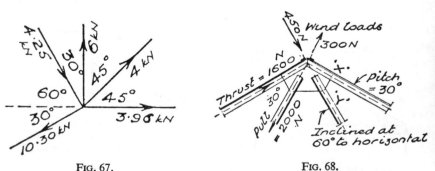

FIG. 67.   FIG. 68.

(8) Verify that the concurrent system of forces given in Fig. 67 is in equilibrium.

(9) Fig. 68 shows the apex joint in a steel roof truss. Find the magnitude and nature of the forces in the members marked 'X' and 'Y' respectively.

(10) The vertical load at a joint in a braced girder is 2 kN (Fig. 69). The forces in members 'AB' and 'BC' are 3 kN and 8 kN respectively,

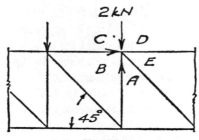

FIG. 69.

acting in the sense indicated by the arrow heads. Determine the forces in members 'DE' and 'EA' respectively. State whether the members are 'struts' or 'ties.'

CHAPTER IV

MOMENTS. PRINCIPLE OF THE LEVER

THE previous chapters have dealt with the translational effect of forces, i.e. their tendency to move bodies from one position to another. A force may, however, have a rotational effect on a body, tending to turn it round some given point which acts as a hinge. In Fig. 70 (a) the force 'F' would turn the disc in a *clockwise* manner of rotation about the hinge 'O.' In Fig. 70 (b) the force would cause an *anticlockwise* rotation about 'O.' To express these turning effects correctly we would say that the disc in Fig. 70 (a) turns as indicated because it is acted upon by a '*clockwise moment.*' Similarly, an '*anticlockwise moment*' is the cause of the rotation indicated in Fig. 70 (b).

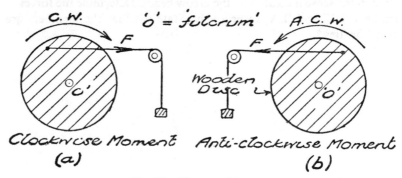

FIG. 70.—TYPES OF MOMENTS.

'Moments' are therefore concerned with rotational effects and, as we will see, play a very important part in structural calculations.

Moment of a Force

In considering the rotational effect of a force on a body we must have in mind some definite point about which the turning tendency is to be measured. The point concerned is usually referred to as the '*fulcrum.*' Thus 'O' is the 'fulcrum' in Fig. 70.

If we wish to open a door, i.e. subject it to a turning moment about the hinge, we must clearly exert a force on it. But a force, however big, applied at the hinge will be of no avail. We must apply the force at some distance from the hinge, and experience shows that the farther we get from the hinge the easier will be the operation of opening the door. This, of course, accounts for the position of the door knob.

Furthermore, it is when we pull at right angles to the door that we get the best turning effect. Even if we do pull at the knob, there will be no tendency for the door to open if the line of action of the pull passes through the hinge (see Fig. 71). The way in which 'distance' enters into the measurement of a 'moment' must therefore be carefully considered.

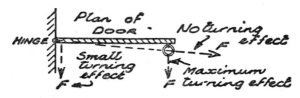

FIG. 71.

Measurement of a Moment.—Two quantities are involved in expressing the value of a '*moment*' or '*turning effect*';

(1) The magnitude of the applied force.

(2) The *perpendicular* distance between the line of action of the force and the point about which turning is being considered, i.e. the fulcrum. This distance is termed the '*arm of the moment.*'

The value of the 'moment' increases directly as these two quantities increase, so that we multiply them together to obtain the actual moment.

Definition: The moment of a force about a given point is the product of

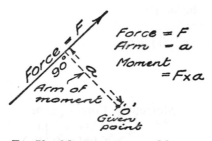

FIG. 72.—MEASUREMENT OF A MOMENT.

the force and the perpendicular distance from the point to the line of action of the force.

$$\text{Moment} = \text{Force} \times \text{arm.}$$

In Fig. 72 the moment of the force 'F' about the point 'O'

$$= \text{Force} \times \text{arm}$$
$$= (F \times a) \text{ units.}$$

Examples of Moments (see Fig. 73)

In each of the examples given 'O' is the point about which the moment is taken. 'C.W.' will be used to denote a *clockwise moment* and 'A.C.W.' will denote an *anticlockwise moment.*

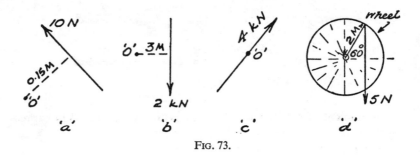

Fig. 73.

Example 'a'

Moment of 10 N force about 'O' $= \text{Force} \times \text{arm}$
$$= 10 \text{ N} \times 0 \cdot 15 \text{ m}$$
$$= 1 \cdot 5 \text{ N m (A.C.W.).}$$

Note.—The unit used to express a 'moment' must contain both 'force' and 'distance' units. To determine whether a moment is 'C.W.' or 'A.C.W.,' imagine the *arm* of the moment to be a crank acted upon by the given force. If the 'crank' tends to turn like the hands of a clock the moment is 'clockwise.' Similarly, an 'anticlockwise' moment may be determined.

Example 'b'

$$\text{Moment} = \text{Force} \times \text{arm}$$
$$= 2 \text{ kN} \times 3 \text{ m}$$
$$= 6 \text{ kN m (C.W.).}$$

Example 'c'

$$\text{Moment} = \text{Force} \times \text{arm}$$
$$= 4\,\text{kN} \times 0$$
$$= 0.$$

The reader should remember that a force has no moment about a point on its own line of action.

Example 'd'

Moment of suspended weight about the axle of the wheel

$$= \text{Force} \times \text{arm}$$
$$= 5\,\text{N} \times (2\cos 60°)\,\text{m}$$
$$= 5\,\text{N} \times 1\,\text{m}$$
$$= 5\,\text{N m (C.W.)}.$$

Note how the *'arm'* is measured in this example.

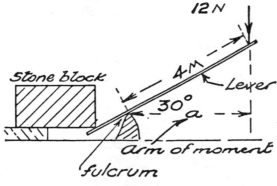

FIG. 74.

Example (see Fig. 74)

Moment of applied vertical force of 12 N about the given fulcrum

$$= \text{Force} \times \text{arm}$$
$$= 12\,\text{N} \times (4\cos 30°)\,\text{m}$$
$$= 41\!\cdot\!57\,\text{Nm (C.W.)}.$$

Example (see Fig. 75)

Moment of reaction at 'A' about fulcrum at 'B'

$$= \text{Force} \times \text{arm}$$
$$= R_A\,\text{kN} \times 5\,\text{m} = (R_A \times 5)\,\text{kN m (C.W.)}.$$

Moment of load 'W_1' kN about 'B' $= (W_1 \times 4)\,\text{kN m (A.C.W.)}$.
Moment of load 'W_2' kN about 'B' $= (W_2 \times 2)\,\text{kN m (A.C.W.)}$.

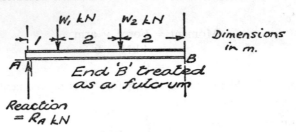

FIG. 75.—APPLICATION IN BEAMS.

Resultant Moment.—When a body is acted upon by several forces, each of which is trying to turn it about a given fulcrum, the net turning effect, i.e. the '*resultant moment*,' is the algebraic sum of all the separate or '*component*' moments. This simply means that we have to take the moment of each force about the fulcrum, add together all those moments which are clockwise and similarly all those which are anticlockwise and finally subtract the smaller of these totals from the larger.

EXAMPLE (i).—*A uniform rule is freely supported at its middle point* (Fig. 76). *Find the manner in which the rule will begin to turn when loaded as indicated in the figure. Calculate the resultant moment about the fulcrum.*

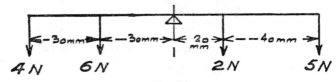

FIG. 76.

A.C.W. Moments	C.W. Moments
4 N × 60 mm = 240 N mm	2 N × 20 mm = 40 N mm
6 N × 30 mm = 180 N mm	5 N × 60 mm = 300 N mm
Total A.C.W. moment = 420 N mm	Total C.W. moment = 340 N mm

The totals indicate that the rule will turn in an anticlockwise manner. The resultant moment about the fulcrum

$$= (420 - 340) \text{ N mm} = 80 \text{ N mm (A.C.W.)}.$$

EXAMPLE (ii).—*The moments disc in Fig. 77 is supported in such a way that there is no tendency to turn when unloaded. Calculate the resultant moment about the fulcrum when the disc is loaded in the manner indicated.*

The tabular method used in the last example is not usually adopted in the solution of 'moments problems.'

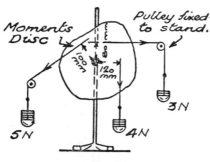

FIG. 77.

Total A.C.W. moment $= 5\,N \times 100\,mm = 500\,N\,mm$
Total C.W. moment $\quad= (3\,N \times 80\,mm) + (4\,N \times 120\,mm)$
$\qquad\qquad\qquad\quad= (240 + 480)\,N\,mm$
$\qquad\qquad\qquad\quad= 720\,N\,mm.$
Resultant moment about fulcrum $= (720 - 500)\,N\,mm$
$\qquad\qquad\qquad\qquad\qquad\quad= 220\,N\,mm\ (C.W.).$

Principle of the Lever

Let us apply the methods used in the last two examples to the case given in Fig. 78.

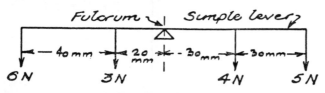

FIG. 78.

Total A.C.W. moment $= [(6 \times 60) + (3 \times 20)]\,N\,mm$
$\qquad\qquad\qquad\quad= (360 \times 60)\,N\,mm = 420\,N\,mm$
Total C.W. moment $\quad= [(4 \times 30) + (5 \times 60)]\,N\,mm$
$\qquad\qquad\qquad\quad= (120 \times 300)\,N\,mm = 420\,N\,mm$
Resultant moment $\quad= (420 - 420)\,N\,mm = zero.$

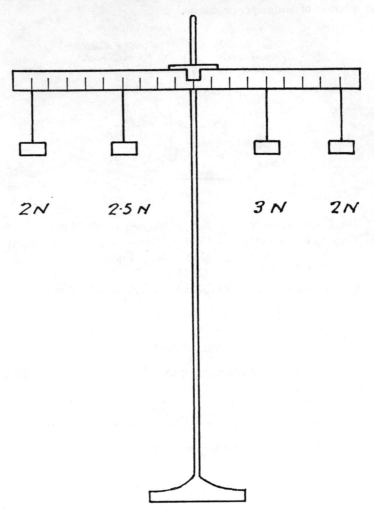

FIG. 79.—PRINCIPLE OF THE LEVER.

The result indicates that the lever will not tend to turn about the fulcrum. This example illustrates a very important law or '*principle*,' which for practical application may be stated as follows:

If a simple lever, capable of turning about a given hinge point or 'fulcrum,' be loaded in such a manner as to keep it in equilibrium, the sum of all the clockwise moments taken about the fulcrum will equal the sum of all the anticlockwise moments.

Just as a body cannot be in positional equilibrium if acted upon by a net resultant force, so it cannot be in rotational equilibrium if a resultant moment act on it.

Experimental Verification of the Principle of the Lever

Fig. 79 shows a typical arrangement of apparatus for verifying the principle in the case of a uniform rule with the fulcrum at the centre. The table on page 51 (Fig. 79A) gives detailed results illustrating the nature of the principle involved.

Expt.	Anticlockwise Moments.				Clockwise Moments.			
	Force.	Arm.	Moment.	Total Moment.	Force.	Arm.	Moment.	Total Moment.
	N	mm	N mm	N mm	N	mm	N mm	N mm
1	0·4	30	12	12	0·2	60	12	12
2	0·5	20	10	10	0·4	25	10	10
3	0·2 0·4	30 40	6 16	— 22	— 0·5	— 44	— 22	— 22
4	— 0·3	— 60	— 18	— 18	0·4 0·2	20 50	8 10	— 18
5 (See Fig. 79)	2·0 2·5	90 40	180 100	— 280	3·0 2·0	40 80	120 160	— 280

FIG. 79A.

EXAMPLES:

In all the following examples, the weight of the 'lever' will be neglected. The method of allowing for the '*self-weight*' of members in such examples will be explained later.

(i) *Calculate the value of 'x' for equilibrium in the case given in Fig. 80 (a).*

$$A.C.W. = C.W.$$
$$(4 \times x) = (2 \times 100)$$
$$4x = 200$$
$$x = 50 \text{ mm.}$$

Care must be exercised to have the '*force*' and '*length*' units consistent throughout. The units may then be omitted from the values in the '*equation of moments.*'

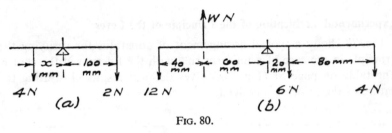

FIG. 80.

(ii) *Find 'W' (Fig. 80 (b)) so that the lever may remain in equilibrium.*

$$A.C.W. = C.W.$$
$$12 \times 100 = (6 \times 20) + (4 \times 100) + (W \times 60)$$
$$1\,200 - 120 - 400 = 60\,W$$
$$60W = 680$$
$$W = 11\tfrac{1}{3} \text{ N.}$$

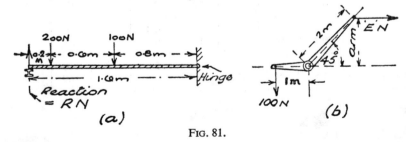

FIG. 81.

(iii) *Fig. 81 (a) shows a trap door freely hinged at one end and resting on a steel beam at the other. Find the reaction exerted by the steel beam for the given loads.*

Taking moments about the hinge:

$$C.W. = A.C.W.$$
$$(R \times 1\cdot6) = (200 \times 1\cdot4) \times (100 \times 0\cdot8)$$
$$1\cdot6R = 280 + 80 = 360$$
$$R = 360/1\cdot6 \text{ N} = 225 \text{ N}.$$

(iv) *Calculate the value of the effort 'E' (Fig. 81 (b))* in order that the applied load of 100 N *may be just balanced.*

In all problems of this type care must be taken to obtain the correct '*arm*' dimension for each moment.

Taking moments about the fulcrum of the cranked lever:

$$C.W. = A.C.W.$$
$$E \times 2 \sin 45° = 100 \times 1$$
$$E \times (2 \times 0\cdot7071) = 100$$
$$E = 100/1\cdot414 = 70\cdot71 \text{ N}.$$

If the effort had been applied at right angles to the lever, the '*moments equation*' would have been:

$$E \times 2 = 100 \times 1$$
$$E = 50 \text{ N}.$$

Readers unfamiliar with trigonometry may obtain the necessary '*moment arms*' by direct measurement off a scale drawing of the given example.

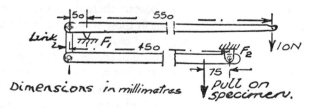

FIG. 82.—COMPOUND LEVER.

(v) *Fig. 82 gives details of a compound lever used in testing specimens in a cement briquette testing machine. Calculate the pull on the specimen corresponding to a test load of 10 N at the end of the lever arm.*

Step 1.—Find the pull in the link by taking moments about fulcrum 'F_1.'

Let 'L' N = pull in link.

$$L \times 50 = 10 \times 550$$
$$\therefore \quad L = 110 \text{ N}.$$

Step 2.—Take moments about fulcrum 'F_2,' using the value of 'L' found in Step 1.

Let 'P' N = pull exerted on specimen.

$$P \times 75 = L \times 450 = 110 \times 450$$

$$\therefore \quad P = \frac{110 \times 450}{75} = 660 \text{ N}.$$

Calculation of Beam Support Reactions

An important application of 'moments' is in the calculation of the '*reactions*' which walls, columns, etc., exert on the beams they support—often the first step in the design of the beam.

Notation.—The evaluation of beam reactions occurs so frequently in structural calculations that it is necessary to adopt a definite system of symbols for the reaction forces.

If 'A' and 'B' be the left-end and right-end reaction points respectively of a beam 'AB,' the left-end reaction is conveniently represented by the symbol 'R_A.' Similarly, 'R_B' would stand for the '*reaction at B.*'

If, as in the case of continuous beams, there are, say, three support points 'A,' 'B' and 'C,' the reactions at these respective points would be 'R_A,' 'R_B' and 'R_C.' Sometimes in the case of loaded frameworks the letters 'A,' 'B,' etc., are used for other purposes. It may be more convenient in these cases to denote the left-end reaction by the symbol 'R_L,' and to use 'R_R' for the right-end reaction. This method is recommended by certain regulations as that to be adopted in all cases of simple beams, etc.

EXAMPLE.—*Calculate the reactions at 'A' and 'B' respectively for the simply supported beam given in Fig.* 83.

We first regard the beam as a simple lever with the fulcrum at 'B' and regard 'R_A' (the reaction at 'A') as an ordinary force acting upwards at 'A.'

Clockwise moment of 'R_A' about 'B' = ($R_A \times 3$) kN m.

The sum of the anticlockwise moments of the loads on the beam about 'B' = $[(6 \times 1\cdot8) + (3 \times 0\cdot6)]$ kN m.

Equating clockwise and anticlockwise moments:

$$R_A \times 3 = (6 \times 1\cdot8) + (3 \times 0\cdot6)$$
$$3R_A = 10\cdot8 + 1\cdot8 = 12\cdot6$$
$$R_A = 4\cdot2 \text{ kN.}$$

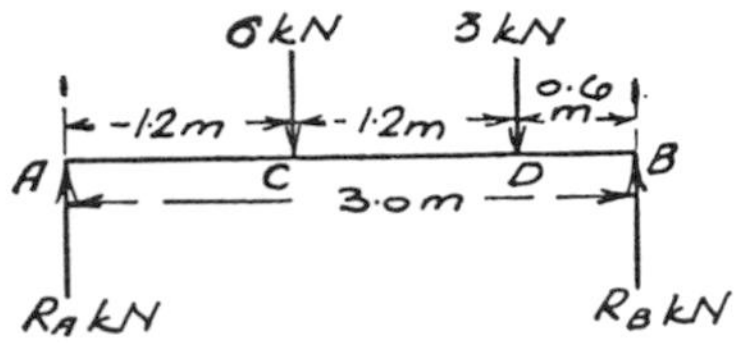

FIG. 83.

The beam is now assumed to have its fulcrum at 'A,' so that 'R_A,' which passes through the fulcrum, will have no moment and therefore will be eliminated from the 'moments equation.'

Anticlockwise moment of 'R_B' about 'A' = $(R_B \times 3)$ kN m. The total clockwise moment of the beam loads about 'A' = $[(3 \times 2\cdot4) + (6 \times 1\cdot2)]$ kN m.

$$\therefore \quad R_B \times 3 = (3 \times 2\cdot4) + (6 \times 1\cdot2)$$
$$3R_B = 7\cdot2 + 7\cdot2 = 14\cdot4$$
$$R_B = 4\cdot8 \text{ kN.}$$

A check on the numerical working is provided by the addition of the two reaction values. Their sum should equal the total load on the beam.

$$R_A + R_B = 4\cdot2 \text{ kN} + 4\cdot8 \text{ kN} = 9 \text{ kN.}$$
Sum of loads $= 6 \text{ kN} + 3 \text{ kN} = 9 \text{ kN.}$

The reader is advised to work out the two reactions independently, as in this example, in order to take advantage of the check on the numerical working. The procedure of obtaining one reaction by subtracting the other from the 'load total' is not recommended to beginners. The method of allowing for the self-weight of a beam is given later.

Fig. 84 shows a form of apparatus devised to demonstrate 'beam support reactions.'

If a single central load be applied, each reaction will equal half the load value.

If a symmetrical system of loading be used, each reaction will equal half the total load value.

The compression spring balances, at the ends of the model beam, record the reactions in any given experimental test.

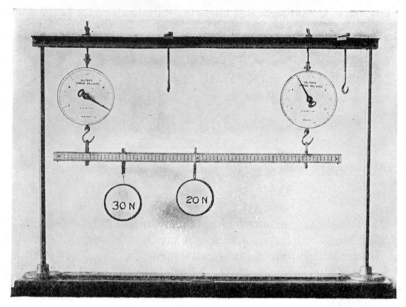

Fig. 84.

It will be noted from the readings of the spring balances that the left-end and right-end reactions are respectively 35N and 15N. Checking these results by the calculation method we have:

$$R_L \times 60 = (20 \times 30) + (30 \times 50) = 600 + 1\,500 = 2\,100$$
$$\therefore R_L = 35N.$$
$$R_R \times 60 = (30 \times 10) + (20 \times 30) = 300 + 600 = 900$$
$$\therefore R_R = 15N.$$

Uniformly Distributed Loads.—Loads are often described in structural calculations as '*so much per unit run,*' e.g. '*2 kN per metre run of beam,*' or more simply '*2 kN per metre.*' In all problems of the present type, i.e. in which the equilibrium of a body is being considered, we may replace such a load system by one concentrated load of equal value at the middle point of the load distribution. The self-weight of a beam may usually be regarded as a uniformly distributed load.

EXAMPLE.—*A 254 mm × 146 mm × 31 kg U.B. is used to carry the loads given in Fig. 85. Calculate the support reactions.*

The total distributed load per metre run = 31 kg (say 300 N) + 7 200 N = 7 500 N or 7·5 kN.

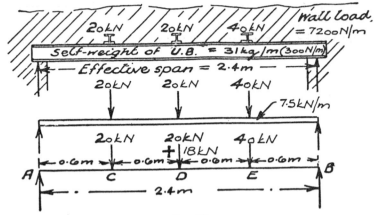

FIG. 85.

The total uniformly distributed load on the beam = 7·5 kN per metre × 2·4 m = 18 kN. We may regard this load as being (*for the present purpose*) concentrated at mid-point of beam.

Moments about B:

$$R_A \times 2·4 = (20 \times 1·8) + (20 \times 1·2) + (40 \times 0·6)$$
$$+ [18 \ (\textit{i.e. unif. dist. load}) \times 1·2]$$
$$= 36 + 24 + 24 + 21·6$$
$$= 105·6$$
$$R_A = 105·6/2·4 \ kN = 44 \ kN.$$

Moments about A:

$$R_B \times 2·4 = (40 \times 1·8) + (20 \times 1·2) + (20 \times 0·6) + (18 \times 1·2)$$
$$= 72 + 24 + 12 + 21·6$$
$$= 129·6$$
$$R_B = 129·6/2·4 = 54 \ kN.$$

If the uniformly distributed load runs for the whole beam length, as in this case, we could find the reactions, in the first instance, for the concentrated loads alone and then add to each reaction half the total uniform load.

EXAMPLE.—*Fig. 86 shows a roof truss carrying an unsymmetrical load system. Calculate the reactions at 'A' and 'B.'*

As has been emphasised previously, the 'arm' to be used in evaluating a moment is necessarily at right angles to the line of action of the force.

We may therefore reduce this example to a simple beam problem as illustrated in Fig. 86.

There will be no need, ordinarily, to actually draw the equivalent beam in such examples.

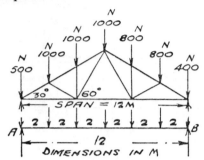

FIG. 86.—REACTIONS FOR A ROOF TRUSS.

Moments about B:

$$R_A \times 12 = (500 \times 12) + (1\,000 \times 10) + (1\,000 \times 8) + (1\,000 \\ \times 6) + (800 \times 4) + (800 \times 2) + (400 \times 0)$$
$$12R_A = 6\,000 + 10\,000 + 8\,000 + 6\,000 + 3\,200 + 1\,600$$
$$= 34\,800$$
$$R_A = 2\,900 \text{ N}.$$

Moments about A:

$$R_B \times 12 = (400 \times 12) + (800 \times 10) + (800 \times 8) + (1\,000 \\ \times 6) + (1\,000 \times 4) + (1\,000 \times 2) + (500 \times 0)$$
$$= 4\,800 + 8\,000 + 6\,400 + 6\,000 + 4\,000 + 2\,000$$
$$12R_B = 31\,200$$
$$R_B = 2\,600 \text{ N}.$$
$$R_A + R_B = 5\,500 \text{ N} = \text{total load on truss.}$$

The self-weight of a truss is proportioned between the joint loads. (See Chapter VIII.)

EXAMPLE.—*Fig. 87 illustrates a beam overhanging its supports. Calculate the support reactions for the loads given.*

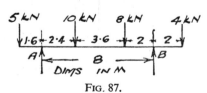

FIG. 87.

In this example we do not take moments about the ends of the beam but about the support points 'A' and 'B' respectively. Considering the fulcrum to be at 'B' it will be observed that the '4 kN load' has a clockwise moment, i.e. the same type of moment as 'R_A' has. These two moments will therefore have to be added.

Moments about B:
$$(R_A \times 8) + (4 \times 2) = (5 \times 9 \cdot 6) + (10 \times 5 \cdot 6) + (8 \times 2)$$
$$8R_A + 8 = 48 + 56 + 16$$
$$8R_A + 8 = 120$$
$$8R_A = 120 - 8 = 112$$
$$R_A = 14 \text{ kN.}$$

Moments about A:
$$(R_B \times 8) + (5 \times 1 \cdot 6) = (4 \times 10) + (8 \times 6) + (10 \times 2 \cdot 4)$$
$$8R_B + 8 = 40 + 48 + 24$$
$$8R_B + 8 = 112$$
$$8R_B = 112 - 8 = 104$$
$$R_B = 13 \text{ kN.}$$
$$R_A + R_B = (14 + 13) \text{ kN} = 27 \text{ kN} = \text{total load on beam.}$$

EXAMPLE.—*Find the reactions at the supports in the case of the beam shown in Fig.* 88.

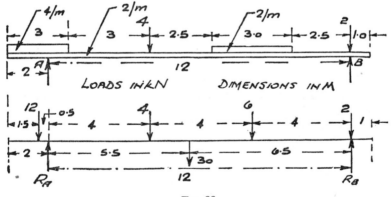

FIG. 88.

Using the rule for replacing uniformly distributed loads by concentrated loads of equal magnitude *for the purpose of beam reaction calculations*, the original load system is reduced to a much simpler form as shown in the lower diagram. When the method is thoroughly understood there should be no actual need to draw a separate equivalent beam diagram.

The '4 kN per metre load,' 3 m long, is partly *to the left* and partly *to the right* of the support 'A.' When replaced by an equivalent load of 12 kN, the load appears at 0·5 m *to the left* of 'A.' There need be no hesitancy in accepting this. If we treated this load as consisting of two separate portions, one in and one out of the centre span, the net result would be exactly as indicated above.

Moments about B:

$$R_A \times 12 = (12 \times 12·5) + (4 \times 8) + (30 \times 6·5) + (6 \times 4)$$
$$+ (2 \times 0)$$
$$= 150 + 32 + 195 + 24$$
$$= 401$$
$$R_A = 33·42 \text{ kN.}$$

Moments about A:

$$(R_B \times 12) + (12 \times 0·5) = (2 \times 12) + (6 \times 8) + (30 \times 5·5)$$
$$+ (4 \times 4)$$
$$12R_B + 6 = 24 + 48 + 165 + 16$$
$$12R_B + 6 = 253$$
$$12R_B = 253 - 6 = 247$$
$$R_B = 20·58 \text{ kN.}$$

$$R_A + R_B = (33·42 + 20·58) \text{ kN} = 54 \text{ kN} = \text{total load on beam.}$$

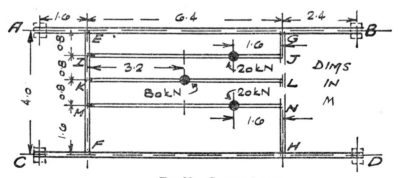

FIG. 89.—COLUMN LOADS.

EXAMPLE.—*Calculate the column loads at 'A,' 'B,' 'C' and 'D' (due to the concentrated loads indicated) in the example of the steel-framed floor given in Fig.* 89.

In such problems the calculated reactions for the secondary beams become the load values for the beams which carry them.

Beam M N: $R_M \times 6\cdot4 = 20 \times 1\cdot6$ $\therefore$ $R_M = 32/6\cdot4\,\text{kN} = 5\,\text{kN}$.
 $R_N \times 6\cdot4 = 20 \times 4\cdot8$ $\therefore$ $R_N = 96/6\cdot4\,\text{kN} = 15\,\text{kN}$.
Beam K L: $R_K = R_L = 40\,\text{kN}$.
Beam I J: $R_I = R_M = 5\,\text{kN}$ (under similar conditions).
 $R_J = R_N = 15\,\text{kN}$ (under similar conditions).

Beam 'EF' takes the left-end reactions from the beams above (see Fig. 90).

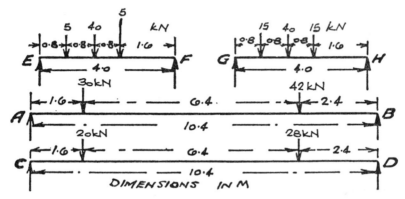

Fig. 90.—Beam Load Diagrams (see Fig. 89).

Beam E F: $R_E \times 4 = (5 \times 3\cdot2) + (40 \times 2\cdot4) + (5 \times 1\cdot6)$
$ = 16 + 96 + 8 = 120$
$ R_E = 30\,\text{kN}.$
$ R_F \times 4 = (5 \times 2\cdot4) + (40 \times 1\cdot6) + (5 \times 0\cdot8)$
$ = 12 + 64 + 4 = 80$
$ R_F = 20\,\text{kN}.$

Beam 'GH' takes the right-end reactions of the beams it supports.

Beam G H: $R_G \times 4 = (15 \times 3\cdot2) + (40 \times 2\cdot4) + (15 \times 1\cdot6)$
$ = 48 + 96 + 24 = 168$
$ R_G = 42\,\text{kN}.$
$ R_H \times 4 = (15 \times 0\cdot8) + (40 \times 1\cdot6) + (15 \times 2\cdot4)$
$ = 12 + 64 + 36 = 112$
$ R_H = 28\,\text{kN}.$

Beams 'AB' and 'CD,' which are supported by the columns, carry the reaction loads transmitted by beams 'EF' and 'GH.'

Beam A B:
$$R_A \times 10\cdot4 = (30 \times 8\cdot8) + (42 \times 2\cdot4)$$
$$= 264 + 100\cdot8 = 364\cdot8$$
$$R_A = 35\cdot1 \text{ kN.}$$
$$R_B \times 10\cdot4 = (42 \times 8) + (30 \times 1\cdot6)$$
$$= 336 + 48 = 384$$
$$R_B = 36\cdot9 \text{ kN.}$$

Beam C D:
$$R_C \times 10\cdot4 = (20 \times 8\cdot8) + (28 \times 2\cdot4)$$
$$= 176 + 67\cdot2 = 243\cdot2$$
$$R_C = 23\cdot4 \text{ kN.}$$
$$R_D \times 10\cdot4 = (28 \times 8) + (20 \times 1\cdot6)$$
$$= 224 + 32 = 256$$
$$R_D = 24\cdot6 \text{ kN.}$$

Summary:

$$R_A = 35\cdot1 \text{ kN}$$
$$R_B = 36\cdot9 \text{ kN}$$
$$R_C = 23\cdot4 \text{ kN}$$
$$R_D = \underline{24\cdot6} \text{ kN}$$
$$\text{Total} = \overline{120\cdot0} \text{ kN}$$

Total load on frame = 120 kN.

Further examples on 'reactions' appear throughout the book.

Moment of a Resultant Force

Theorem: The resultant moment of a system of forces about any given point in the plane of the forces is equal to the moment of the resultant of the system about the same point.

In Fig. 91 there is shown a system of forces 'F₁,' 'F₂,' etc., with 'R,' the resultant of the system, represented in its correct

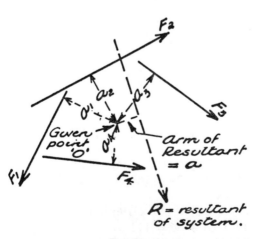

FIG. 91.

position in space. The point 'O' is any selected point in the plane of the forces. In this case 'R' has a clockwise moment about the point 'O.' Regarding clockwise moments as *'positive,'* the theorem expressed in the form of an equation of moments becomes:

$$R \times a = (F_2 \times a_2) + (F_3 \times a_3) - (F_1 \times a_1) - (F_4 \times a_4).$$

This theorem is of great assistance in the solution of structural problems.

EXAMPLE.—*Verify the theorem above for the case of two intersecting forces, 7 N and 8 N respectively, whose lines of action contain an angle of 60° (see Fig. 92).*

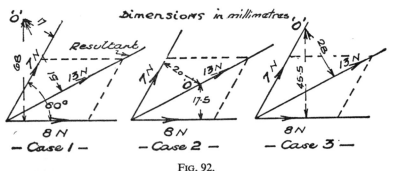

FIG. 92.

The reader should carry through this exercise to a fairly large scale. In addition to verifying the theorem, it provides excellent practice in *'taking moments.'*

Three positions are taken for the point 'O.' The parallelogram of forces gives the resultant in its correct position in space. The 'arms' of the various moments are measured in actual millimetres on the drawing paper. The following results illustrate the significance of the theorem.

Case 1.—Moments about 'O'

7 N force: Force × arm = 7 N × 17 mm = 119 N mm A.C.W.
8 N force: Force × arm = 8 N × 68 mm = 544 N mm A.C.W.

                      Resultant moment (by addition) = 663 N mm A.C.W.

13 N force (Resultant):

         Force × arm = 13 N × 51 mm = 663 N mm A.C.W.
∴ **Sum of component moments = moment of resultant.**

Case 2.—Moments about 'O'

7 *N force:* Force × arm = 7 N × 20 mm = 140 N mm C.W.
8 *N force:* Force × arm = 8 N × 17·5 mm = 140 N mm A.C.W.

Resultant moment (by algebraic addition) = 0

13 *N force (Resultant):*

Force × arm = 13 N × 0 mm = 0 N mm.
∴ **Resultant moment = moment of resultant.**

Case 3.—Moments about 'O'

7 *N force:* Force × arm = 7 N × 0 mm = 0 N mm
8 *N force:* Force × arm = 8 N × 45·5 mm = 364 N mm A.C.W.

Resultant moment (by addition) = 364 N mm A.C.W.

13 *N force (Resultant):*

Force × arm = 13 N × 28 mm = 364 N mm A.C.W.
∴ **Resultant moment = moment of resultant.**

In each case, therefore, the theorem has been verified. Exercise 12 at the end of the chapter involves the verification of the theorem for a system of four concurrent forces.

The theorem is extensively employed in Chapter VI.

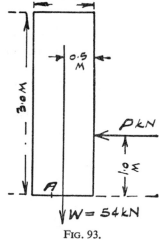

FIG. 93.

EXAMPLE.—*Fig. 93 shows a masonry pier weighing 54 kN. A horizontal force 'P kN' acts at 1·0 m from the base of the pier. It is desired that the resultant of 'P' and the weight (54 kN) shall pass through a given point 'A,' 0·167 m from the centre of the base. Find the corresponding value of 'P.'*

If the resultant of 'P' and 'W' passes through 'A,' it will have no moment about 'A.' Hence by the theorem above, the resultant moment of 'P' and 'W' will be zero about 'A.' i.e. the moment of 'P' must be *equal and opposite to that of 'W.'*

∴ P × 1 (A.C.W.) = W × 0·167 (C.W.)
P × 1 = 54 × 0·167
∴ P = 9 kN.

This example illustrates a typical 'retaining wall problem.' Retaining walls are considered in detail in Chapter XV.

EXERCISES 4

(1) Write down the moment of the given force about the fulcrum 'O' in each of the cases shown in Fig. 94.

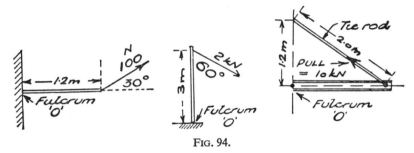

FIG. 94.

(2) A nut requires an applied moment of 15 N m in order to turn it. Calculate the necessary effort if applied as in (a) and (b) respectively, in Fig. 95.

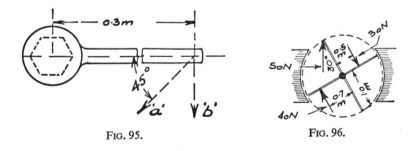

FIG. 95.　　　　FIG. 96.

(3) A swing door is pushed by three people in the manner indicated in the diagram shown in Fig. 96. Calculate the resultant turning moment on the door.

(4) Fig. 97 gives a diagram of the lever arm of a timber beam-testing machine. Assuming the lever arm to be balanced by the compensating weight when the travelling weight is at the zero of the scale, calculate the length in millimetres of the scale graduations, which represent 1 kN pull on specimen.

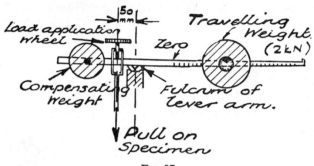

FIG. 97.

(5) A plank, resting on two scaffold poles, is lashed to one and is free at the other (Fig. 98). Calculate the pull on the lashing when the plank carries the loads indicated. Neglect self-weight of plank.

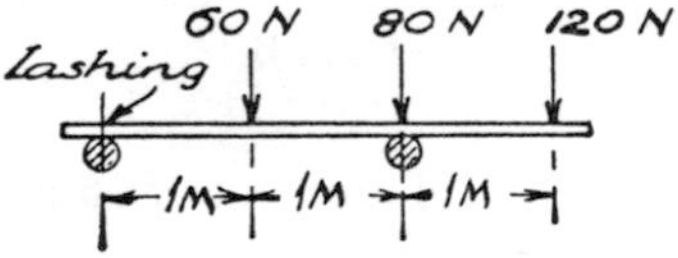

FIG. 98.

(6) Fig. 99 shows a north-light roof truss carrying wind loads. Find the value of the vertical reaction at the left end of the truss. (Take moments about right end of truss.)

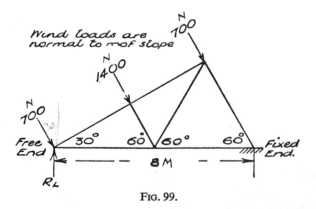

FIG. 99.

(7) Calculate the support reactions (for the loads indicated) for each of the beams shown in Fig. 100.

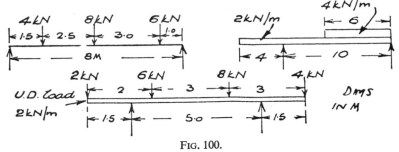

FIG. 100.

(8) Find the support reactions for the steel beam given in Fig. 101. The density of the brickwork is 20 kN/m³. The wall is 250 mm thick. Neglect the self-weight of beam.

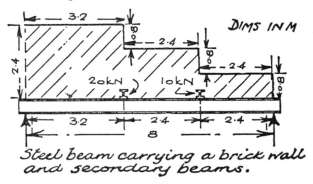

Steel beam carrying a brick wall and secondary beams.

FIG. 101.

(9) A roof truss carries rafter and tie loads as shown in Fig. 102. Calculate the reactions at the supports.

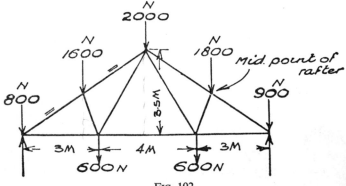

FIG. 102.

(10) Fig. 103 shows a girder 'AB' along which a weight of 20 kN is able to travel. The subsoil under the concrete at end 'A' must not (due to the travelling weight alone) be subjected to a higher pressure than

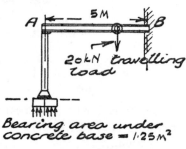

FIG. 103.

12 kN/m². Calculate the minimum permissible distance between the weight and the centre line of the column.

(11) Find the resultant moment *about the point 'P'* of all the forces

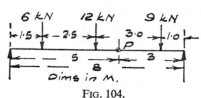

FIG. 104.

which act on the beam (Fig. 104) to the left of 'P.' Similarly find the resultant moment of all the forces which act to the right of 'P.' In which respect do these two resultant moments (i) agree, (ii) differ?

Repeat the exercise, taking any other point on the beam.

(See page 197 for an application of this exercise.)

(12) Show that the resultant of the co-planar system of concurrent forces given in Fig. 105 will have the same moment about point 'A' as

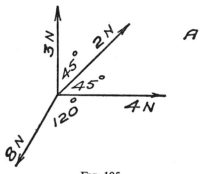

FIG. 105.

the algebraic sum of the moments of the four forces (taken separately) about point 'A.' Select a suitable position for point 'A,' in the approximate position shown in figure. Repeat, taking point 'A' on one of the forces.

(13) Calculate the load carried by each of the columns at 'A,' 'B,' 'C,' and 'D' respectively (Fig. 106), due to the given point load of 60 kN on the secondary beam.

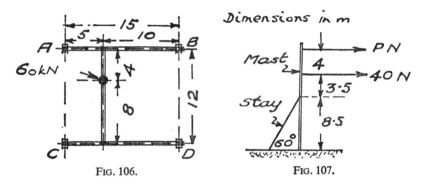

FIG. 106.   FIG. 107.

(14) Fig. 107 shows a mast supporting two horizontal pulls. The mast is kept vertical by a back stay. Assuming the maximum safe load for the stay to be 400 N, find the highest permissible value of 'P.'

The bottom of the mast may be considered to be a pin-jointed connection.

CHAPTER V

PARALLEL FORCE SYSTEMS. COUPLES. GENERAL CONDITIONS OF EQUILIBRIUM

PARALLEL force systems may be divided into two classes:

(i) *Like parallel systems*, in which all the forces act in the same direction.

(ii) *Unlike parallel systems*, in which some of the forces act in the opposite direction to the others. (See Fig. 108.)

Like   *Unlike*

FIG. 108.—PARALLEL FORCE SYSTEMS.

Like parallel systems are the more common in structural problems, but there is one case of an unlike parallel system which is of especial importance. The theorem given on page 62 with respect to resultant moments is also true for parallel force systems. This may be verified by means of the experimental apparatus shown in Fig. 109. (See also Fig. 136.)

If we reverse the direction of the equilibrant of the system, as found in the experiment, we have a parallel force system with its resultant. The resultant will be found to be equal in magnitude to the sum of the forces, act in a direction parallel to the forces, and have its line of action

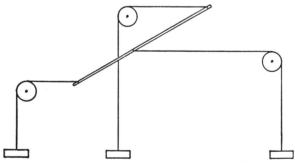

FIG. 109.—PARALLEL FORCES.

in such a position as to comply with the provisions of the theorem referred to.

Like Parallel Systems

Two Like Parallel Forces

EXAMPLE.—*Find the magnitude, direction and position of the resultant of the two like parallel forces given in Fig.* 110.

Magnitude of resultant

$$= \text{sum of forces}$$
$$= 12 \text{ N} + 8 \text{ N} = 20 \text{ N}.$$

Direction: Vertically downwards.

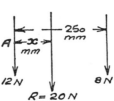

FIG. 110.

Position: Let 'x' mm = distance from point 'A' (on the line of action of the '12 N' force).

Moments about 'A':

Moment of resultant = $20 \text{ N} \times x \text{ mm} = 20x$ N mm.
Sum of moments of components = $(12 \times 0) + (8 \text{ N} \times 250 \text{ mm})$
$$= (0 + 2\,000) \text{ N mm} = 2\,000 \text{ N mm}.$$
$$\therefore \quad 20x = 2\,000$$
$$x = 100 \text{ mm}.$$

It will be noted that the line of action of the resultant force divides the distance between the forces *inversely* as the forces.

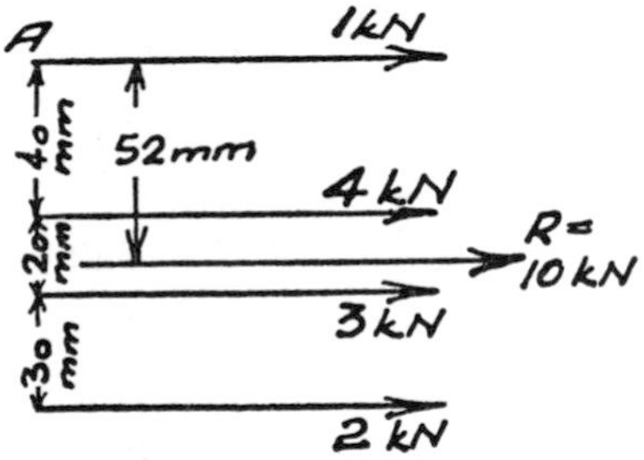

Thus $\dfrac{100 \text{ mm}}{150 \text{ mm}} = \dfrac{8 \text{ N}}{12 \text{ N}}$.

This rule is used in the graphical problem explained on page 105.

Several Like Parallel Forces

EXAMPLE.—*Find the resultant of the like parallel system shown in Fig.* 111.

FIG. 111.

Magnitude of resultant = $(2 + 3 + 4 + 1)$ kN
$$= 10 \text{ kN}.$$

Direction: The resultant will act horizontally towards the right.

Position: Let 'x' mm = distance from point 'A' on the '1 kN' force.

Moments about 'A':

$$10 \times x = (2 \times 90) + (3 \times 60) + (4 \times 40)$$
$$= 180 + 180 + 160 = 520$$
$$x = 52 \text{ mm.}$$

Note that in writing down the equations we are not equating clockwise to anticlockwise moments, as in the principle of the lever, but merely expressing the equality of resultant moments, obtained in two different ways.

EXAMPLE.—*Fig.* 112 *shows a roof truss carrying wind loads on the windward side. Find the resultant wind thrust on the roof slope.*

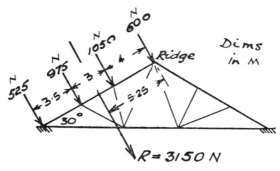

FIG. 112.

Magnitude of resultant = (525 + 975 + 1 050 + 600) N
= 3 150 N.

Direction: Normal to roof slope, acting inwards.

Position: Let 'x' m = distance, from the ridge, at which the resultant acts.

Taking moments about the ridge:

$$3\,150x = (600 \times 0) + (1\,050 \times 4) + (975 \times 7) + (525 \times 10{\cdot}5)$$
$$= 0 + 4\,200 + 6\,825 + 5\,512{\cdot}5$$
$$= 16\,537{\cdot}5$$

$$x = \frac{16\,537{\cdot}5}{3\,150} \text{ m} = 5{\cdot}25 \text{ m.}$$

Unlike Parallel Systems

Two Unlike Parallel Forces

EXAMPLE.—*A dock gate,with water on both sides, is subject to the two horizontal thrusts shown in Fig. 113. Find the resultant water thrust on the gate.*

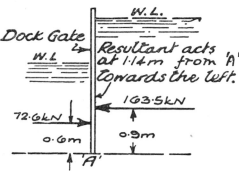

FIG. 113.

Magnitude of resultant = the 'algebraic' sum of the forces
$$= (163 \cdot 5 - 72 \cdot 6) \text{ kN}$$
$$= 90 \cdot 9 \text{ kN.}$$

Direction: The resultant acts towards the left, at right angles to the gate.

Position: Let 'x' m = distance from bottom of gate ('A').
$$90 \cdot 9 \, x = [(163 \cdot 5 \times 0 \cdot 9) - (72 \cdot 6 \times 0 \cdot 6)]$$
$$\text{A.C.W.} = \quad \text{A.C.W.} \quad - \quad \text{C.W.}$$

Note that the kind of moment the resultant produces (A.C.W. in this case) is given the positive sign.

$$90 \cdot 9 \, x = (147 \cdot 15 - 43 \cdot 56) = 103 \cdot 59$$
$$x = 1 \cdot 14 \text{ m.}$$

In the case of two unlike and unequal parallel forces, the resultant does not act between the forces, but beyond the greater force.

Several Unlike Parallel Forces

EXAMPLE.—*Find the resultant of the unlike parallel system given in Fig. 114.*

Magnitude of resultant = algebraic sum of forces
$$= (5 + 6 - 2 - 4 - 1)\,\text{N}$$
$$= 4\,\text{N}.$$

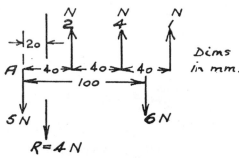

FIG. 114.

Direction: Vertically downwards.

Position: Let 'x' mm = distance from the point 'A' (on the '5 N' force) measured towards the right.

$$(4 \times x) = (6 \times 100) - (2 \times 40) - (4 \times 80) - (1 \times 120)$$
$$= 600 - 80 - 320 - 120$$
$$4\,x = 80$$
$$x = 20\ \text{mm}.$$

Couples

If, in such an example as the one solved above, *the sum of the forces acting upwards equalled the sum of those acting downwards,* two possibilities would ensue:

(i) If the upward and downward resultant forces acted in the same vertical line, the system would be in equilibrium.

(ii) If they acted out of line the net effect of the system would be one of *rotation.* A system of forces, such as that suggested in (ii), is known as a '*couple.*'

Definition: A couple consists of two equal, unlike parallel forces, acting out of line.

The effect of a couple on a body is one of pure rotation.

Moment of a Couple.—The 'moment of a couple' measures the actual turning tendency the couple has on the body to which it is applied. It is obtained by multiplying one of the forces by the perpendicular distance between them (Fig. 115).

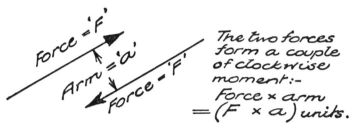

FIG. 115.

Moment of couple = Force × arm of couple.

A couple has a '*clockwise*' or an '*anticlockwise*' moment according to the manner in which it tends to cause rotation (Fig. 116).

FIG. 116.—TYPES OF COUPLES.

Practical Examples of Couples.—Whenever we measure the turning effect a force has on a body, as in the simple lever problems in Chapter IV, we are really finding the moment of a couple. The reaction at the point of support of the body, i.e. the fulcrum, provides the 'equal and unlike parallel force' to form the couple with the net applied force.

Example (i).—Fig. 117 shows a force applied to the handle of a door. The hinge provides a reaction which forms a couple with the applied force. The door opens under the moment of this couple.

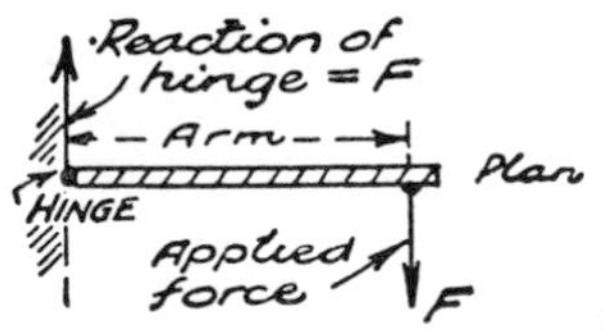

Example (ii).—A very important application of couples occurs in beam problems (see Chapter XII). When a beam deflects, the fibres of the beam material become stretched or compressed according to their position in the beam. These fibres act like tiny springs, and they resist deformation. At a beam section, therefore, we have, acting across the section, a series of thrusts and pulls. The resultant thrust and the resultant pull form a couple which resists the bending of the beam at the section (see Fig. 118). The moment of this couple is termed the '*moment of resistance*' at the section (see page 239).

FIG. 117.

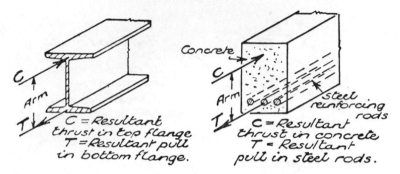

FIG. 118.—COUPLES IN BEAMS.

Important Properties of Couples.—There are a number of 'theorems' enunciated with respect to the subject of '*couples*.' These have applications in beam problems and in column problems (e.g. in '*eccentric loading*'), and will be referred to, as necessary, later. The reader will at this stage be able to appreciate the following facts about the nature of 'couples.'

(i) A couple cannot be balanced by a single force. To bring to equilibrium a body acted upon by a couple, it is necessary to apply to the body another couple of *equal* and *opposite* moment (see page 239).

(ii) A couple has the same moment about any point in the plane of the forces. The value of the moment is that of the couple itself.

Consider the cases shown in Fig. 119. The resultant turning effect or 'moment' about the point 'O' in each case equals the algebraic sum of the two moments of the respective forces 'F' forming the couple.

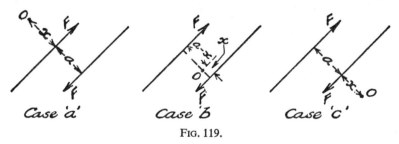

FIG. 119.

Case (*a*). Resultant moment about 'O' = $F \times (a + x) - F \times x$
$$= Fa + Fx - Fx$$
$$= Fa \text{ (clockwise)}$$
$$= \textit{moment of couple.}$$

Case (*b*). Resultant moment about 'O' = Fx + F(a − x)
$$= Fx + Fa − Fx$$
$$= Fa \text{ (clockwise)}$$
$$= \textit{moment of couple.}$$

Case (*c*). Resultant moment about 'O' = F(a + x) − Fx
$$= Fa + Fx − Fx$$
$$= Fa \text{ (clockwise)}$$
$$= \textit{moment of couple.}$$

An important application of this theorem occurs later in this chapter in connection with the general conditions of equilibrium.

EXAMPLE.—*A vertical rod is hinged at the bottom (Fig. 120). Calculate the turning moment on the rod (about the hinge) for the given system of loads.*

Apparently the dimensions given are insufficient as the position of the hinge is not defined. The force system, however, reduces to a couple, and all we require to find is the moment of the couple (*theorem* (ii), page 76).

To illustrate the principles of parallel force systems this problem will be solved in two ways.

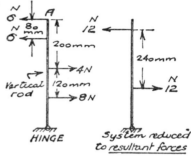

FIG. 120.

(i) *By finding the 'force' and 'arm' of the couple to which the system reduces.*

The resultant force to the left is (6 + 6) N = 12 N, acting in position shown.

The resultant force to the right is (8 + 4) N = 12 N, acting in position shown.

The arm of the couple is the distance between the lines of action of these two forces = 240 mm. (The reader should check this figure).

$$\text{Moment of couple} = \text{Force} \times \text{arm}$$
$$= 12 \text{ N} \times 240 \text{ mm}$$
$$= 2\,880 \text{ N mm (A.C.W.).}$$

(ii) *By taking moments of the forces in the system about 'any' convenient point.*

The point 'A' (see Fig. 120) is chosen.

Resultant moment about 'A'

$$= [(4 \times 200) + (8 \times 320) - (6 \times 80)] \text{ N mm}$$
$$= (800 + 2\,560 - 480) \text{ N mm}$$
$$= 2\,880 \text{ N mm (A.C.W.).}$$

The latter method is clearly the easier to employ.

The moment of the couple is therefore 2 880 N mm (A.C.W.), and this is the required turning effect of the force system about the hinge.

EXAMPLE (ii).—*A rigid T-shaped member is hinged at the point 'O'* (Fig. 121). *Assuming a load of* 10 *kN to be applied as shown, calculate the necessary values of the forces 'C' and 'T' respectively in order to maintain the number in equilibrium.* (*Neglect the self-weight of the lever.*)

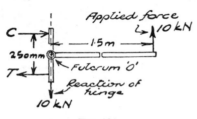

FIG. 121.

For vertical equilibrium it is necessary for the hinge at 'O' to exert a force of '10 kN' on the member, vertically downwards. The two forces of '10 kN' constitute a couple of anticlockwise moment: 'force × arm' = 10 kN × 1·5 m = 15 kN m.

To maintain equilibrium, 'C' and 'T' must form a couple of clockwise moment = 15 kN m.

The forces 'C' and 'T' are therefore equal and

$$C \text{ (or T)} \times 250 \text{ mm} = (15 \times 1\,000) \text{ kN mm}$$

$$\therefore \quad C = T = \frac{15 \times 1\,000}{250} \text{ kN} = 60 \text{ kN.}$$

The reader should compare this problem with the beam problem discussed on page 242.

Conditions of Equilibrium for a Non-concurrent Force System

Consider the force system shown in Fig. 122. If we resolve both forces, horizontally, by the cosine rule, we get 'F cos θ' acting towards the left and 'F cos θ' acting towards the right. There is no net horizontal force, therefore the system is in '*horizontal equilibrium.*'

Resolving vertically we get 'F sin θ' up and 'F sin θ' down. The system is therefore also in '*vertical equilibrium.*'

We can see that the given system is not in '*complete equilibrium*' as it forms a couple, and would cause rotation.

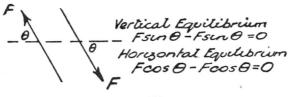

Vertical Equilibrium
$F\sin\theta - F\sin\theta = 0$
Horizontal Equilibrium
$F\cos\theta - F\cos\theta = 0$

FIG. 122.

If we were given a more complicated system of non-concurrent forces we could not decide so easily whether it reduced to a 'resultant force,' or to a 'couple' or was, in fact, 'in equilibrium.' To show that a non-concurrent system of forces is in equilibrium we have to resolve the forces into their vertical and horizontal components and show that the total horizontal component equals zero and that the total vertical component equals zero. If such were the case the system would not reduce to a single resultant force. But we still have to show that the system does not reduce to a couple. To do this, we may select *any* convenient point in the plane of the forces, and having taken the moments of all the forces about this point, verify that the total clockwise moment equals the total anticlockwise moment.

The position of the point about which moments are taken is immaterial because a couple has the same moment about any point in the plane of the forces. If, therefore, we get zero moment about any convenient selected point, there is no possibility of the system reducing to a couple.

For a co-planar system of forces to maintain a body in equilibrium there must be (i) no resultant force acting on the body and (ii) no resultant moment about any point in the plane of the forces.

Experimental Verification of the Laws of Equilibrium

A suitable form of apparatus is shown in Fig. 123. The weights suspended over the pulleys should be fairly large in order that the weight of the thin disc of cardboard may be neglected. A space diagram of four non-concurrent forces in equilibrium is obtained. Each force is resolved into its respective vertical and horizontal components. These are then tabulated in the manner indicated in Fig. 123A. The convention of signs adopted for force direction is indicated.

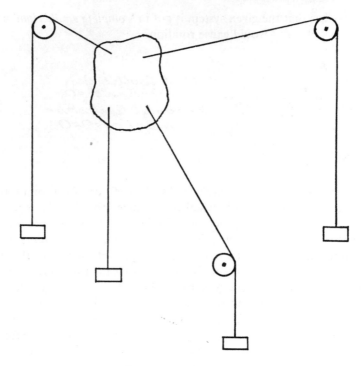

SPACE DIAGRAM.

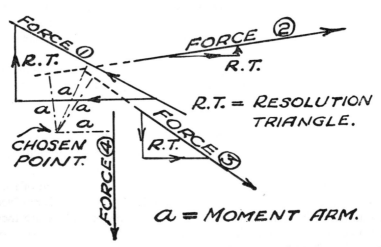

FIG. 123 (see also Fig. 123A).—LAWS OF EQUILIBRIUM.

A suitable point is chosen and the moment of each force is taken about it. Moments are deemed to be 'plus' or 'minus' according as to whether they are 'clockwise' or 'anticlockwise' respectively. The moment values are tabulated as shown in the figure. The totals given at the foot of the table verify the conditions of equilibrium.

TABLE OF COMPONENTS.

FORCE.	HORIZONTAL COMP.		VERTICAL COMP.		ARM.	MOMENT.	
N	+ N	− N	+ N	− N	MM	+Nmm	−Nmm
1	4		3·50	1·90		95		380
2	2	1·95		0·26		70	140	
3	2	1·55			1·16	82·5	165	
4	1				1·00	75	75	
TOTALS		3·50	3·50	2·16	2·16		380	380

FIG. 123A (see Fig. 123).

Principle of Moments

In the problems we have hitherto taken, illustrating the *'principle of the lever,'* the structural unit concerned had a definite fulcrum about which turning was considered. The 'principle of moments' is much wider in its application. We may now imagine the fulcrum to be *anywhere* in space in the plane of the forces, not necessarily on the member concerned. The choice of a fulcrum about which to take moments is just a question as to which is the best position from the point of view of assisting in the solution of the problem in hand. If several of the unknown forces in a question pass through one common point, this point is usually a suitable one about which to take moments, as these unknown forces will then not appear in the *'moments equation.'* We may have to choose several points before solving all the unknowns, but, of course, the other conditions of equilibrium are also utilised as desirable. Care must be taken to ensure that all the forces in the problem are considered in taking moments, including any reactions whose lines of action do not pass through the chosen '*fulcrum.*'

Summary of 'Equilibrium Conditions'

Concurrent Forces

(i) The algebraic sum of all the horizontal components of the forces must equal zero, i.e. the total sum to the *left* must equal the total sum to the *right*.

(ii) The algebraic sum of all the vertical components of the forces must equal zero, i.e. the total sum *upwards* must equal the total sum *downwards*.

Non-concurrent Forces

The two conditions given for concurrent forces must be satisfied and, in addition, the algebraic sum of all the moments taken about any point in the plane of the forces must equal zero, i.e. the sum of all the clockwise moments must equal the sum of all the anticlockwise moments.

Examples on 'Conditions of Equilibrium'

(i) *A bracket (the self-weight of which may be neglected) is hinged in the manner indicated in Fig. 124. Calculate the value of the horizontal reaction at the upper hinge and the magnitude and direction of the reaction at the lower hinge.*

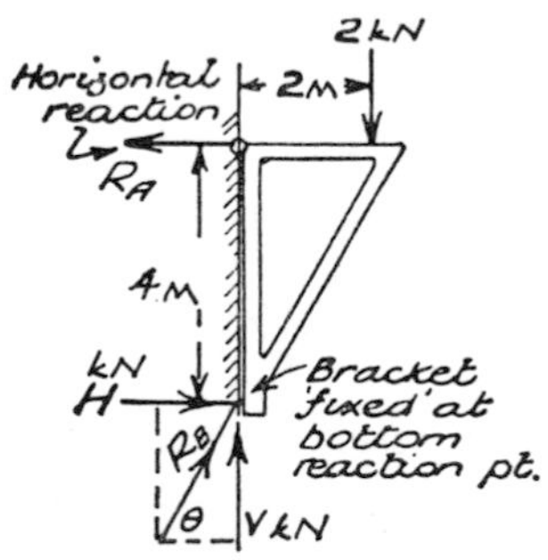

FIG. 124.

Let 'V' kN and 'H' kN be the vertical and horizontal components respectively of the reaction at the lower hinge, and let 'R_A' = horizontal reaction at the upper hinge.

Expressing the condition for *vertical* equilibrium.

$$V \text{ kN (up)} = 2 \text{ kN (down)}$$
$$\therefore \quad V = 2 \text{ kN.}$$

The condition for *horizontal* equilibrium gives:

$$R_A \text{ (left)} = H \text{ (right)}$$

Taking moments about the lower hinge point—chosen because it gets rid of the completely unknown lower hinge reaction from the 'moments equation':

$$(2 \text{ kN} \times 2 \text{ m}) \text{ C.W.} = (R_A \times 4 \text{ m}) \text{ A.C.W.}$$
$$\therefore \quad R_A = 1 \text{ kN}$$

But $H = R_A$. $\qquad \therefore \quad H = 1 \text{ kN}.$

Having found 'H' and 'V,' we can find 'R_B,' the total reaction at the lower hinge, by the parallelogram of forces:

$$R_B{}^2 = H^2 + V^2$$
$$= 1^2 + 2^2 = 5$$
$$\therefore \quad R_B = \sqrt{5} \text{ kN} = 2{\cdot}24 \text{ kN}.$$

If 'θ' be the angle made by the lower reaction with the horizontal.

$$\tan \theta = V/H = 2/1 = 2$$
$$\therefore \quad \theta = 63° \, 26'.$$

(ii) *Calculate the left-end and right-end reactions respectively for the roof truss (given in Fig. 125), due to the 600 N resultant wind load. The left-end reaction may be assumed to be vertical.*

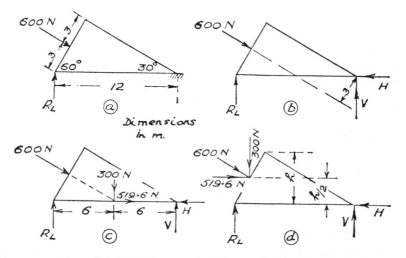

FIG. 125.—REACTIONS BY VARIOUS METHODS.

This problem will be solved in various ways in order to illustrate different methods of attack in such questions.

First Method (Figs. '*a*' and '*b*').

Taking moments about the right-end reaction point:

$$R_L \times 12 = 600 \times 3$$
$$\therefore \quad R_L = 150 \text{ N.}$$

Replace the right-end reaction by components 'V' and 'H.'
The vertical component of the 600 N load $= 600 \cos 60° = 300$ N.

For vertical equilibrium: $R_L + V = 300$.
$\therefore \quad 150 + V = 300. \quad \therefore \quad V = 150$ N.

For horizontal equilibrium: $H = 600 \cos 30°$.
$\therefore \quad H = 600 \times 0.866 = 519.6$ N.

$$R_R = \sqrt{150^2 + 519.6^2} = 540.8 \text{ N.}$$

Let 'θ' = angle 'R_R' makes with the horizontal.

$$\tan \theta = \frac{150}{519.6} = 0.2887$$

$$\theta = 16° \, 6'.$$

Second Method (Fig. '*c*')

The 600 N force may be resolved into its vertical and horizontal components at any convenient point in its line of action. A simple solution of the problem may be effected by resolving the force at the point where its line of action cuts the bottom tie.

$$R_L = V = \frac{300 \text{ N}}{2} = 150 \text{ N, as the vertical component '300 N' acts}$$

at the mid-point of the tie.

For horizontal equilibrium, $H = 600 \times \cos 30° = 519.6$ N.
R_R is found, as in first method, from its components 'V' and 'H.'

Third Method (Fig. '*d*')

Resolve the '600 N' load into its horizontal and vertical components at the point of application of the load on the rafter.

Let 'h' m = height of truss
$h = 6 \sin 60° = 6 \times 0.866 = 5.196$.

Taking moments about right-end reaction point:

$$(R_L \times 12) + \left(519 \cdot 6 \times \frac{5 \cdot 196}{2}\right) = 300 \times 10 \cdot 5$$

$$12R_L = 3\,150 - 1\,350 = 1\,800$$
$$R_L = 150 \text{ N.}$$
$$\text{V (right-end reaction)} = 300 - R_L = 300 - 150 = 150 \text{ N.}$$

$H = 519 \cdot 6$ N (horizontal equilibrium). The right-end reaction is compounded of 'H' and 'V' as before.

The results by all the three methods agree. The best method to employ in any given case will depend upon the particular type of problem.

Method 2 gave the quickest solution in this example, because the line of action of the '600 N' load cut the bottom tie in a very convenient point, i.e. at centre of span.

A final check of the problem will be made by the graphical method shown in Fig. 126.

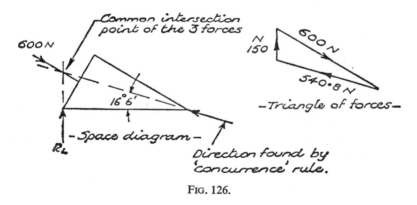

FIG. 126.

Further examples on the application of the conditions of equilibrium will be found throughout the book.

EXERCISES 5

(1) Find the magnitude, direction and position of the resultant, in each of the two cases of *like* parallel force systems shown in Fig. 127.

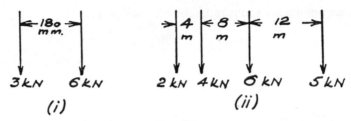

FIG. 127.

(2) In Fig. 128 are given two *unlike* parallel force systems. Find, completely, the resultant of each of these respective systems.

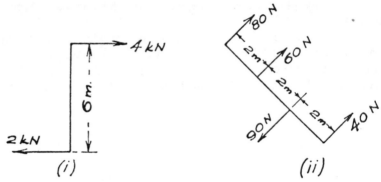

FIG. 128.

(3) Find the resultant of the vertical load system carried by the simply supported beam shown in Fig. 129.

Calculate the support reactions for the beam (i) by taking the loads as given, (ii) by assuming the beam to carry the single resultant load calculated.

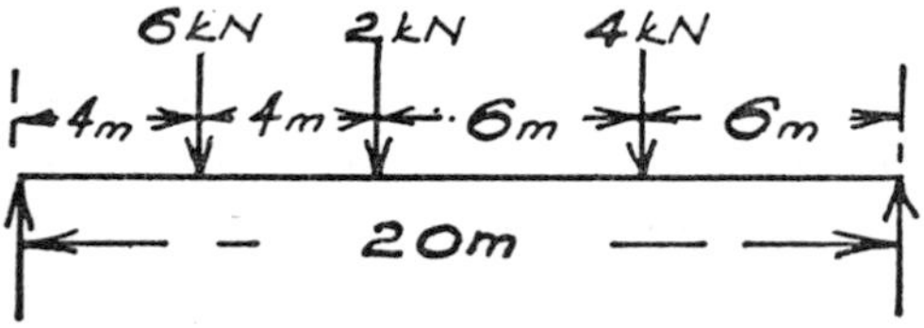

FIG. 129.

(4) A roof truss is subjected to the system of wind '*suction*' loads shown in Fig. 130. Find the resultant wind force on the truss due to suction effect.

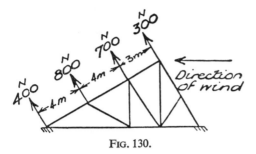

FIG. 130.

(5) In designing a compound grillage to support several columns in line, it is desirable to have the resultant column load at the centre of length of the grillage. Assuming the grillage to extend 0·6 m beyond the centre line of the left-hand column (Fig. 131), find the necessary overall length of the grillage.

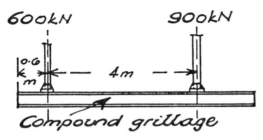

FIG. 131.

(6) A system of moving loads (Fig. 132) crosses a beam of 20 m span. How far is the resultant of the load system from the right end of the beam when the left-hand wheel of the system is 5 m from the left end of the beam?

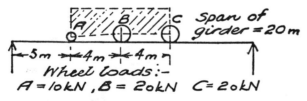

FIG. 132.

(7) Find the moment of the couple to which the system of forces given in Fig. 133 reduces (i) by finding the *force* and *arm* of the couple, (ii) by taking moments of the given forces about any convenient point.

s.m.(m)—4

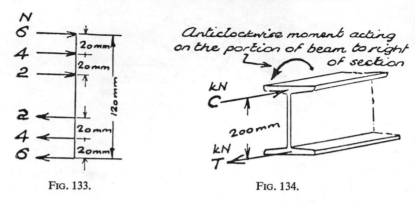

FIG. 133.

FIG. 134.

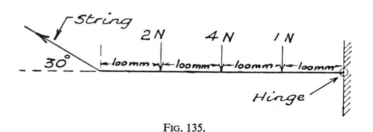

FIG. 135.

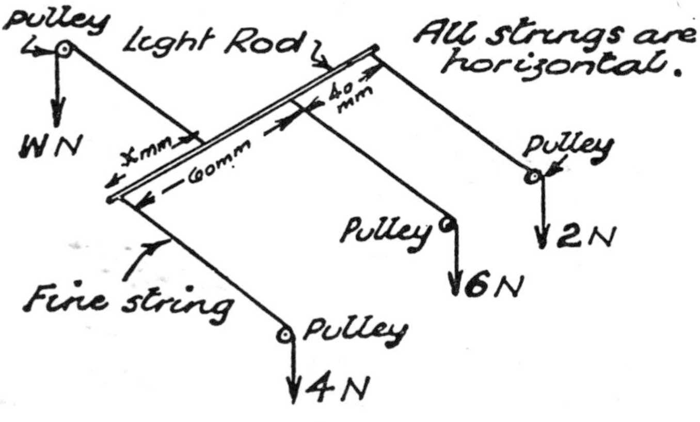

FIG. 136.

(8) If the beam section (Fig. 134) is subjected to a couple of anti-clockwise moment 6 000 kN mm, calculate the value of 'C' (the thrust in the top flange) and 'T' (the pull in the bottom flange).

(9) Obtain the two reactions in the ladder example given on page 37, by calculation method.

(10) A rod, whose weight may be neglected, is held in a horizontal position by a string in the manner indicated in Fig. 135.

(i) *Calculate* the pull in the string and also the direction and magnitude of the hinge reaction. (ii) Check the calculated values by a graphical method.

(11) Find the values of 'W' and 'x' respectively (Fig. 136) for the light rod shown to remain in equilibrium.

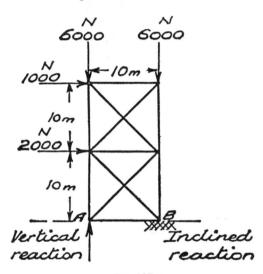

FIG. 137.

(12) A vertical steel frame is subjected to the load system given in Fig. 137. Calculate the reactions at 'A' and 'B' respectively.

CHAPTER VI

CENTRE OF GRAVITY

Fig. 138 shows a body divided up into very small portions. Each of these portions is subjected to a vertical pull due to the attractive force of gravity. All the pulls together constitute a *like parallel system* of forces. This system will have a resultant, equal in magnitude to the sum of the forces, acting vertically downwards in a definite line of action. In Fig. 138 the same body is shown turned from one position to another. The component pulls in the two cases will be in a relatively different arrange-

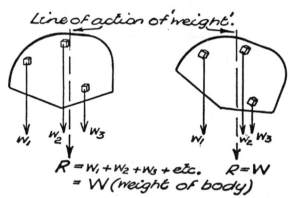

$$R = W_1 + W_2 + W_3 + \text{etc.} \qquad R = W$$
$$= W \text{ (weight of body)}$$

Fig. 138.—Centre of Gravity.

ment amongst themselves, so that while their resultant will remain the same in 'magnitude' it will act along a new 'line of action.' Although we could obtain many such lines of action by successively turning the body round, we would find that they would all pass through one common point in, or near, the body. To this point is given the name '*centre of gravity*.' Each body has, therefore, one centre of gravity.

Definition: The centre of gravity (C.G.) of a body is that point in space through which the resultant pull of the earth, i.e. the 'weight' of the body, acts, for all possible positions of the body.

The C.G. of a body is not necessarily in the material of which the body is composed.

Position of Centre of Gravity

Regular Geometrical Solids.—If a solid has a *geometrical centre,* this point will be the C.G., assuming the solid to be of uniform density throughout. If a solid has a central axis, the C.G. will lie somewhere along this axis. The exact position in the case of certain solids, such as a right cone, may be found by a formula. The proofs of some of these formulae are rather beyond the scope of the book and they will be assumed. Fig. 139 indicates the C.G. positions for a few simple bodies.

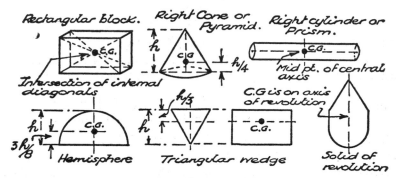

FIG. 139.—STANDARD CASES.

Two Equal Bodies

It should be noted that in the case of a body composed of *two* simpler bodies (whose C.G. positions, we know), the C.G. will be somewhere along the line joining the C.G.'s of the two respective component bodies. If we imagine the body in such a position that the C.G. of one of the

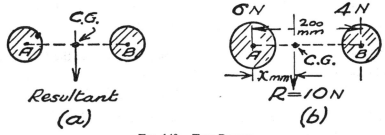

FIG. 140.—TWO BODIES.

component bodies is vertically above that of the other, it will be clear that, as the earth's pull will pass through the two C.G.'s, the C.G. of the body must be on the line joining them.

In Fig. 140 (*a*) we have to find the position of the resultant of two equal parallel forces. The C.G. is the mid-point of 'AB.'

Two Unequal Bodies.—In Fig. 140 (*b*) we have to find the point in which the resultant 'force' cuts the line 'AB.'

Resultant of system = W = 6 N + 4 N = 10 N.

Let '*x*' = distance of resultant from 'A.'

Moments about 'A':

$$(10 \times x) = (6 \times 0) + (4 \times 200)$$
$$= 0 + 800 = 800$$
$$\therefore \quad x = 80 \text{ mm.}$$

The C.G. of the two bodies is 80 mm from 'A.'

Rule.—In the case of two bodies, to obtain the C.G., divide the distance between their respective C.G.'s *inversely* as the bodies.

In the example, $4 \text{ N}/6 \text{ N} = \dfrac{80 \text{ mm}}{(200 - 80) \text{ mm}} = \dfrac{80 \text{ mm}}{120 \text{ mm}} = \dfrac{2}{3}$.

The rule is applied in the example on page 105. In working examples, the 'moments' method given previously is the easier.

Several Unequal Bodies.—The bodies shown in Fig. 141 have their C.G.'s in one straight line. The methods of parallel force systems are again applied.

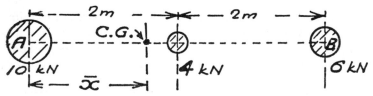

FIG. 141.—SEVERAL UNEQUAL BODIES.

The symbol '$\bar{x}$' ('*x bar*') is usually employed to denote the *horizontal* distance of a C.G. from an axis of reference.

$$W = (10 + 4 + 6) \text{ kN} = 20 \text{ kN.}$$

Moments about 'A': Let $\bar{x}$ = C.G. distance from 'A.'

$$20 \times \bar{x} = (10 \times 0) + (4 \times 2) + (6 \times 4)$$
$$20\bar{x} = 32$$
$$\bar{x} = 1 \cdot 6 \text{ m.}$$

The C.G. is on the line 'AB,' at a point 1·6 m from 'A.'

Several Bodies whose C.G.'s are not in One Straight Line

The C.G.'s of the three bodies shown in Fig. 142 are assumed to be in one plane. Suitable axes 'OX' and 'OY' must be chosen to which to relate the C.G. of the system.

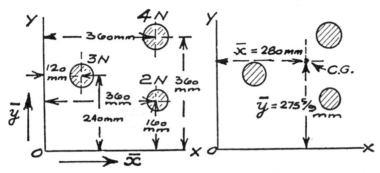

FIG. 142.—GENERAL CASE OF UNEQUAL BODIES.

Let $\bar{x}$ = distance of C.G. *from* axis 'OY.'
$\bar{y}$ = distance of C.G. *from* axis 'OX.'
Total mass = $(3 + 4 + 2)\,N = 9\,N$.

Moments about axis 'OY' to find $\bar{x}$.
In this case we multiply each body by its distance from 'OY.'
$$9\bar{x} = (3 \times 120) + (4 \times 360) + (2 \times 360)$$
$$= 360 + 1\,440 + 720 = 2\,520$$
$$\bar{x} = 280 \text{ mm.}$$

Moments about axis 'OX' to find $\bar{y}$.
$$9\bar{y} = (3 \times 240) + (4 \times 360) + (2 \times 160)$$
$$= 720 + 1\,440 + 320 = 2\,480$$
$$\bar{y} = 275\tfrac{5}{9} \text{ mm.}$$

The C.G. is shown in position in Fig. 142.

C.G. of Composite Bodies

Provided a body be of uniform density, *volumes* may be used instead of *masses* in calculating C.G. positions. If the terms in the previously given *'moments equations'* were divided by the density of the material each term would represent a *'volume moment.'*

EXAMPLE.—*Calculate the position of the C.G. of the given solid* (*Fig. 143*).

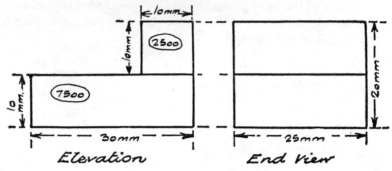

FIG. 143.—COMPOSITE SOLID.

Total volume of solid $= [(10 \times 30 \times 25) + (10 \times 10 \times 25)]$ mm³
$= (7\,500 + 2\,500)$ mm³
$= 10\,000$ mm³.

It is sometimes helpful to indicate, on each component part, the volume of the particular portion, so as to avoid error in taking moments.

Let $\bar{x} =$ distance of C.G. from left face of solid.
    $\bar{y} =$ distance of C.G. from bottom face of solid.
$10\,000\,\bar{x} =$ the total sum of '*each volume multiplied by the distance of its own C.G. from the left face*'
$= (7\,500 \times 15) + (2\,500 \times 25)$
$= 112\,500 + 62\,500$
$= 175\,000$

$$\bar{x} = \frac{175\,000}{10\,000} = 17 \cdot 5 \text{ mm.}$$

$10\,000\,\bar{y} =$ the total sum of '*each volume multiplied by the distance of its own C.G. from the bottom face*'
$= (7\,500 \times 5) + (2\,500 \times 15)$
$= 37\,500 + 37\,500$
$= 75\,000$

$$\bar{y} = \frac{75\,000}{10\,000} = 7 \cdot 5 \text{ mm.}$$

The solid is of uniform thickness ($= 25$ mm), hence the C.G. is 12·5 mm from the front face. The distance from front (or back) face in any given case could be found by taking moments about that face, using the procedure indicated above.

EXAMPLE.—*A masonry pillar has the detail shown in Fig.* 144. *Find the position of the C.G.*

The C.G. will lie somewhere along the vertical central axis of the pillar.

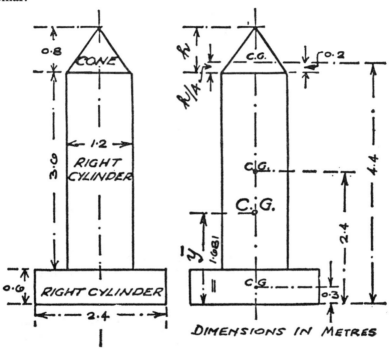

FIG. 144.—COMPOSITE REGULAR SOLID.

Let $\bar{y}$ = height of C.G. above base.

$$\text{Volume of cone} = \frac{\pi r^2 h}{3} = \frac{\pi \times 0.6^2 + 0.8}{3} \text{ m}^3 = 0.096\pi \text{ m}^3$$

$$\text{Volume of right cylinder (upper)} = \pi r^2 h = \pi \times 0.6^2 \times 3.6 \text{ m}^3$$
$$= 1.296\pi \text{ m}^3$$

$$\text{Volume of right cylinder (lower)} = \pi r^2 h = \pi \times 1.2^2 \times 0.6 \text{ m}^3$$
$$= 0.864\pi \text{ m}^3$$

$$\text{Total volume} = (0.096\pi + 1.296\pi + 0.864\pi) \text{ m}^3$$
$$= 2.256\pi \text{ m}^3.$$

S.M.(M)—4*

$$2 \cdot 256\pi \times \bar{y} = (0 \cdot 096\pi \times 4 \cdot 4) + (1 \cdot 296\pi \times 2 \cdot 4) + (0 \cdot 864\pi \times 0 \cdot 3)$$
$$2 \cdot 256\bar{y} = 0 \cdot 4224 + 3 \cdot 1104 + 0 \cdot 2592$$
$$= 3 \cdot 792$$
$$\bar{y} = 1 \cdot 681 \text{ m.}$$

C.G. is 1·681 m above bottom of pillar.

C.G. of a Plane Figure

A figure, having no mass, will not be affected by gravitational force, so that the term '*centre of gravity*' is not strictly applicable in this case. In order to get a physical interpretation of 'C.G.' in the case of '*areas*' and '*sections*,' we may imagine the figure to represent the outline of an extremely thin slice of material, so thin that the C.G. may be regarded as being practically on the surface. We may therefore employ the general methods used in the case of solids, and take '*moments of areas*' just as we took '*moments of volumes*.' The term '*centroid*' is sometimes used instead of 'centre of gravity.'

The determination of the C.G. of an area is frequently required in structural calculations. The methods of solution given should be thoroughly mastered by the reader.

Figures with Axes of Symmetry.—If a figure has an axis of symmetry, the C.G. will be somewhere on that axis (Fig. 145). If there are two axes

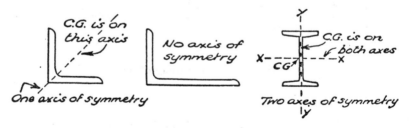

FIG. 145.—STRUCTURAL SECTIONS.

of symmetry, their point of intersection will be the required C.G. Symmetrical axes may be used to find the C.G. of a rectangle, but it is easier to find the point by drawing in the diagonals (see Fig. 146 (*a*)).

Parallelogram.—In the case of the parallelogram (Fig. 146 (*b*)) the thin strips indicated may be regarded as being rectangles. The line 'AB' which passes through the respective C.G.'s of all the horizontal strips will clearly pass through the C.G. of the parallelogram. Similarly the C.G. must be on the line 'C.D.' The quickest method again in this case will be to draw-in the diagonals of the parallelogram.

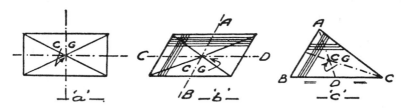

FIG. 146.—PLANE SECTIONS.

Triangle.—Consider thin strips of the triangle (Fig. 146 (c)), parallel to the base 'BC.' The mid-point of each strip will lie on a straight line joining the apex 'A' to the mid-point of 'BC.' This is a geometrical theorem, the line 'AD' being termed a *median* of the triangle 'ABC.' The median 'AD' passes through the C.G. of each strip and will therefore contain the C.G. of the triangle. Similarly the other two medians of the triangle must pass through its C.G. We may define the C.G. of a triangle as being the intersection point of any two of its medians.

Right-angled Triangle.—It is shown in geometry that the common intersection point of the three medians of *any* triangle divides each median into two parts having a ratio of 1 to 2 (see Fig. 147). This fact

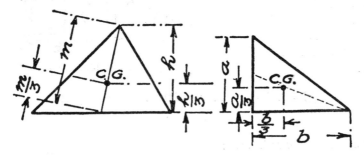

FIG. 147.—C.G. OF A TRIANGLE.

is very useful in dealing with right-angled triangles, and is especially applicable to structural problems when extended, by the principle of similar triangles, to the form shown in Fig. 147.

Other Geometrical Figures.—Structural calculations sometimes involve the consideration of the C.G. position in such figures as *semicircles, parabolae,* etc. The reader is referred to structural handbooks for a complete list of such cases. Fig. 148 indicates a few important C.G. positions.

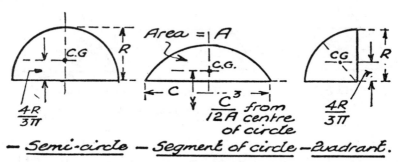

FIG. 148.—STANDARD CASES.

Centre of Gravity of Composite Figures

The general method of treament of practical structural sections is exemplified in the diagram given in Fig. 149.

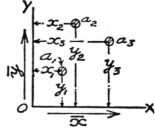

The areas 'a_1,' 'a_2,' 'a_3,' are assumed to represent component parts of a given figure or structural section. 'OX' and 'OY' are two convenient axes of reference. In a practical example it would be usual to regard the bottom edge of the figure and the left edge as forming these respective axes.

FIG. 149.—GENERAL CASE.

Let 'A' = total area of the component parts.
$$A = a_1 + a_2 + a_3.$$

Moments about axis 'OY':

$$A\bar{x} = a_1x_1 + a_2x_2 + a_3x_3$$
$$= \sum ax \text{ (i.e. \textit{the sum of all such quantities as '}a \times x\text{')}}$$

$$\therefore \bar{x} = \frac{\sum ax}{A}.$$

Moments about axis 'OX':

$$A\bar{y} = a_1y_1 + a_2y_2 + a_3y_3$$
$$= \sum ay$$
$$\therefore \bar{y} = \frac{\sum ay}{A}.$$

In simple words the procedure is as follows:

Divide up the given figure into suitable simple figures. Take each of these component areas and multiply by its own C.G. distance from the selected axis of reference. Add together all such '*area moments.*' Equate the sum to the product '*total area of figure multiplied by the required C.G. distance from the given axis.*' Hence determine the C.G. distance.

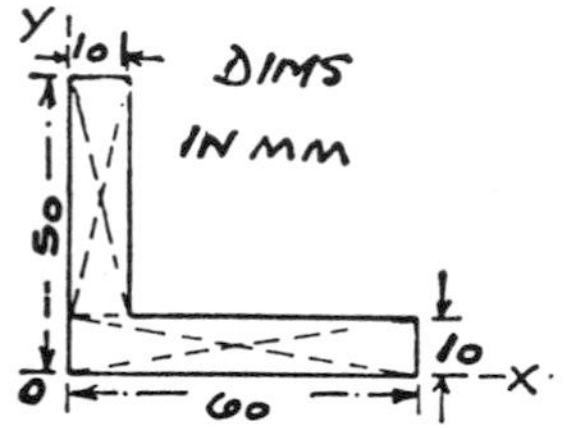

EXAMPLE (i).—*Find the C.G. position in the case of the unequal angle section shown in Fig. 150.*

In this case we divide the figure into two rectangles.

FIG. 150.—UNEQUAL ANGLE.

Component areas: upper rectangle = 40 mm × 10 mm =   400 mm²
lower rectangle = 60 mm × 10 mm =   600 mm²
Total area = 1 000 mm².

Axis 'OX' is taken to coincide with the bottom edge of the section and 'OY' is assumed to coincide with the left edge. It is not necessary to draw-in these axes provided the meanings of '$\bar{x}$' and '$\bar{y}$' respectively are clearly stated.

Moments about 'OY' (left edge of section):

Moment of total area = sum of moments of component areas

$$1\,000\bar{x} = (400 \times 5) + (600 \times 30)$$
$$= 2\,000 + 18\,000 = 20\,000$$
$$\bar{x} = \frac{20\,000}{1\,000} = 20 \text{ mm.}$$

Moments about 'OX' (bottom of section):

Moment of total area = sum of moments of component areas

$$1\,000\bar{y} = (400 \times 30) + (600 \times 5)$$
$$= 12\,000 + 3\,000 = 15\,000$$
$$\bar{y} = \frac{15\,000}{1\,000} = 15 \text{ mm.}$$

C.G. is 20 mm from left edge of figure and 15 mm from the bottom.

(ii) *Determine the position of the C.G. of the given Z-section (Fig. 151).*

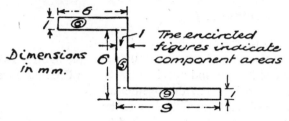

FIG. 151.—Z-SECTION.

Let $\bar{x}$ = distance of C.G. from left edge of section.

$\bar{y}$ = distance of C.G. from bottom edge of section.

Total area of section $= [(6 \times 1) + (5 \times 1) + (9 \times 1)] \text{ mm}^2$
$= (6 + 5 + 9) \text{ mm}^2 = 20 \text{ mm}^2.$

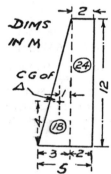

FIG. 152.—RETAINING-WALL SECTION.

$20\bar{x} = (6 \times 3) + (5 \times 5{\cdot}5) + (9 \times 9{\cdot}5)$
$= 18 + 27{\cdot}5 + 85{\cdot}5 = 131$

$$\bar{x} = \frac{131}{20} \text{ mm} = 6{\cdot}55 \text{ mm.}$$

$20\bar{y} = (6 \times 6{\cdot}5) + (5 \times 3{\cdot}5) + (9 \times 0{\cdot}5)$
$= 39 + 17{\cdot}5 + 4{\cdot}5 = 61$

$$\bar{y} = \frac{61}{20} \text{ mm} = 3{\cdot}05 \text{ mm.}$$

(iii) *Fig. 152 shows the section of a retaining wall. Calculate the distances of the C.G. from the base and from the vertical back of the wall respectively.*

Divide the section into a rectangle and a triangle.

Total area of section $= [(12 \times 2) + (\tfrac{1}{2} \times 3 \times 12)] \text{ m}^2$
$= (24 + 18) \text{ m}^2 = 42 \text{ m}^2.$

Let $\bar{x}$ = distance of C.G. from back of wall.

$\bar{y}$ = distance of C.G. from base of wall.

Moments about back of wall:

$$42\bar{x} = (24 \times 1) + (18 \times 3).$$

[The C.G. of the triangle is $(\frac{1}{3} \times 3 \text{ m})$, from the right-angle corner, so that, from the back of the wall, it is $(\frac{1}{3} \times 3 \text{ m}) + 2 \text{ m} = 3 \text{ m}.]$

$$\therefore \quad 42\bar{x} = 24 + 54 = 78$$

$$\bar{x} = \frac{78}{42} \text{ m} = 1\tfrac{6}{7} \text{ m}.$$

Moments about bottom of wall:

$$42\bar{y} = (24 \times 6) + (18 \times \tfrac{12}{3})$$
$$= 144 + 72 = 216$$

$$\bar{y} = \frac{216}{42} \text{ m} = 5\tfrac{1}{7} \text{ m}.$$

Allowance for Rivet Holes in Structural Sections

In cases in which rivet holes have to be deducted, it is convenient to treat the section, first of all, as being '*gross*' (i.e. having no holes), and then to assume the rivet holes to be negative superimposed areas.

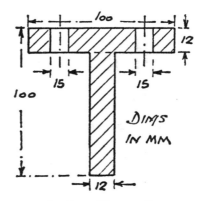

In the example shown in Fig. 153, the gross section is symmetrical about a vertical axis. The rivet holes are also symmetrical about this axis, so that the C.G. must lie somewhere on this axis.

Let '$\bar{y}$' be depth of C.G. below top of section.

Total gross area = $[(100 \times 12) + (88 \times 12)]$ mm².

Area of rivet holes = $(2 \times 15 \times 12)$ mm².

FIG. 153.—RIVET HOLES IN SECTION.

Net area = $(1\ 200 + 1\ 056 - 360)$ mm².

Moments about top edge of section:

$$(1\ 200 = 1\ 056 - 360)\bar{y} = (1\ 200 \times 6) + (1\ 056 \times 56) - (360 \times 6)$$
$$1\ 896\bar{y} = 7\ 200 + 59\ 136 - 2\ 160$$
$$= 64\ 176$$
$$\bar{y} = 33 \cdot 85 \text{ mm}.$$

EXAMPLE.—*Find the centre of gravity of the rivet group given in Fig. 154.*

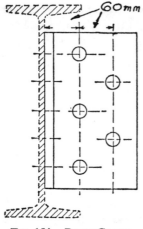

FIG. 154.—RIVET GROUP.

The determination of the C.G. in such a case is the first step in the calculation of the strength of the connection.

The C.G. will lie on a horizontal line through the centre of the middle rivet in the first row (principle of symmetry).

Let $\bar{x}$ = distance of C.G. from left edge of cleat.

Let 'A' mm² = area of one rivet.

Total area = 5A mm².

$$\therefore \quad 5A\bar{x} = (3A \times 60) + (2A \times 120)$$

$$5A\bar{x} = 180A + 240A$$

$$= 420A$$

$$\therefore \bar{x} = \frac{420}{5} = 84 \text{ mm.}$$

EXAMPLE.—*Determine the depth beneath the top of section, of the gravity of the given girder section (Fig. 155). Make no allowance for rivet holes.*

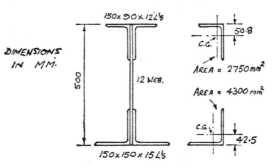

FIG. 155.—BUILT-UP GIRDER SECTION.

To solve this problem we require to know certain 'properties' of the 'British Standard Sections' involved. We must know the 'sectional area' and the position of the C.G. of the 'angle' sections given. These properties are shown in Fig. 155, having been found by reference to 'section property tables.' Such tables are to be found in any of the handbooks issued by steel firms.

Total area of girder section

$$= [(2 \times 2\,750) + (2 \times 4\,300) + (500 \times 12)]\ mm^2$$
$$= (5\,500 + 8\,600 + 6\,000)\ mm^2$$
$$= 20\,100\ mm^2.$$

Let '$\bar{y}$' = distance of C.G. below top of section.

Moments about top of section:

$$20\,100\bar{y} = (5\,500 \times 50\cdot8) + (8\,600 \times 457\cdot5) + (6\,000 \times 250)$$
$$= 279\,400 + 3\,934\,500 + 1\,500\,000$$
$$= 5\,713\,900$$
$$\bar{y} = 284\cdot3\ mm.$$

EXAMPLE.—*The diagram given in Fig. 156 represents the end of a building which is subjected to a uniform wind pressure of 750 N per square metre of area. Calculate the overturning moment, due to the wind, about the base 'AB.'*

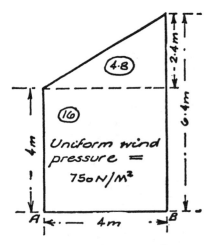

FIG. 156.—UNIFORM WIND LOAD.

The wind pressure being uniform, the resultant wind force will act at the 'centre of gravity' of the area.

$$Total\ area = \left[(4 \times 4) + \left(\frac{4 \times 2\cdot4}{2} \right) \right] m^2$$
$$= (16 + 4\cdot8)\ m^2 = 20\cdot8\ m^2.$$

Let '$\bar{y}$' be height of C.G. above base 'AB.'

$$20 \cdot 8 \bar{y} = (16 \times 2) + (4 \cdot 8 \times 4 \cdot 8)$$
$$= (32 + 23 \cdot 04)$$
$$= 55 \cdot 04$$
$$\bar{y} = 55 \cdot 04 / 20 \cdot 8 \text{ m} = 2 \cdot 646 \text{ m}.$$

The resultant wind load $= 20 \cdot 8 \text{ m}^2 \times 750 \text{ N/m}^2$
$$= 15\ 600 \text{ N}.$$

Overturning moment about base $=$ Force $\times$ arm
$$= 15\ 600 \text{ N} \times 2 \cdot 646 \text{ m}$$
$$= 41\ 280 \text{ N m}.$$

Note: We could have dealt with each component area separately.

Graphical Methods for Determining the C.G. of a Plane Figure

A graphical method applicable to any plane figure is explained in the next chapter. Certain figures may be dealt with by special forms of graphical solution, based upon principles already considered.

Quadrilateral

Join 'A' to 'C,' forming two triangles 'ABC' and 'ADC,' whose respective C.G.'s are 'G_1' and 'G_2.' Now join 'B' to 'D,' forming the two triangles 'BAD' and 'BCD.' The C.G.'s of these two triangles are

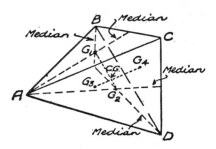

FIG. 157.—QUADRILATERAL.

'G_3' and 'G_4' respectively (the medians used to fix these two points are omitted from the diagram shown, in order to avoid confusion of lines). Join '$G_3 G_4$.' The C.G. of the quadrilateral will be the intersection point of '$G_1 G_2$' and '$G_3 G_4$,' as it must lie on both these lines.

Fig. 158 illustrates the application of the same principle of '*sub-dividing a given figure into two simpler figures in two different ways.*'

Alternative Method.—Divide the given figure into two simpler figures and find 'G$_1$' and 'G$_2$,' their centres of gravity. Join 'G$_1$G$_2$.' Just as in the case of two solids (page 92), we must find the point which divides 'G$_1$G$_2$' *inversely* as the respective areas. In the example of the acute-angle section given in Fig. 158, 'G$_2$X' is set off in any suitable direction.

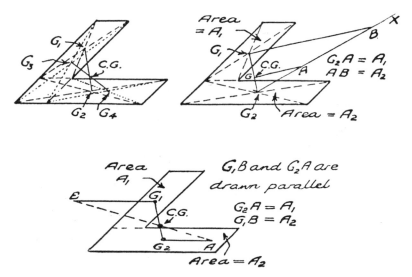

FIG. 158.—ALTERNATIVE GRAPHICAL METHODS.

Along 'G$_2$X,' 'G$_2$A' is marked off to a convenient area scale to represent area 'A$_1$' and 'AB' is similarly marked off to represent area 'A$_2$.' 'B' is joined to 'G$_1$' and 'AG' is drawn parallel to 'BG$_1$.' By principles of geometry, 'G' divides G$_1$G$_2$ in the required ratio, and will be, therefore, the C.G. of the given section.

The third method shown is based upon the same principles as the second method.

Retaining-wall Section.—The figure given in Fig. 159 represents the type section for a 'gravity' retaining wall. To determine the stability of such a wall it is necessary to find the C.G. of the section (see Chapter XV). The methods already explained may be employed, but for a four-sided figure with a pair of opposite sides parallel, a simpler procedure is usually adopted.

Draw AA' and BB' each equal to the base DC, and CC' and DD'

each equal to the top AB. Join A'C' and B'D'. The point of intersection of these two transversals will be the C.G. of the given wall section. The C.G. will also be in the line 'MN,' where 'M' and 'N' are the mid-points of the top and base respectively.

Alternatively, if 'a' = length of base and 'b' = length of top, we may find 'G' by the formula:

$$NG = \frac{a + 2b}{3(a + b)} \times NM.$$

Proof.—The proof of the foregoing methods is an interesting application of the general principles of C.G. determination. The figure is divided into two triangles.

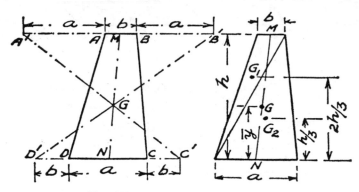

FIG. 159.—RETAINING-WALL SECTION.

Let $\bar{y}$ = *height* of C.G. above base (Fig. 159).

Total area $= \left(\dfrac{ah}{2} + \dfrac{bh}{2}\right) = \dfrac{h}{2}(a + b).$

Taking moments about the base:

$$\frac{h}{2}(a + b) \times \bar{y} = \left(\frac{ah}{2} \times \frac{h}{3}\right) + \left(\frac{bh}{2} \times \frac{2h}{3}\right)$$

$$= h^2\left(\frac{a}{6} + \frac{2b}{6}\right)$$

$$\therefore \bar{y} = \frac{2h}{a + b}\left(\frac{a + 2b}{6}\right) = \frac{a + 2b}{3(a + b)} \times h.$$

But $\bar{y}:h = $ NG:NM by the principle of similar triangles,

$$\therefore \quad NG = \frac{a + 2b}{(3a + b)} \times NM.$$

Further, considering the graphical construction, $\triangle$s GND' and GMB' are similar.

$$\therefore \qquad \frac{NG}{GM} = \frac{ND'}{MB'} = \frac{b + a/2}{a + b/2} = \frac{2b + a}{2a + b}$$

$$\therefore \quad \frac{NG}{GM + NG} = \frac{2b + a}{(2a + b) + (2b + a)} = \frac{a + 2b}{3(a + b)}$$

$$\therefore \qquad \frac{NG}{NM} = \frac{a + 2b}{3(a + b)}.$$

The geometrical construction used therefore gives the correct position for the C.G.

Beam-reaction Problems Involving 'C.G.'

EXAMPLE (i).—*Calculate the support reactions for the beam shown in Fig. 160. The wall is uniform in thickness. The self-weight of the beam may be neglected.*

The beam is regarded as having a single-point load acting through the C.G. of the wall and equal in magnitude to the total weight of the wall.

If a scale drawing is made, the C.G. of the wall may be found graphically.

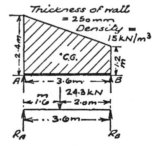

FIG. 160.—BEAM REACTIONS.

By formula: distance of C.G. of wall from support at 'A'

$$= \frac{a + 2b}{3(a + b)} \times l = \frac{2\cdot4 + (2 \times 1\cdot2)}{3(2\cdot4 + 1\cdot2)} \times 3\cdot6 \text{ m}$$

$$= \frac{4\cdot8}{10\cdot8} \times 3\cdot6 = 1\cdot6 \text{ m}$$

$$\text{Total weight of wall} = \text{volume} \times \text{density}$$
$$= [\tfrac{1}{2}(2\cdot4 + 1\cdot2) \times 3\cdot6] \times 0\cdot25 \times 15 \text{ kN}$$
$$= 24\cdot3 \text{ kN}$$
$$R_A \times 3\cdot6 = 24\cdot3 \times 2\cdot0 = 48\cdot6$$
$$\therefore \quad R_A = 13\cdot5 \text{ kN}$$
$$R_B \times 3\cdot6 = 24\cdot3 \times 1\cdot6 = 38\cdot88$$
$$\therefore \quad R_B = 10\cdot8 \text{ kN}$$
$$R_A + R_B = (13\cdot5 + 10\cdot8) \text{ kN} = 24\cdot3 \text{ kN}$$

We could have treated the wall section as being composed of a rectangle (3·6 m × 1·2 m) plus a triangle and had two equivalent beam loads.

EXAMPLE (ii).—*Fig. 161 shows a beam carrying a wall of varying height but of uniform thickness 0·3 m. Taking the particulars given, obtain the support reactions due to this load.*

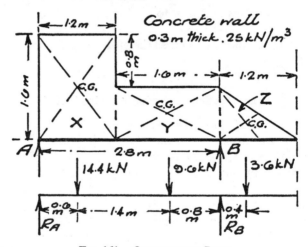

FIG. 161.—OVERHANGING BEAM.

We could, as in the last example, find the C.G. of the whole wall and reduce the problem to that of an overhanging beam carrying a single-point load. It is easier, in this case, to divide the wall up into convenient portions.

$$\text{Load X} = (1\cdot6 \times 1\cdot2 \times 0\cdot3 \times 25) \text{ kN} = 14\cdot4 \text{ kN}$$
$$\text{Y} = (0\cdot8 \times 1\cdot6 \times 0\cdot3 \times 25) \text{ kN} = 9\cdot6 \text{ kN}$$
$$\text{Z} = (0\cdot4 \times 1\cdot2 \times 0\cdot3 \times 25) \text{ kN} = 3\cdot6 \text{ kN}$$
$$\text{Total load} = 27\cdot6 \text{ kN}$$

These loads are regarded as acting through their respective C.G.'s.

$$(R_A \times 2\cdot8) + (3\cdot6 \times 0\cdot4) = (14\cdot4 \times 2\cdot2) + (9\cdot6 \times 0\cdot8)$$
$$\therefore \quad R_A \times 2\cdot8 = 31\cdot68 + 7\cdot68 - 1\cdot44 = 37\cdot92$$
$$R_A = 13\cdot543 \text{ kN}$$
$$R_B \times 2\cdot8 = (14\cdot4 \times 0\cdot6) + (9\cdot6 \times 2\cdot0) + (3\cdot6 \times 3\cdot2)$$
$$= 8\cdot64 + 19\cdot2 + 11\cdot52$$
$$= 39\cdot36$$
$$\therefore \quad R_B = 14\cdot057 \text{ kN}$$
$$R_A + R_B = (13\cdot543 + 14\cdot057) \text{ kN}$$
$$= 27\cdot6 \text{ kN.}$$

Eccentrically Loaded Columns.—Loads which do not lie on the symmetrical axes 'XX' and 'YY' of a column section are termed *'eccentric loads.'* One method of dealing with an eccentric load system (in certain circumstances) is to reduce the system to one resultant load acting at the C.G. of the system.

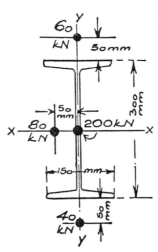

EXAMPLE.—*Calculate the eccentricities, with respect to the 'principal' axes 'XX' and 'YY,' of the resultant column load for the case shown in Fig.* 162.

WE require to find the position of the C.G. of the four given loads expressing '$\bar{x}$' and '$\bar{y}$' with respect to axes 'YY' and 'XX' respectively, as axes of reference.

Moments about YY (distances to left, positive):

FIG. 162.—ECCENTRICALLY LOADED COLUMN.

$$(60 + 200 + 40 + 80)\bar{x} = (60 \times 0) + (200 \times 0) + (40 \times 0) + (80 \times 50)$$
$$380\bar{x} = 4\,000$$
$$\bar{x} = 4\,000/380 \text{ mm} = 10\cdot53 \text{ mm.}$$

Moments about XX (distances upwards, positive):

$$380\bar{y} = (60 \times 200) + (200 \times 0) - (40 \times 200) + (80 \times 0)$$
$$= 12\,000 - 8\,000 = 4\,000$$
$$\bar{y} = 4\,000/380 \text{ mm} = 10\cdot53 \text{ mm.}$$

The resultant load is 10·53 mm eccentric with respect to both the principal axes of the section. In practical problems the eccentricity with respect to the 'XX' axis is sometimes denoted by the symbol 'ex.' Similarly 'ey' would represent the eccentricity with respect to the 'YY' axis. Thus $\bar{x} = ey$ and $\bar{y} = ex$ in the foregoing example.

Experimental Determination of C.G. Position

If a section has a very irregular outline, made up of portions of curves and straight lines, it may be difficult to determine the C.G. by calculation or by graphics. If the C.G. has been fixed by either of these methods, it may be useful to check the accuracy of the result by an experimental method. The given section should be drawn either full size or to a convenient scale, on a piece of stiff cardboard, or sheet metal, of uni-

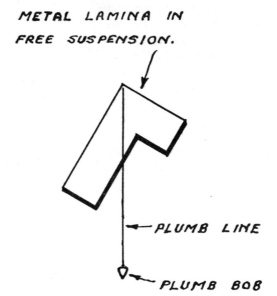

FIG. 163.—EXPERIMENTAL DETERMINATION OF CENTRE OF GRAVITY.

form thickness. Three small holes must be made near the edge of the template from which, in turn, the template is suspended on a smooth pin (see Fig. 163). At each suspension the direction of the plumb line is marked on the template. The intersection of two of these markings gives the required C.G. position, the third line provides a check on accuracy. The reader should use this simple method for verifying the C.G. position in the case of a 'triangle' or a 'semi-circle,' etc.

Application of C.G. in Reinforced Concrete Beam Section

Readers interested in reinforced concrete theory should note the example illustrated in Fig. 164. To reduce the composite concrete and

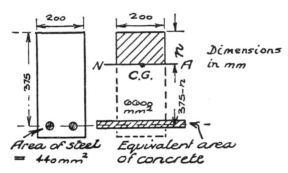

FIG. 164.—REINFORCED CONCRETE BEAM SECTION.

steel section to an equivalent '*all-concrete*' section, the steel area must be multiplied by a constant usually taken as '15.' The 440 mm² of steel thus becomes (440 × 15) mm² = 6 600 mm² of equivalent concrete. In order to calculate the strength in bending of a beam of the given section it is first of all necessary to find an axis 'NA' (see page 240) such that the C.G. of the concrete area above it, taken in conjunction with the '*equivalent concrete area*' below it, shall lie on the axis. All other concrete below 'NA' is ignored.

Let n mm = depth of the axis 'NA' from top of section. Moment of area above 'NA' must equal moment of area below 'NA,' both moments being taken about 'NA.'

$$\therefore \quad (200 \times n \times n/2) = 6\,600 \times (375 - n)$$
$$\therefore \quad 100n^2 = 2\,475\,000 - 6\,600n$$
$$n^2 + 66n - 24\,750 = 0.$$

$n = 128$ satisfies this quadratic equation. The C.G. of the '*equivalent all-concrete section*' is therefore 128 mm below the top of the section. The position of 'NA' is thus determined.

EXERCISES 6

(1) Calculate the position of the centre of gravity of the four masses shown in Fig. 165.

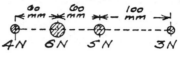

FIG. 165.

(2) A solid of uniform density is composed of three rectangular blocks (Fig. 166). Calculate the distance of the C.G. of the solid from (*a*) the left face, (*b*) the bottom.

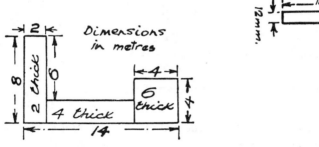

FIG. 166.

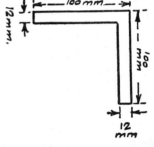

FIG. 167.

(3) Find the C.G. of the equal-angle section given in Fig. 167. Check the position by a graphical method.

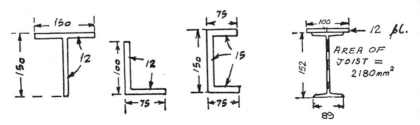

ALL DIMENSIONS IN MM

FIG. 168.

(4) Determine the centre of gravity of each of the sections given in Fig. 168.

(5) The girder section (Fig. 169) has two 25 mm diameter bolt holes in the bottom flange. Find the C.G. of the net section.

(6) A retaining-wall section has a vertical back. The widths at the top and bottom of the wall are, respectively, 3 m and 9 m. The height of the wall is 20 m. Calculate the distance of the C.G. of the wall section from the back of the wall. Check the distance by a graphical method.

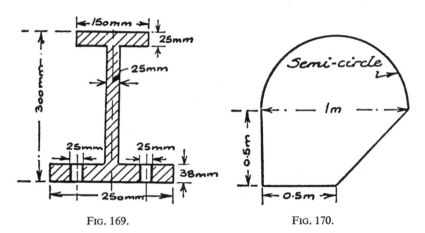

FIG. 169.   FIG. 170.

(7) Fig. 170 gives the outline of a masonry slab of uniform thickness, which has to be hoisted by a crane. Find the best position at which to fix a bolt for the crane chain in order that the slab shall remain horizontal during lifting.

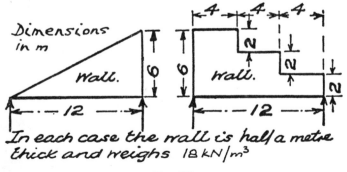

FIG. 171.

(8) Find the beam-support reactions in each of the cases given in Fig. 171. Neglect the self-weight of the beams.

(9) Determine the centre of gravity of the three column loads indicated in Fig. 172.

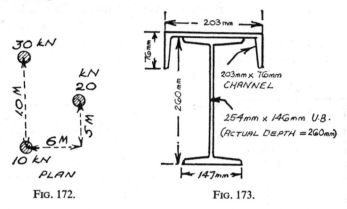

FIG. 172.

FIG. 173.

(10) Find the centre of gravity of the structural section shown in Fig. 173. The following *properties* are taken from section tables:

Sectional area of 254 mm × 146 mm U.B. = 5 500 mm².

Sectional area of 203 mm × 76 mm B.S.C. = 3 034 mm².

Thickness of web of channel = 7·1 mm.

Distance of C.G. of channel section from back of channel web = 21·3 mm.

(11) A casting has a square hole in the position indicated in Fig. 174.

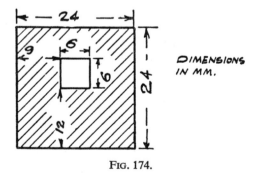

DIMENSIONS IN MM.

FIG. 174.

The material is of uniform thickness and density. Find the C.G. of the casting.

(12) Fig. 175 shows a reinforced concrete retaining wall with the

retained earth resting on the base. Calculate the distance, from the left edge of base, of the centre of gravity of the combined earth and wall.

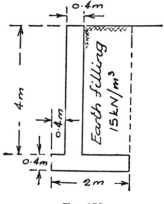

FIG. 175.

The average wall density $= 24$ kN/m³, and the earth weighs 15 kN/m³. (Take 1 m length of wall, in plan, and reduce to 'weights' before taking moments.)

CHAPTER VII

THE LINK POLYGON

THE *'link polygon'* construction is used for the solution of a number of types of structural problems. These range from simple beam-reaction problems to problems involving bending moments and beam deflections. An important application occurs in masonry arch design, the line of thrust in the arch ring being determined by a link polygon. It is clear, therefore, that the principles underlying this graphical construction should be thoroughly understood.

Resultant of a Co-planar Force System

Consider the three forces 'AB,' 'BC' and 'CD,' shown in Fig. 176. If we represent these forces by the vector lines *'ab,' 'bc'* and *'cd'* respec-

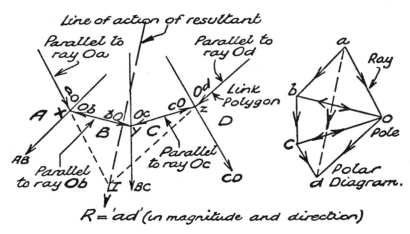

FIG. 176.—RESULTANT OF CO-PLANAR SYSTEM.

tively, the vector line *'ad'* will represent the resultant of the three forces in *magnitude and direction*. We do not know, however, where it acts in space in relationship to the forces in the system. To ascertain this it is necessary to proceed as indicated on page 117.

Procedure

Choose *any* point 'O' adjacent to the load line '*abcd*' and draw in the '*rays*' '*Oa*,' '*Ob*,' '*Oc*' and '*Od*.' At any convenient point 'X' in the line of action of force 'AB' and in the space 'A,' draw a '*link*' parallel to '*Oa*.' Draw link 'XY' (in space B) parallel to the ray '*Ob*,' and so on as indicated in Fig. 176. Produce the links in the spaces 'A' and 'D' to intersect in the point 'I.' The resultant of the given force system will pass through the point 'I.'

Proof.—The triangles formed by the '*rays*' in the '*polar diagram*' are really 'resolution triangles' for the forces in the system. Because all the triangles have a common apex point 'O' (the '*pole*') they achieve the 'resolution' in such a manner as to lead to the rule indicated above for the position of point 'I.' At the point 'X,' the force 'AB' is resolved into two components '*aO*' and '*Ob*' by means of the resolution triangle '*abO*.' At the point 'Y,' by resolution triangle '*bcO*,' the force 'BC' is replaced by its two components '*bO*' and '*Oc*.' Similarly at 'Z,' the components '*cO*' and '*Od*' are substituted for force 'CD.' But the components in space 'B' cancel out each other (they are equal in *magnitude* and *opposite* in direction). Similarly in space 'C' the components balance each other. The original system is therefore reduced to two forces, viz. force '*aO*' in space 'A' and force '*Od*' in space D.' The resultant of these two forces passes through 'I,' their point of intersection. The resultant of forces 'AB,' 'BC' and 'CD' must therefore pass through 'I.'

In the practical employment of the link polygon the arrow heads and the letters, as shown on the 'links' in Fig. 176, should be omitted. The arrow heads on the 'rays' are also omitted.

EXAMPLE (i).—*Find the resultant of the like parallel system of forces given in Fig. 177, by the link polygon method. Check the position of the resultant by the usual method of moments.*

The graphical solution is shown in Fig. 177. The reader should draw out the diagrams to the following scales: 25 mm = 2 m and 25 mm = 4kN.

Calculation of resultant.—The magnitude of the resultant = $(4 + 8 + 6 + 2)$ kN = 20 kN.

Let '*x*' m = distance at which the resultant acts from the 4 kN force.

Moment of resultant = sum of moments of components.

$$20 \times x = (4 \times 0) + (8 \times 4) + (6 \times 10) + (2 \times 14)$$
$$= 0 + 32 + 60 + 28$$
$$= 120$$
$$x = 6 \text{ m (as in graphical method).}$$

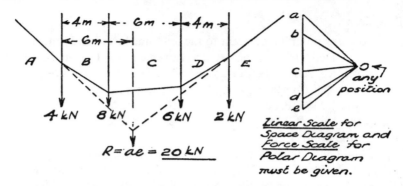

FIG. 177.—RESULTANT OF A PARALLEL SYSTEM.

EXAMPLE (ii).—*The loads shown acting on the roof truss in Fig. 178 are due to wind acting from the right. Find, by link polygon construction, the position of the resultant wind thrust on the truss.*

The load line is drawn, to a suitable force scale, *parallel to the wind loads.* The pole 'O' is chosen in any convenient position and the link polygon constructed in the usual way. The resultant wind load is a force

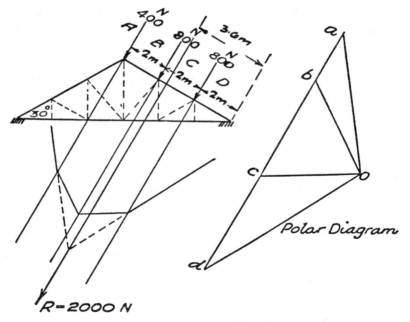

FIG. 178.— ᴇSULTANT WIND LOAD.

of 2 000 N, acting at 3·6 m from the right-end reaction point and in a
direction normal to the roof slope.

Experimental Demonstration of the Link Polygon

The apparatus shown in Fig. 179 may be used to illustrate the nature
of the link polygon. The fine string is suitably loaded so as to hang in
segments in front of a wall board. The string directions are transferred
to the board and the magnitudes of the weights noted. 'AB,' 'BC' and
'CD' form the load system.

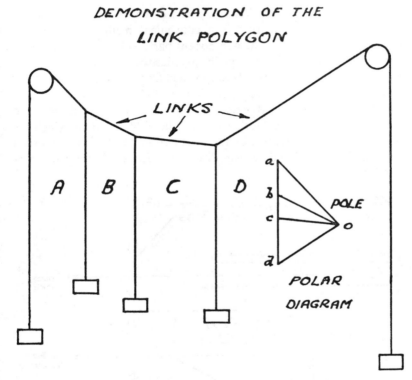

FIG. 179.—THE LINK POLYGON.

If these three loads be set out as a '*load line*' and lines be drawn
parallel to the string segments, from the appropriate points, the lines
will all intersect in one common point—the '*pole*' in the usual construction. The lengths of the top and bottom rays in the built-up '*polar*

diagram' will be found to represent the magnitudes of the suspended weights at the left and right ends of the string respectively. If spring balances were inserted in the string segments, their readings would correspond to the force values given by the corresponding '*rays*' in the polar diagram.

The experimental apparatus clearly demonstrates the 'vector diagram' nature of the 'polar diagram,' and the fact that the '*links*' represent lines of action of component forces.

Graphical Solution of Centre of Gravity Problems

When a body or an 'area' is composed of several component parts the link polygon construction is a convenient method of graphical solution. We may imagine the component parts as being subjected to a system of parallel 'pulls,' acting in any desired direction. Using the procedure previously described we can then fix the line of action of the resultant 'pull' for the chosen direction. The C.G. will lie somewhere along this line of action. By selecting another direction and repeating the method for this case, another line will be obtained along which the C.G. must lie. The C.G. will be the intersection point of any two such lines. The horizontal and vertical directions are usually the most convenient to take.

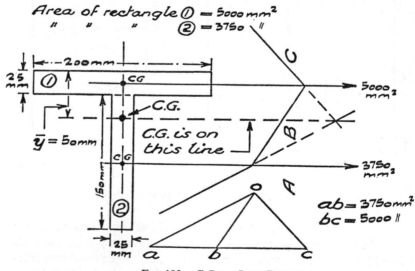

FIG. 180.—C.G. BY LINK POLYGON.

EXAMPLE.—*Find, by graphical construction, the height above base of the C.G. of the T-section given in Fig.* 180 (page 120).

The horizontal direction is chosen for the 'pulls' in this case.

Draw the section to a scale of $\frac{3}{4}$ fullsize. Divide up into two rectangles, marked (1) and (2) respectively in Fig. 180.

$$\text{Area of rectangle (1)} = 200 \times 25 = 5\,000 \text{ mm}^2.$$
$$\text{Area of rectangle (2)} = 150 \times 25 = 3\,750 \text{ mm}^2.$$

Set off '*ab*' to represent 'AB' (= 3 750 mm^2) and '*bc*' to represent 'BC' (= 5 000 mm^2). The area scale should be about 10 mm = 1 000 mm^2. Choose any pole 'O' (above or below the line '*ac*') and construct the polar and link polygons.

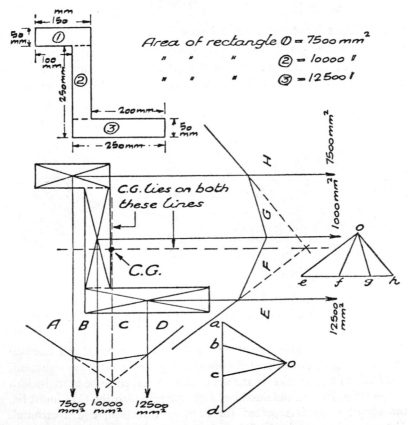

FIG. 181.—C.G. OF STRUCTURAL SECTION.

The C.G. is on the vertical axis of symmetry of the section and is 50 mm below the top of the section.

EXAMPLE.—*By means of the link polygon method, determine the C.G. of the Z-section given in Fig. 181.*

The section is divided into three rectangles:

$$\text{Area of rectangle (1)} = 150 \times 50 = 7\ 500 \text{ mm}^2$$
$$\text{Area of rectangle (2)} = 200 \times 50 = 10\ 000 \text{ mm}^2$$
$$\text{Area of rectangle (3)} = 250 \times 50 = 12\ 500 \text{ mm}^2$$

The complete graphical solution is given in Fig. 181. The C.G. is $154\frac{1}{6}$ mm from the left edge of the section and $129\frac{1}{6}$ mm from the bottom.

Recommended scales: Linear: $\frac{3}{4}$ full size.
Area: 10 mm = 2 000 mm².

Centre of Gravity of an Irregular Figure

As shown in Fig. 182, the link polygon construction may be employed to determine the C.G. of an irregular figure. The area is divided up into thin strips, the number of strips taken being governed by the nature of

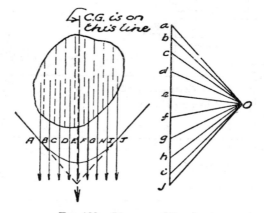

FIG. 182.—IRREGULAR FIGURE.

the outline of the figure. The area of each strip is computed and assumed to 'act' vertically downwards through the C.G. of the strip, as indicated. The C.G. of a strip may, in the usual case, be taken to be at mid-width of the strip. Having obtained one line, along which the C.G. must lie, the construction is repeated for another direction, e.g. horizontal. The second line obtained will intersect the first line at the required C.G.

Beam Support Reactions by Link Polygon

Procedure (see Fig. 183).—Draw the beam to a convenient scale and indicate the lines of action of the respective loads. If the beam should carry a distributed load system, the load must be divided up into a number of portions and the weight of each portion regarded as acting through its own centre of gravity. The number of portions chosen will be decided by the nature of the loading.

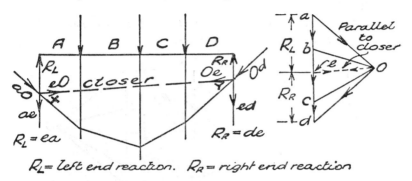

FIG. 183.—BEAM-SUPPORT REACTIONS.

A polar diagram is drawn for the loads and the link polygon constructed therefrom exactly in the manner previously described. The two outer links (in spaces 'A' and 'D') are not produced in this case to intersect. The points 'X' and 'Y' are found, these being the points in which the two outer links cut the reaction lines, respectively at the left end and right end of the beam. Points 'X' and 'Y' are joined, forming a line to which the name '*closer*' is given.

From the pole 'O' in the polar diagram a line 'Oe' is drawn, parallel to the 'closer.' The vector line 'ea' will represent the *left-end reaction* and 'de' will represent the *right-end reaction*, both, of course, to the force scale of the polar diagram.

Proof.—As previously shown on page 117, the force system 'AB,' 'BC' and 'CD' is equivalent to the two forces 'aO' (in space 'A') and 'Od' (in space 'D').

At point 'X,' by means of the resolution triangle 'aOe' resolve force 'aO' into force 'ae' (vertically downwards) and force 'eO' (along the closer). Similarly at point 'Y', by means of resolution triangle 'Ode,' resolve force 'Od' into force 'ed' (vertically downwards) and force 'Oe' along the closer. The two forces along the closer, being '*equal and opposite*, cancel out and we are left with force '*ae*' (at the left-end reaction point) and force '*ed*' (at the right-end reaction point). To balance these forces the support at the left end must exert an upward force equal to '*ea*' and, at the right end, the support must provide a reaction equal to '*de*.' The reader will note that in the preliminary drawing-in of the links in the appropriate spaces the reaction lines are completely ignored. The reaction lines do not form the boundary lines of a '*space*,' and enter into the construction only at the point when their intersection with the outer links is being considered.

EXAMPLE.—*Find the support reactions for the beam given in Fig.* 184 (i) *by a graphical method,* (ii) *by calculation.*

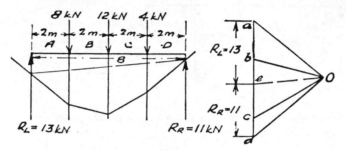

FIG. 184.—EXAMPLE ON REACTIONS.

In Fig. 184 the load line '*abcd*' is drawn and any pole 'O' chosen. The link polygon is constructed, the closer line drawn in and a line 'O*e*' is drawn parallel to the closer from the pole 'O.' In this case the line '*de*' scales 11 kN (= R_R) and '*ea*' scales 13 kN (= R_L).

It does not matter at what point in space 'A' the link polygon is commenced; and if the link in the space 'D' should come above the beam line, produce the right-end reaction line upwards to obtain the point of intersection for the closer.

Suggested scales: Linear: 50 mm = 1 m.
　　　　　　　　 Force:　50 mm = 8 kN.

Calculation of reactions:
$$R_L \times 8 = (8 \times 6) + (12 \times 4) + (4 \times 2)$$
$$= 48 + 48 + 8 = 104$$
$$R_L = 104/8 \text{ kN} = 13 \text{ kN}.$$
$$R_R \times 8 = (4 \times 6) + (12 \times 4) + (8 \times 2)$$
$$= 24 + 48 + 16 = 88$$
$$R_R = 88/8 \text{ kN} = 11 \text{ kN}.$$

EXAMPLE.—*Apply the link polygon construction to the example shown in Fig. 185, in which the support reactions are required for a beam which overhangs at both ends.*

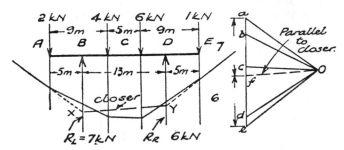

FIG. 185.—OVERHANGING BEAM.

Proceed as in the last example. When drawing the link in space 'B,' ignore altogether the reaction line of 'R_L.' Similarly, disregard the reaction line for 'R_R' when drawing the link in space 'D.' Produce the links in spaces 'A' and 'E' until they cut their respective reaction lines (as 'X' and 'Y'). Join 'XY' to form the 'closer' and proceed as in last example.

$$R_L = fa = 7 \text{ kN.}$$
$$R_R = ef = 6 \text{ kN.}$$

Suggested scales: Linear: 20 mm = 1 m.
Force: 12·5 mm = 1 kN.

The reader should check these reactions by the method shown on page 59.

Reactions for a Roof Truss

When a frame, such as a roof truss, carries loads inclined to the vertical, at least one reaction must be inclined. The practical treatment of reactions for trusses with inclined loads is considered in Chapter VIII.

In the example given in Fig. 186, the left-end reaction (R_L) is assumed to be vertical. With this assumption the value of 'R_L' and the direction and value of 'R_R' may be determined.

EXAMPLE.—*Find, by link polygon method, the support reactions for the roof truss shown in Fig. 186. The truss carries wind loads acting inwards, normally to the roof slope.*

Having drawn the load line '*bcdef*,' a pole 'O' is chosen in any position and the polar diagram completed in the usual manner. (The letter 'A' is placed in the space between the support reactions to correspond with the procedure adopted in Chapter VIII.) As the direction of 'R$_R$' is unknown, it is *necessary to begin the link polygon at the right-end reaction point*. If this were not done it would be impossible to ascertain the point in which the link in space 'F' (i.e. the link parallel to 'O*f*')

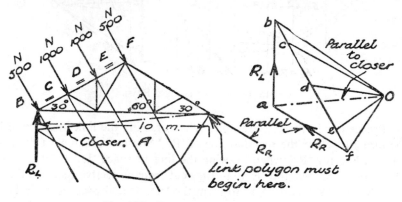

FIG. 186.—REACTIONS FOR WIND LOADS.

would cut the reaction line, as the direction of the reaction cannot at this stage be drawn-in. The only point we are certain will be on the reaction line is the *right-end reaction point*, hence we arrange the link in space 'F' to go through this point and construct the link polygon by working towards the left. The intersection of the link in space 'B' with the left-end reaction is determinable, hence the 'closer' may be constructed.

The left-end reaction being vertical in this case, a vertical line is drawn downwards from '*b*' in the polar diagram, to cut a line drawn from 'O,' parallel to the closer, in the point '*a*.' Point '*f*,' in the load line, is joined to '*a*.'

'*fa*' will represent the right-end reaction. The left-end reaction will be represented by the vector line '*ab*.' The proof is similar to that given on page 123.

The magnitudes of both reactions will be found to be 1 732 N.

Suggested scales: 20 mm = 1 m and 50 mm = 1 000 N.

Further examples illustrating the employment of the link polygon occur throughout the book.

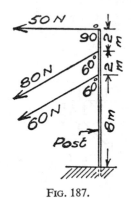

FIG. 187.

(1) A vertical post 12 m high is pulled by three wires as shown in Fig. 187. Find, by means of a link polygon, the resultant pull on the post.

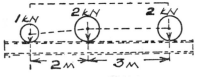

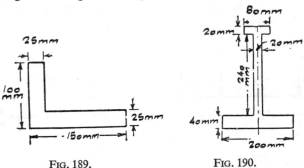

FIG. 188.

(2) The wheel loads given in Fig. 188 are transmitted to a girder by the carriage of a travelling crane. Find (i) by link polygon, (ii) by calculation method, the distance, from left-hand wheel, at which the resultant wheel load acts.

(3) Find, by link polygon construction, the centre of gravity of the unequal angle section given in Fig. 189.

FIG. 189.

FIG. 190.

(4) Verify, by means of a link polygon, that the C.G. of the girder section shown in Fig. 190 is $96\frac{2}{3}$ mm above the bottom of the section.

(5) In order to test the stability of the retaining wall represented by the section given in Fig. 191, it is necessary to determine the distance of the C.G. of the section from the vertical back of the wall. Find this distance by the link polygon method.

(If a 'line of action' through the centre of gravity of one component area be coincident with that through another, the two areas may be regarded as being one combined area for purposes of the link polygon construction.)

s.m.(m)—5*

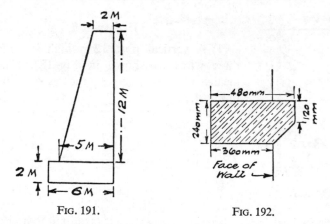

FIG. 191.   FIG. 192.

(6) Fig. 192 shows the section of a stone corbel built into a brick wall. Show, by link polygon construction, that the centre of gravity of the corbel section is $133\frac{1}{3}$ mm from the face of the wall.

(Divide up the section into simple component figures, e.g. two rectangles and a triangle.)

(7) Find the beam support reactions for each of the beam examples given in Fig. 193 (i) by link polygon, (ii) by calculation.

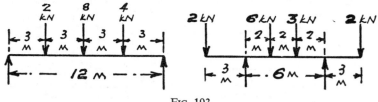

FIG. 193.

(8) Obtain, graphically, the support reactions for the given braced girder (Fig. 194).

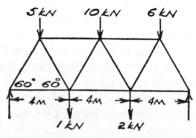

FIG. 194.

(The 'panel' loads shown may be regarded as acting (in their respective vertical lines of action) on a simple beam 12 m long.)

(9) The roof truss given in Fig. 195 carries four inclined wind loads (acting at 90° to roof slope). Find the value of each support reaction by means of a link polygon.

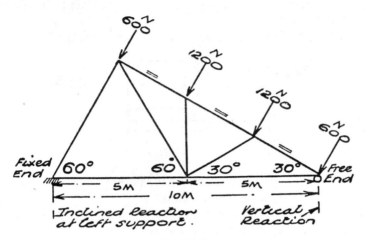

FIG. 195.

(Commence the link polygon at the fixed end, i.e. the left-end reaction point. Carefully watch the correspondence between 'space' letters and polar diagram 'ray' letters. The ray corresponding to the space letter to the left of the left-hand rafter will not be required to be considered, as the link to be drawn parallel to it passes through the reaction point and need not actually be drawn in.)

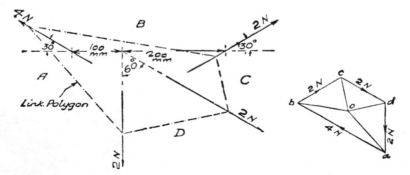

FIG. 196.—NON-CONCURRENT SYSTEM IN EQUILIBRIUM.

(10) A law of equilibrium states '*the link polygon for a system of non-concurrent forces in equilibrium is a closed diagram.*'

Fig. 196 shows a non-concurrent system of four forces in equilibrium, viz. 4 N, 2 N, 2 N and 2 N respectively, with the corresponding force polygon. Draw out a space diagram according to the data given. Taking a '*pole*' in the approximate position shown, construct a link polygon and verify the law given above.

(An example of such a link polygon is shown in broken lines.)

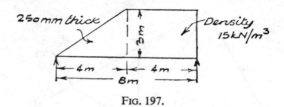

FIG. 197.

(11) Find the support reactions for the beam given in Fig. 197 (i) by link polygon, (ii) by calculation method.

(12) Check the position of the resultant wind load for the example given in Fig. 112 by means of a link polygon.

CHAPTER VIII

CONSTRUCTION OF A STRESS OR FORCE DIAGRAM FOR A LOADED FRAME

Introduction.—Fig. 198 shows an experimental roof truss. When a load is placed on the hook at the apex of the truss it will be observed that the various members forming the frame alter in length—the inclined members become shorter and the horizontal string member is extended.

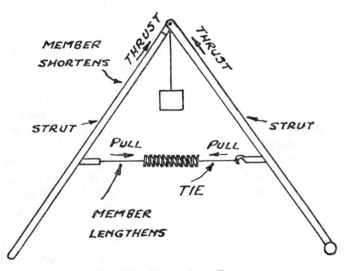

FIG. 198.—MODEL ROOF TRUSS.

Alteration in length is just what members of a practical frame, such as a steel or timber roof truss or a lattice girder, undergo when external load is applied to the frame. We may regard the members of a frame as being very strong springs, and as such they tend to resist any attempt at deformation. Members which are extended are said to be in *tension*. They exert pulls on their end connections and are termed '*ties.*' Those members which are shortened are in 'compression' and are termed 'struts.' A 'strut' exerts a thrust at both its ends (Fig. 199).

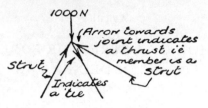

FIG. 199.—STRUTS AND TIES.

The purpose of drawing a *stress diagram* is to determine which members of a loaded frame are 'struts' and which are 'ties,' and also to ascertain the magnitude of the force in each member. A better name for a 'stress diagram' would be *'force diagram.'* This name is sometimes applied, but 'stress diagram' is distinctive of the type of force diagram now under consideration.

Assumptions made.—It is assumed that each member pulls or pushes in the direction of its length. If we draw a diagram showing the centre lines of the members forming the frame (the *'frame'* or *'space'* diagram) we may regard these lines as indicating the lines of action of the various member forces. For the typical practical frame, in which the joints are not constructed with a view to possessing special rigidity, this is a reasonable assumption.

It is also assumed that the loading may be reduced to a system of point loads, acting at the joints or *'nodes'* of the frame. In the case of a roof truss the 'purlins,' which actually transmit the roof load to the truss, are situated at, or near, the joints. Loads applied between joints tend to cause bending in the members, and special calculations may have to be made to allow for this. For the purpose of stress-diagram construction such loads are proportioned between the adjacent joints in the manner of beam support reaction calculations.

Calculation of Joint Loads for a Roof Truss

In examination problems the joint loads are usually given. The examples following illustrate the methods of obtaining these loads. It is assumed that the purlins transmit the roof load to the truss at the joints.

EXAMPLE (i).—*Calculate the joint loads for the roof truss given in*

Fig. 200. The roof covering is composed of the material indicated in the following list of weights:

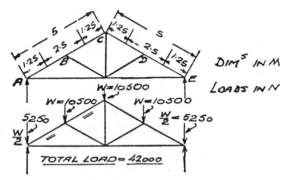

FIG. 200.—JOINT LOADS FOR TRUSS.

Purlins and ridge:	50 N *per sq. metre of roof area.*
Common rafters:	80 N *per sq. metre of roof area.*
Roof boarding:	150 N *per sq. metre of roof area.*
Felt:	50 N *per sq. metre of roof area.*
Slating laths:	30 N *per sq. metre of roof area.*
Slates:	450 N *per sq. metre of roof area.*
Possible snow (say):	150 N *per sq. metre of roof area.*
	960 N *per sq. metre of roof area.*

The trusses are spaced at 4 m centres.

Each truss has to support a roof area of $(5 + 5)\,\text{m} \times 4\,\text{m} = 40\,\text{m}^2$. The total roof load for one truss is therefore $(40 \times 960)\,\text{N} = 38\,400\,\text{N}$. The self-weight of the truss is added in to the roof-covering load, and for this a provisional estimate must be made. Experience of previous similar calculations will indicate the allowance to be made.

The reader is referred to text-books on building construction, etc., for formulae and graphs which will also assist in arriving at a reasonable figure for the self-weight of a particular type of truss.

The truss will be assumed to weigh 3 600 N in the present example.

Total load carried by one truss $= (38\,400 + 3\,600)\,\text{N}$
$$= 42\,000\,\text{N}.$$

We have now to divide up this load proportionately between the joints.

Area of roof for joint 'A' = $1 \cdot 25$ m × 4 m = 5 m^2
Area of roof for joint 'B' = $2 \cdot 5$ m × 4 m = $10 \cdot 0$ m^2

The loads at 'B', 'C' and 'D' will be equal and each will be twice that at 'A' or 'E'.

$$\text{Let 'W' N} = \text{load at 'B', 'C' or 'D'.}$$
$$\therefore \quad \text{W/2 N} = \text{load at 'A' or 'E'.}$$
$$\therefore \quad 3W + 2(W/2) = 42\ 000$$
$$\therefore \quad W = 10\ 500 \text{ N.}$$

EXAMPLE (ii).—*Assuming the data given in Example* (i), *calculate the joint loads for the truss given in Fig.* 201. *The trusses are at 4 m centres.*

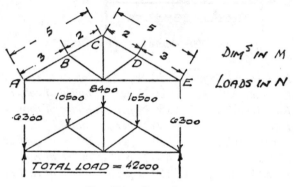

FIG. 201.—JOINT LOADS.

As in previous example, the total load carried by the truss = 42 000 N.

Area of roof for joint 'A' = $1 \cdot 5$ × 4   = $6 \cdot 0$ m^2
Area of roof for joint 'B' = $(1 \cdot 5 + 1 \cdot 0)$ × 4 = $10 \cdot 0$ m^2
Area of roof for joint 'C' = $(1 \cdot 0 + 1 \cdot 0)$ × 4 = $8 \cdot 0$ m^2
Area of roof for joint 'D' = (as 'B')   = $10 \cdot 0$ m^2
Area of roof for joint 'E' = (as 'A')   = $6 \cdot 0$ m^2
                                                $40 \cdot 0$ m^2

The load at any particular joint will be proportional to the area of roof associated with it.

$\therefore$ Joint load at 'A' = joint load at 'E'

$$= \frac{6}{40} \times 42\ 000 \text{ N} = 6\ 300 \text{ N.}$$

∴ Joint load at 'B' = joint load at 'D'

$$= \frac{10}{40} \times 42\,000 \text{ N} = 10\,500 \text{ N}.$$

Joint load at 'C' $= \frac{8}{40} \times 42\,000 \text{ N} = 8\,400 \text{ N}.$

Total load $= [(2 \times 6\,300) + (2 \times 10\,500) + 8\,400]\text{ N} = 42\,000 \text{ N}.$

Wind Loads on Roof Trusses.—The magnitude and treatment of wind pressure in the design of a structure depend upon the nature of the structure. In the case of roof trusses the following points should be noted:

(i) The wind may blow from the left or from the right on to the truss. The two cases may have to be considered independently in order to arrive at the worst case for any particular member of the truss.

(ii) It is the wind load acting at right angles to the roof slope which has to be considered. Empirical formulae are sometimes used in order to derive a practical value for this component pressure from a given horizontal wind pressure. Some building regulations give the necessary normal wind pressure directly.

(iii) In general, it will be found that with slopes up to about 35°, the windward side of a roof truss will be subjected to *'negative wind loads,'* i.e. loads acting outwards at right angles to the roof slope. Trusses with slopes steeper than 35° will generally be subjected to *'positive wind loads,'* i.e. loads acting inwards at right angles to the roof slope. On the leeward side, the wind loads will always be *'negative.'*

(iv) As in the case of dead loads, all wind loads are actually transmitted to the roof truss through the purlins. The usual example will therefore have both the dead loads and the wind loads acting at the joints of the truss.

An example illustrating the calculation of winds loads will be found on page 151.

Nature of a Stress Diagram

A stress diagram is an amalgamation into one diagram of a number of force polygons. Consider the example of the northlight roof truss given in Fig. 202. For the loading given, the support reactions are: 1 050 N (left end) and 1 150 N (right end).

The triangle of forces *'abe'* is drawn for the three forces acting at the

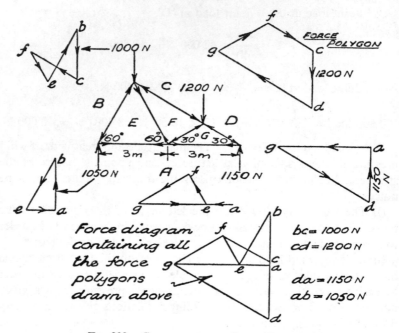

FIG. 202.—COMPOSITION OF A STRESS DIAGRAM.

left-end reaction point. We find that member 'BE' is a 'strut,' i.e. it exerts a thrust at the joint. The magnitude of the thrust may be found by scaling the vector line '*be*.' Member 'BE' will also exert this thrust at the apex joint of the truss. This fact enables us to construct the vector line '*eb*' (see top left-hand diagram), and knowing that '*bc*' must scale 1 000 N (the external load at the apex), we are enabled to complete the force polygon for the four forces at the apex joint. By taking the various joints in a proper sequence—and using the results obtained at one joint to aid the solution of the next—the force polygon for each joint of the given truss has been constructed, as indicated in Fig. 202. We have thus completely 'solved' the truss.

Now examine the lower diagram shown in the figure. The reader should have no difficulty in tracing out, in this diagram, each of the five force polygons previously drawn for the respective five joints of the truss. Such a diagram—a composition of a number of force polygons—is termed a '*stress diagram.*' We have now to consider how to construct a stress diagram directly, without having to draw a number of separate force polygons.

Rules for Constructing a Stress Diagram

Taking the typical example given in Fig. 202, the steps in the procedure are as follows:

(1) Draw a scale diagram of the frame, indicating the loads acting at the various points.

(2) Obtain the support reactions and mark these clearly at their respective positions in the frame (or 'space') diagram.

(3) Letter-up the spaces by Bow's notation. Throughout the book, the letter 'A' will be allocated to the space between the two reactions. The letter 'B' will occupy the first space at the left end of the truss and lettering will proceed round the frame in a clockwise manner, completing first the external spaces formed by the loads.

(4) Draw the *'load line bcd'* to a suitable force scale. Insert the letter *'a'* on this line by marking off *'da'* to represent '1 150 N' (i.e. the right-end reaction). (The completed load line represents the *'force polygon'* for the external load system.)

(5) Consider the truss joint by joint. First take the left-end reaction joint. From *'b'* (in the 'load line') draw *'be'* parallel to 'BE' and from *'a'* draw *'ae'* parallel to 'EA.' (In order to get the intersection point *'e'* we have to draw from the point *'a'* towards the left.)

(6) Having completed the portion of the stress diagram for a given joint, determine whether the members concerned are *'struts'* or *'ties'*. (The rules for this are given later.)

(7) Take the next convenient joint in the frame—the apex joint in the given example. Considering the forces round the joint in clockwise order, we have, 'BC,' 'CF,' 'FE' and 'EB'. Draw *'cf'* and *'ef'* from *'c'* and *'e'* respectively parallel to forces 'CF' and 'FE.' This fixes the point *'f'* in the stress diagram. The forces in clockwise order round the central joint in the bottom tie are 'AE,' 'EF,' 'FG' and 'GA.' 'AE' and 'EF' are already represented in the stress diagram, so we draw *'fg'* parallel to 'FG' until it meets *'ag'* (parallel to 'GA') in the point *'g.'* Thus the point *'g'* is fixed. The rafter joint at which the '1 200 N' load acts and the right-end reaction joint are similarly dealt with. The accuracy of work in the construction of a stress diagram is checked by the fact that a *'stress diagram must close.'* In the example taken, the accuracy will be checked by ascertaining whether the vector line drawn from *'d'* parallel to member 'DG' passes through the point *'g,'* which has been already fixed in the stress diagram.

(8) The stress diagram being now complete, it is usual to draw up a

table, as indicated in Fig. 203, giving the force in each member. The force in any given member will be obtained by scaling (to the force scale of the 'load line') the corresponding vector line in the stress diagram.

Determination of 'Struts' and 'Ties'

The method will be explained by taking one particular joint in the previous example, i.e. the apex joint of the truss.

Take the member letters in *clockwise order* round the joint, thus: 'CF,' 'FE' and 'EB.' Considering, firstly, the member 'CF,' examine the stress diagram and ascertain in which direction you have to proceed along the appropriate vector line '*cf*' in order to obtain the same sequence of letters. This is clearly '*c*' to '*f*' (*not* '*f*' to '*c*'), i.e. upwards towards the left. Place an arrow *at the apex joint on the member* '*CF*' in this direction. The arrow is pointing towards the joint, indicating that the member is pushing (see Fig. 202). Member 'CF' is therefore a '*strut*.' For member 'FE,' the order of letters indicates that we must proceed from '*f*' to '*e*' in the stress diagram, i.e. downwards towards the right. The arrow on 'FE' at the apex joint therefore denotes a 'pull,' i.e. member 'FE' is a '*tie*.' Similarly, the nature of the force in any member at any joint may be determined.

General Remarks.—The following points should be carefully noted:

(i) Concentrate on one joint only at a time. See that all the 'letters' found are placed in the stress diagram.

(ii) Place the arrow heads on the frame diagram before you leave a given joint. The arrow heads should be placed on the members fairly near the joint considered, to avoid possible error.

(iii) If you are held up at a joint because there are too many 'letters' unfixed in the stress diagram, it is probably because you have missed a joint which will provide you with the necessary letters.

The reader is strongly advised to draw the stress diagram for the example given in Fig. 202.

Suggested scales:

20 mm = 1 m for the frame diagram, and
50 mm = 1000 N for the stress diagram.

Types of Stress Diagrams

1. Truss with Symmetrical Loads

In this case the two support reactions are equal and the point '*a*' is the mid-point of the load line. Fig. 203 shows the stress diagram for a

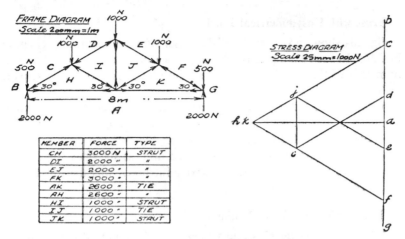

FRAME DIAGRAM
Scale 200mm = 1m

STRESS DIAGRAM
Scale 25mm = 1000N

MEMBER	FORCE	TYPE
CH	3000 N	STRUT
DI	2000 "	"
EJ	2000 "	"
FK	3000 "	"
AK	2600 "	TIE
AH	2600 "	"
HI	1000 "	STRUT
IJ	1000 "	TIE
JK	1000 "	STRUT

FIG. 203.—TRUSS WITH SYMMETRICAL LOADS.

king-post roof truss. The reader should develop this diagram himself, following the rules given on page 137.

A suitable table for recording the force in each member is shown in the figure. In the case of a symmetrical truss with symmetrical loads the stress diagram will be a symmetrical figure about a horizontal line through the point 'a.' It is advisable in examination work, if time permits, to complete the whole stress diagram even though half the diagram would be sufficient to obtain all the member forces. The 'closure' check may thereby be applied.

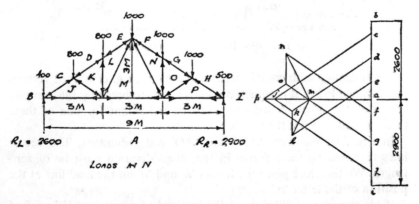

FIG. 204.—UNSYMMETRICAL LOADS.

2. Truss with Unsymmetrical Loads

Fig. 204 shows an example of this case. Another example is given in Fig. 205. In such examples the support reactions must first be obtained. By means of the reaction values, the point '*a*' is fixed (see rule 4, page 137).

Note that the letter 'A' is associated with all the bottom horizontal members of the truss. The reader should find no difficulty in tracing the development of the stress diagram. In Fig. 204 the three joints on the left-hand rafter were first considered and, as the apex joint could not then be proceeded with, the next joint taken was the lower joint 'JKLMA.'

3. Truss with Redundant Members

A '*redundant member*' is one which could be removed from a frame without, theoretically, causing its collapse, the remaining members being capable of maintaining static equilibrium—provided the load system remain unaltered. Distortion of the frame, on a change of the load system, might make such a member a *practical* necessity. Figs. 205 and 206 illustrate typical cases of redundant members.

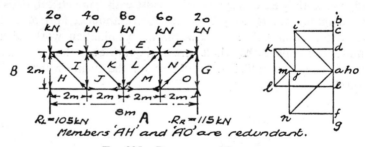

FIG. 205.—REDUNDANT MEMBERS.

When dealing with complicated load systems, e.g. combined vertical and inclined loads, it is sometimes convenient to break up into simpler component systems. A member may be redundant for one of these load systems and not for another.

In Fig. 205, members 'AH' and 'AO' are redundant. The vector lines representing these forces in the stress diagram must be of zero length. We therefore place the letters '*h*' and '*o*' on the load line at the position of the letter '*a*.'

If we consider equilibrium at the left end of member 'AH,' we find

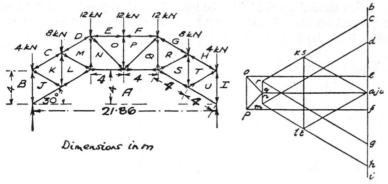

FIG. 206.—FOOT BRIDGE TRUSS.

that the forces in member 'BH' and reaction 'AB' are both vertical.
Therefore for horizontal equilibrium member 'AH' cannot be exerting
any force. Similarly, in Fig. 206, the force in member 'AJ' must be zero.
If member 'AJ' did exert a force, it would have a horizontal component.
This is impossible, as there is no other horizontal force at the left-end
reaction point to balance it. Member 'AU' is also redundant.

4. Truss with Loads on the Bottom Tie

It is convenient in this case (Fig. 207) to place the letter 'A' in the
space nearest to the left-end reaction. Note that in this example there
are several letters beneath the bottom tie. In setting out the '*load line*'

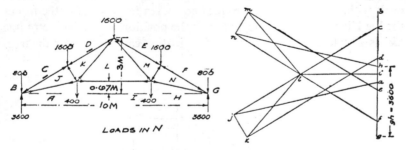

FIG. 207.—LOADS ON BOTTOM TIE.

the vector lines must be drawn with due regard to the direction of the
forces concerned. Thus '*gh*' is measured upwards (because the reaction
'GH' acts upwards), and '*hi*' is plotted downwards (because load 'HI'
acts downwards).

Having fixed the point '*a*' in the load line, the vector line '*ab*' should be verified—it should scale 3 600 N. The setting out of the load line having been checked, the stress diagram is constructed in the usual way.

5. Truss with Unsymmetrical Loads Solved with the Aid of a Link Polygon

The position of the point '*a*' in the load line (Fig. 208) is fixed by drawing in the closer line in the link polygon and drawing '*Oa*' parallel to the closer (see page 123).

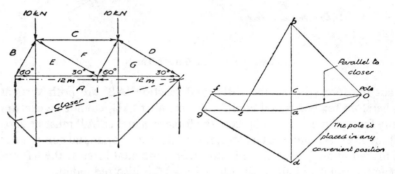

FIG. 208.—REACTIONS BY LINK POLYGON.

6. Cantilever Truss

In trusses of the type in Fig. 209 we may proceed with the stress diagram without first obtaining the reactions. The load line '*bcdefg*' having been drawn, the joint at the extreme right end of the truss is first dealt with. The stress diagram does not begin to grow from the top of the load line in this example, but from the position where the vector line '*de*' is situated. It will be noted that, in this case, the stress diagram

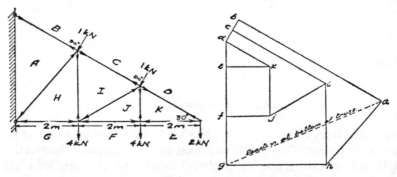

FIG. 209.—CANTILEVER TRUSS.

develops to the right of the load line. If the truss has a vertical member connecting the reaction points, we require to know the direction of one of the reactions before this member can be solved.

7. Truss Overhanging its Supports

In Fig. 210, the truss overhangs the supports at both ends. The reader should construct the stress diagram to the following scales: 20 mm = 1 m and 10 mm = 1 kN.

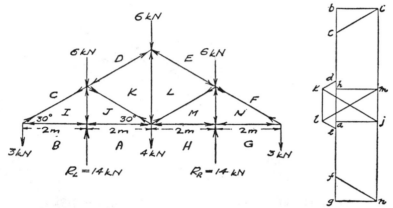

FIG. 210.—OVERHANGING TRUSS.

8. Truss Solved by Means of a Calculated Member

It would not be possible to draw a complete stress diagram for the truss given in Fig. 211A by the usual methods. When joint 'DEPONM' was being considered, it would be found that more than two members

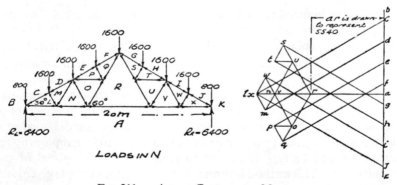

FIG. 211A.—AID OF CALCULATED MEMBER.

were unknown. This would hold up the construction of the stress diagram. There are various methods of getting over the difficulty. A good method in such cases is to calculate the force in a suitable member of the truss. In this case the tie 'AR' is a suitable member.

If the tie were removed, the truss would collapse by turning about the apex joint as a fulcrum (see Fig. 211B). The tie member must therefore be pulling with just sufficient force to maintain the left-hand portion of the truss in equilibrium.

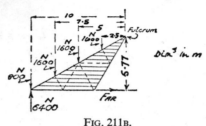

FIG. 211B.

Let 'F' N be the pull in the tie member 'AR.'
Taking moments about apex of truss:

$$(F \times 5{\cdot}77) + (1\ 600 \times 2{\cdot}5) + (1\ 600 \times 5) + (1\ 600 \times 7{\cdot}5)$$
$$+ (800 \times 10) = 6\ 400 \times 10$$
$$5{\cdot}77F + 4\ 000 + 8\ 000 + 12\ 000 + 8\ 000 = 64\ 000$$

$$5{\cdot}77F = 64\ 000 - 32\ 000 = 32\ 000$$

$$F = \frac{32\ 000}{5{\cdot}77} = 5\ 540\ N.$$

The force in member 'AR' is therefore 5 540 N. The vector line 'ar,' which represents this force in the stress diagram, must scale 5 540 N. This fixes the point 'r' in the stress diagram. The joint 'ANORA' can now be completed and the remainder of the stress diagram finished off in the usual way.

Alternative Solution.—A method known as the '*reversal of diagonal*' method may be employed to get over the difficulty of the completion of the stress diagram in this case. Member 'PO,' which is a diagonal of a quadrilateral formed by four members, is assumed to be temporarily removed and to be replaced so as to form the other diagonal of the quadrilateral. This method is further explained in the example given later (see Fig. 217).

Wind Load Stress Diagrams

Determination of Reactions

When a frame is subjected to loads which are inclined to the vertical, it is not possible for both reactions to be vertical. One, or both, of the reactions must contribute a horizontal force to balance the horizontal effect of the inclined loads. If one end of the truss is so supported as to allow practical freedom for horizontal movement—so that it may be regarded as being supported by a roller bearing—it is usual to assume that the reaction at this end is vertical. Such freedom is sometimes necessary, in order to allow for expansion due to temperature change. In this case the reaction at the '*fixed*' end must supply the necessary horizontal force to prevent the lateral movement of the truss.

If both ends are *similarly fixed* on rigid supports, it is a usual assumption that each reaction is parallel to the resultant force acting on the truss.

In the case of a roof truss supported by steel columns (of approximately equal stiffnesses), it is usual to assume that the horizontal components of the reactions are equal.

Regulations differ in their requirements in respect to the combination of *dead* and *wind* loads. It will be clear that the actual design load for any given member of the frame will be the algebraic sum of the loads produced independently (and simultaneously) by the dead load and by the wind loads. Some regulations require the 'negative' wind loads ('*suction loads*') to be taken without the accompaniment of the positive wind loads. When the various loads are treated independently careful tabulation will be necessary in order to arrive at the maximum possible load for any particular member.

The general principles of stress-diagram construction for wind loads are illustrated in the examples which follow.

*Examples of Roof Trusses Supporting Wind Loads

Positive Wind Loads Only

In this case (Fig. 212) we can determine the direction of the right-end reaction by the rule on page 37. A line is drawn from '*e*,' in the load line, parallel to the ascertained direction to meet a vertical line (representing the direction of the free-end reaction) drawn from '*b*.' The lines meet

*More advanced students should consult C.P. 3, Chapter V.

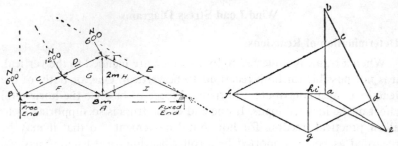

FIG. 212.—POSITIVE WIND LOADS.

in the point '*a*.' The stress diagram is then readily constructed. Member 'HI' is redundant.

Positive Wind Loads and Dead Loads Combined

The example given in Fig. 213 illustrates a method of dealing with dead loads and positive wind loads acting simultaneously. The wind is assumed to act from the left. The wind loads and dead loads, acting at the respective joints on the left slope of the truss, are combined into resultant single loads by parallelograms of forces. In this case the direction of the fixed-end reaction is determined by the construction of a *link polygon.*

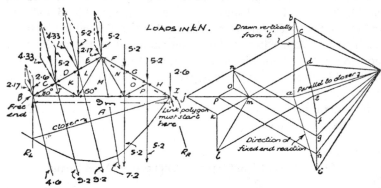

FIG. 213.—POSITIVE WIND AND DEAD LOADS.

To avoid the difficulty of having to find where the link in space 'H' cuts the fixed-end reaction line of action, it is necessary to commence drawing the link polygon from the fixed-end reaction point (which, of course, will be on the reaction). As we start drawing the links from this point, the link corresponding to space 'I' is not utilised. The reader should very carefully check that the links are drawn in their proper spaces in these examples. The reaction problem of this example was previously dealt with on page 126.

Negative Wind Loads and Dead Loads Combined

The example given in Fig. 214 assumes wind acting from the left, so that the negative wind loads act outwards on the right slope of the truss. Instead of drawing complete parallelograms of forces, the negative wind

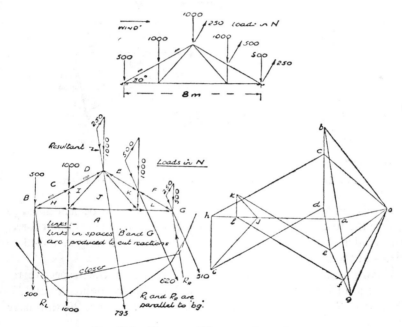

FIG. 214.—NEGATIVE WIND AND DEAD LOADS.

loads and dead loads have been combined by the drawing of one vector line at the end of the other (see page 10). The resultant force on the truss is represented by '*bg*' and each reaction is assumed to be parallel to '*bg*' in this example.

There is no need to start the link polygon at any particular reaction point in this case, as the directions of the reactions can be drawn-in on the frame diagram. The point '*a*' is on the line '*bg*.'

It will be noted that the links in spaces 'B' and 'G' respectively have to be produced till they cut the reaction lines in order that the closer line may be drawn-in.

Positive Wind Loads, Negative Wind Loads and Dead Loads Combined

In the example illustrated in Fig. 215 the wind is taken as acting from the right. Both reactions are assumed to be parallel to the resultant thrust on the truss.

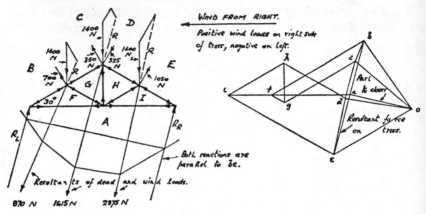

FIG. 215.—POSITIVE WIND, NEGATIVE WIND AND DEAD LOADS.

In this example the *successive vector line* method has been adopted to reduce the forces acting, at any particular joint, to one resultant force. The reader should particularly notice the compounding of the three forces at the apex joint of the truss.

Having drawn the load line '*bcde*' to represent the respective resultant joint loads 870 N, 1 615 N and 2 375 N, the points '*e*' and '*b*' are joined. '*be*' represents the resultant force on the truss. We thus have the direction of each reaction fixed.

No loads were given for the eaves joints of the truss in this case. Such loads are sometimes omitted in problems on stress-diagram construction on the supposition that the common rafters between the roof trusses transmit their bottom end loads, not to purlins carried by the truss, but to wall plates.

Example Involving Equal Horizontal Reactions

The roof truss shown in Fig. 216 is supported on steel columns of equal stiffnesses. The particular point to note in such problems is the effect, on the values of the reactions, of the overturning moment caused by the combined horizontal components of the positive and negative wind loads. The lower diagram in Fig. 216 shows the load system reduced to resultant vertical and horizontal components. (The reader should check these components graphically or by the '*cosine*' rule).

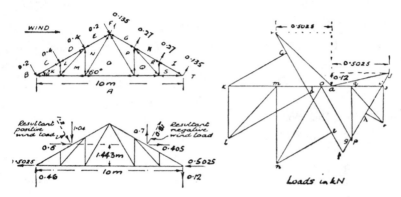

FIG. 216.—EQUAL HORIZONTAL REACTIONS.

Vertical Reactions (*due to vertical loads only*)

$$R_L \times 10 = (1{\cdot}04 \times 7{\cdot}5) - (0{\cdot}7 \times 2{\cdot}5) \text{ [Note the minus sign.]}$$
$$= 7{\cdot}0 - 1{\cdot}75$$
$$= 6{\cdot}05$$

$$R_L = \frac{6{\cdot}05}{10} \text{ kN} = 0{\cdot}605 \text{ kN } (upwards).$$

$$R_R \times 10 = (1{\cdot}04 \times 2{\cdot}5) - (0{\cdot}7 \times 7{\cdot}5)$$
$$= 2{\cdot}6 - 5{\cdot}25$$
$$= -2{\cdot}65$$

$$R_R = -\frac{2{\cdot}65}{10} \text{kN} = -0{\cdot}265 \text{ kN } (downwards).$$

Vertical Reactions (*due to overturning moment*)

Total horizontal force acting on truss

$$= (0{\cdot}6 + 0{\cdot}405) = 1{\cdot}005 \text{ kN}$$
$$\text{Height of truss} = 5 \tan 30° = 2{\cdot}886 \text{ m}.$$

$$\text{Arm of overturning moment} = \frac{2{\cdot}886}{2} = 1{\cdot}443 \text{ m}.$$

Taking moments about right-end reaction point:

$$R_L \times 10 = 1{\cdot}005 \times 1{\cdot}443 = 1{\cdot}45$$

$$\therefore \quad R_L = \frac{1{\cdot}45}{10} \, kN = 0{\cdot}145 \, kN \; (downwards).$$

Moments about left-end reaction point:

$$R_R \times 10 = 1{\cdot}005 \times 1{\cdot}443 = 1{\cdot}45$$

$$\therefore \quad R_R = \frac{1{\cdot}45}{10} \, kN = 0{\cdot}145 \, kN \; (upwards).$$

Net Vertical Reactions

R_L = 0·605 kN (upwards) − 0·145 kN (downwards)
   = 0·46 kN (upwards).
R_R = 0·265 kN (downwards) − 0·145 kN (upwards)
   = 0·12 kN (downwards).

Horizontal Reactions

The horizontal reactions being equal at both supports, the value of

$$\text{each reaction} = \frac{0{\cdot}6 + 0{\cdot}405}{2} \, kN = \frac{1{\cdot}005}{2} \, kN = 0{\cdot}5025 \, kN.$$

Construction of Stress Diagram

The load line 'bcdefghij' is constructed in the manner indicated in Fig. 216. From 'j' a vector line is drawn horizontally towards the left to represent 0·5025 kN, the horizontal component of the right-end reaction. At the end of this line a vector line is drawn vertically downwards to represent 0·12 kN, the vertical (downward) component of the right-end reaction. We thus get 'ja' as the vector line representing the total right-end reaction. By joining 'a' to 'b' the complete polygon of forces, for the external load system acting on the truss, is formed. The stress diagram is constructed in the usual way off this polygonal diagram.

Calculation of Wind Joint Loads

The principles of the calculations for wind loads are the same as for dead loads. As an example we will take the truss illustrated in Fig. 213. The dead load was taken as 1 kN per m² of roof area. The wind load was assumed to be 0·833 kN per m² inwards on the windward side. The trusses were assumed to be at 3 m centres.

Length of principal rafter $= \dfrac{4\cdot5}{\cos 30°} = 5\cdot196$ m.

Area of roof slope on each side $= (5\cdot196 \times 3)$ m²
$$= 15\cdot6 \text{ m}^2.$$

Dead load: at 1 kN per sq. metre, the total on each roof slope $=$ $(1\cdot0 \times 15\cdot6)$ kN $= 15\cdot6$ kN.

Full joint load $= \dfrac{15\cdot6}{3}$ kN $= 5\cdot2$ kN.

∴   Half joint load $= 2\cdot6$ kN.

Positive wind load: at 0·833 kN per sq. metre, total load on windward roof slope $= (0\cdot833 \times 15\cdot6)$ kN $= 13\cdot0$ kN.

Full joint load $= \dfrac{13\cdot0}{3}$ kN $= 4\cdot33$ kN.

∴   Half joint load $= 2\cdot17$ kN.

French Truss by 'Reversal of Diagonal' Method

In the example illustrated in Fig. 211A, the truss carried dead loads only and was solved by the calculation of the force in one of the members of the truss.

In Fig. 217 we have a similar type of truss with positive wind and dead loads. Both reactions are assumed to be parallel to the resultant force on the truss. The stress diagram can be constructed until the joint 'NMDEPO' is reached. The procedure then is as follows:

Remove member 'PO' and replace by the diagonal member shown by a broken line. The amended space notation is shown in the separate small inset diagram.

The original member 'PQ' becomes redundant and is ignored. Its redundancy can be verified by considering equilibrium conditions at its lower end. Joint DE ① NM may now be completed and the point ① fixed.

The stress diagram may be developed by taking the joint 'EF ② ①,' the temporary member ① ② being represented by the broken line in ① ② the stress diagram.

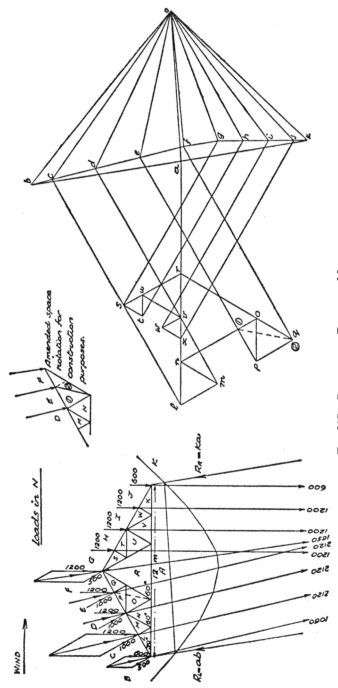

FIG. 217.—REVERSAL OF DIAGONAL METHOD.

The letter 'q' is substituted for the figure ②, the diagonal member 'PO' is replaced in its correct position and the stress diagram completed. The justification of the method lies in the fact that the reversal of the diagonal 'PO' does not affect the loads carried by the members of the truss *outside the quadrilateral* of which it is the diagonal. This means that 'FQ' is correctly represented by 'F②,' i.e. 'Fq' in the stress diagram.

EXERCISES 8

(The reader is strongly advised to draw-out the stress diagrams already dealt with in this chapter.)

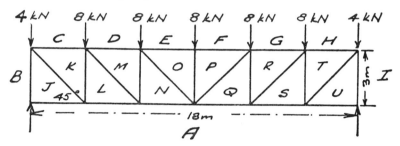

FIG. 218.

(1) Construct the stress diagram for the symmetrically loaded braced girder given in Fig. 218. Tabulate the forces in the various members, differentiating between struts and ties.

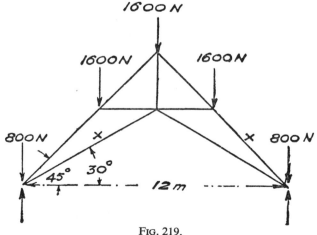

FIG. 219.

(2) Fig. 219 shows a form of roof truss which provides good head-room at mid-span. Draw the stress diagram for the truss. Determine the force in each of the two members marked by a cross and state whether they are respectively in tension or compression.

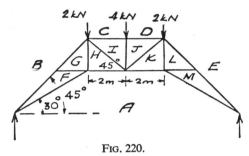

Fig. 220.

(3) Determine the forces in members 'AF,' 'IJ' and 'JK' respectively, of the loaded frame given in Fig. 220, by the construction of the complete stress diagram.

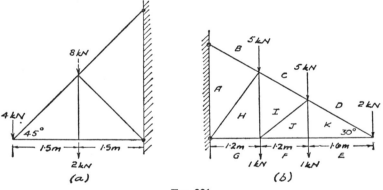

Fig. 221.

(4) Draw the stress diagrams for the given cantilever trusses (Fig. 221 (a) and Fig. 221 (b)). Tabulate the various member forces (stating whether the members are 'struts' or 'ties') in the case of Fig. 221 (a).

(5) Calculate the support reactions for the unsymmetrically loaded truss given in Fig. 222. Construct the stress diagram for the truss and determine the force in the inclined member indicated by a cross.

(6) The reactions for the overhanging roof truss given in Fig. 223 have been evaluated and the struts and ties determined. The ties are denoted by a negative sign and the struts by a positive sign. Check the

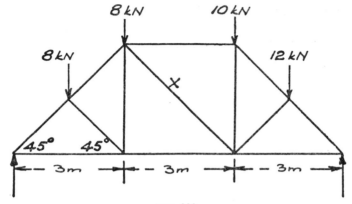

FIG. 222.

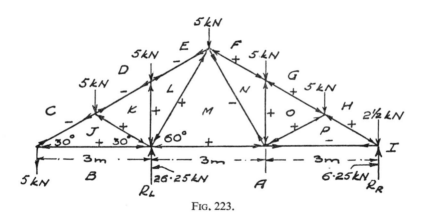

FIG. 223.

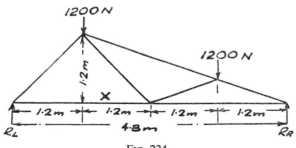

FIG. 224.

reaction values and verify the correctness of the strut and tie indications by constructing a complete stress diagram for the truss.

(7) Construct the stress diagram for the roof truss given in Fig. 224. Find the force in the member indicated by a cross. State whether the member is a 'strut' or a 'tie'.

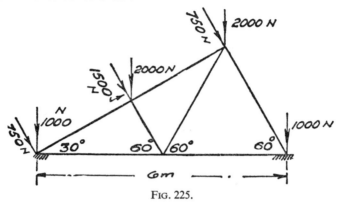

FIG. 225.

(8) Draw the stress diagram for the truss shown in Fig. 225. The truss carries dead loads and positive wind loads. The reactions are to be assumed to be parallel to the resultant force on the truss.

(9) The king-post truss (Fig. 226) carries dead loads and positive wind loads. The wind load is to be taken at 0·4 kN per sq. metre of roof

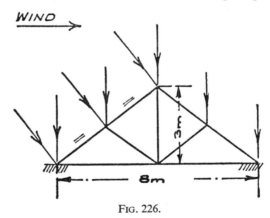

FIG. 226.

srea, and the total dead load at 1·0 kN per sq. metre of roof area. Calculate the respective joint loads for the truss, assuming the trusses to be apaced at 3 m centres.

(10) Draw the stress diagram for the braced girder given in Fig. 227. Indicate struts and ties on your frame diagram by writing alongside the members a plus sign for a strut and a minus sign for a tie.

Give the force in each of the inclined members of the girder.

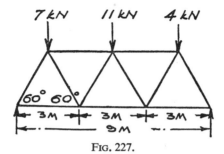

FIG. 227.

(11) Using the 'reversal of diagonal' method, construct the stress diagram for the truss shown in Fig. 211A.

CHAPTER IX

STRESS AND STRAIN. YOUNG'S MODULUS

Nature of Stress

A STRESS diagram of the type considered in the last chapter enables us to find the force or '*total stress*' in any given member of a loaded frame. We now have to investigate the implication of the term '*stress*' as normally employed in structural calculations.

Member 'NA' in the frame shown in Fig. 218 carries a tensile load of 32 kN. Is the member unsafely loaded? We cannot answer this question until the net cross-sectional area of the member is known. It is not the load in the member that ultimately matters but the *intensity* of the loading, i.e. how much the member has to carry for every square unit (usually taken as 'sq. millimetre') of sectional area. Assuming that the net sectional area of the member referred to equals 250 mm², the load carried by one sq. millimetre is 128 N. This is quite safe—assuming the material of the member to be mild steel. The intensity of loading expressed as '*so many units of load per unit of sectional area*' is termed the *stress* in the material of the given member.

Uniform Stress.—In cases in which the load carried by a member may be regarded as being uniformly distributed over a given cross-sectional area of the member, the stress at that section may be obtained by dividing the total load by the area of the section (Fig. 228).

$$\text{Stress} = \frac{\text{Load}}{\text{Area}}.$$

Units for Expressing Stress Values

Stress values for structural steelwork, reinforced concrete and timber are usually expressed in newtons per sq. millimetre (N/mm²). Masonry calculations and foundation pressure problems frequently involve the units 'N/m²' and 'kN/m²'.

Non-uniform Stress.—If the load distribution over a beam section be not uniform, the value obtained by dividing the 'total load' by the 'area of section' would merely give an *average* stress value. The structural

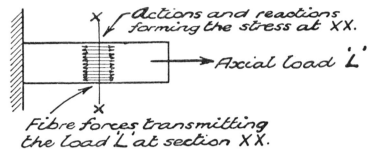

FIG. 228.—NATURE OF STRESS.

designer does not think in terms of 'average stress' values, but has to consider the intensity of stress in the most heavily loaded fibres of the member he is designing. In some cases the formula used for determining this maximum stress may not look much like the basic expression *load/area*, but the nature of stress remains the same, i.e. it is an '*intensity*' or '*rate*' of loading.

If a tiny beam fibre has a stress intensity of 3 newtons per sq. millimetre, the implication is that if the sectional area of the fibre were enlarged up to one sq. millimetre and the load were increased in the same proportion, the sq. millimetre would be carrying a load of 3 newtons.

In this chapter we will confine our attention to uniform stress.

Three Types of Simple Stress

The three basic types of stress are (i) *tensile*, (ii) *compressive*, and (iii) *shear*.

Tensile Stress.—This stress occurs in the fibres of a member which is subjected to a pull.

In Fig. 229, 'A' sq. millimetres of cross-sectional area at section 'XX' have to carry 'L' newtons

∴   each sq. millimetre carries L/A newtons
∴   *tensile stress* at 'XX' = L/A newtons/mm².

Compressive Stress.—When a member transmits a thrust the material of the member is subjected to this form of stress (Fig. 230).

Compressive stress at section 'XX' = $\dfrac{\text{Load}}{\text{Area}} = \dfrac{L}{A}$ newtons/mm².

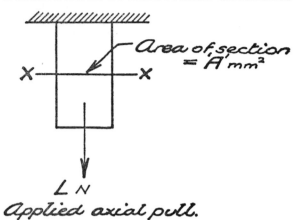

Area of section = A' mm²

L N

Applied axial pull.

FIG. 229.—TENSILE STRESS.

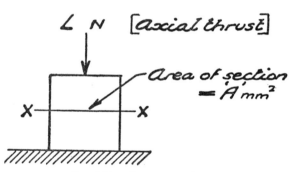

L N [axial thrust]

Area of section = A' mm²

FIG. 230.—COMPRESSIVE STRESS.

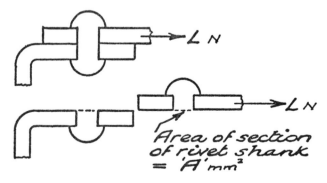

L N

L N

Area of section of rivet shank = A' mm²

FIG. 231.—SHEAR STRESS.

Shear Stress.—This is the type of stress that occurs in the steel of a rivet which is loaded as indicated in Fig. 231. When one portion of a member tends to slide over another portion at a certain plane in the material, the fibres at that plane are said to be in '*shear.*' Beam webs are subjected to 'shear stress.'

If the rivet shank has a sectional area of 'A' mm² and the applied shear load be 'L' newtons, the shear stress in the rivet steel will be 'L/A' newtons/mm².

EXAMPLES:

(i) *A steel tie-member has a solid rectangular section,* 100 mm × 20 mm. *It carries an axial (i.e. centrally applied) load of* 210 kN. *Calculate the stress in the steel.*

$$\text{Tensile stress} = \frac{\text{Load}}{\text{Area}} = \frac{210\ \text{kN}}{(100 \times 20)\ \text{mm}^2}$$

$$= \frac{210 \times 1\ 000}{2\ 000}\ \text{N/mm}^2 = 105\ \text{N/mm}^2.$$

(ii) *A short concrete column of square section,* 250 mm × 250 mm, *carries an axial load which induces a compressive stress of* 4·2 N/mm² *in the concrete. Calculate the value of the load.*

$$\text{Compressive stress} = \frac{\text{Load}}{\text{Area}}$$

$$\text{Let L N} = \text{load}$$

$$4·2 = \frac{\text{L}}{62\ 500}$$

Each rivet provides a shear area of 452mm²

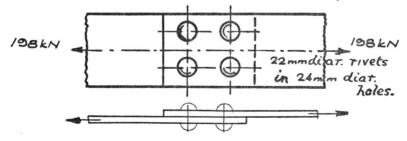

198kN   198kN   22mm diar. rivets in 24mm diar. holes.

FIG. 232.—LAP JOINT.

(It is usual to leave the units out in such equations, but great care must be taken to employ the same units on both sides of the equation.)

$$\therefore \quad L = (4 \cdot 2 \times 62\ 500)\ N$$
$$= 262 \cdot 5\ kN.$$

(iii) *Calculate the shear stress in the rivets in the lap-joint shown in Fig. 232.*

The total shear area provided is 4 times the sectional area of one rivet.

$$\text{Gross area of one rivet} = \frac{\pi d^2}{4} = \frac{\pi \times 24^2}{4}\ mm^2$$

$$= 452\ mm^2$$
$$\therefore \quad \text{total shear area} = (4 \times 452)\ mm^2$$
$$= 1\ 808\ mm^2$$

$$\text{Shear stress} = \frac{\text{Shear load}}{\text{Area under shear}}$$

$$= \frac{198 \times 1\ 000}{1\ 808}\ N/mm^2$$

$$= 109 \cdot 5\ N/mm^2.$$

Working Stresses

The maximum safe value for the stress in the material of a practical structural member depends upon several factors. It will depend upon the nature of the material. It will also depend upon whether the stress is 'tensile,' 'compressive' or 'shear.' The actual manner in which the member is employed in the structure will affect the maximum permissible stress, e.g. a 'long' compression member should not be so highly stressed as a 'short' one.

The nature of the loading, e.g. 'live' or 'dead,' will also be a determining factor.

For cases commonly occurring in structural design, regulations issued by local authorities or professional institutions specify the working stresses which should be employed.

Ultimate Stress.—Working stresses are usually fixed from the results of practical tests to destruction, carried out by means of suitable testing machines. The tests give *ultimate stress* values from which the safe working stresses are derived.

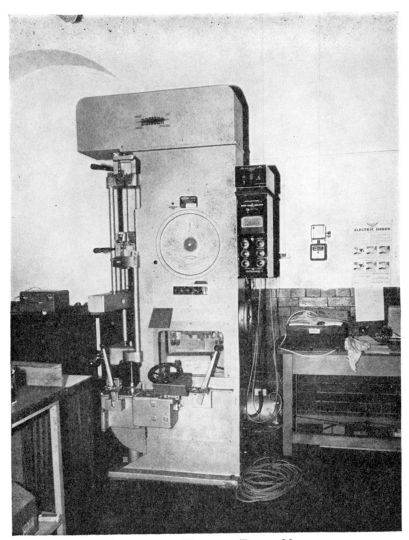

Fig. 233.—Denison Universal Testing Machine.
Figs. 233–7 are reproduced by courtesy of Denison-Avery Ltd.

Fig. 233 shows a Denison Universal Testing Machine.

In Fig. 234 a steel specimen is shown in position during a tensile test to destruction. The 'waisting' at yield is clearly seen. The two clamps on the specimen are 200 mm apart, this being the usual gauge length for measurements of elongation.

Fig. 234.—Steel Tension Test.

Fig. 235.—Shear Test on Steel Specimen.

Fig. 235 shows a specimen of steel, of circular section, being tested for ultimate shear stress.

Fig. 236 shows a tensile test on a glued timber specimen. Off-set

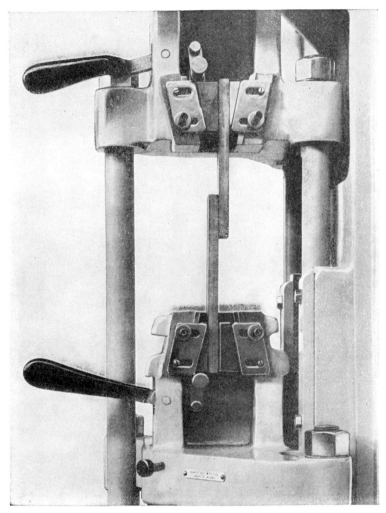

FIG. 236.—TENSION TEST ON TIMBER SPECIMEN.

wedges are used in this test. Timber specimens are tested along the grain and across the grain.

In Fig. 237 another form of timber test is illustrated. The object of

FIG. 237.—CLEAVAGE TEST ON TIMBER SPECIMEN.

this test is to determine the '*cleavage*' strength of timber of a given type.

Concrete is tested for compressive strength. The concrete to be tested is poured into moulds in the form of cubes, usually 150 mm cubes. After 'curing' for a definite period, say 28 days, the concrete cubes are tested to destruction. Fig. 238 shows a cube after having been tested to destruction.* B.S. 1881 gives full details of the manner in which the cubes shall be manufactured, cured and tested.

Readers who are interested in the testing of materials should consult the B.S. (British Standards) which are concerned with the particular materials in question (see Appendix I).

FIG. 238.—COMPRESSION TEST ON CONCRETE CUBE.

'B.S. No. 4360' deals with the testing of 'structural steel.' The testing of cement is the subject of 'B.S. No. 12.' The methods of testing small clear specimens of timber are dealt with in 'B.S. No. 373.' These, and other similar standards, may be obtained from the British Standards Institution, 2 Park Street, London, W1A 2BS.

Factor of Safety.—A *'factor of safety'* is a number which is divided into an 'ultimate stress' in order to obtain a suitable 'working stress.' The value of the 'factor of safety' is such as to provide an appropriate margin of safety in the employment of the derived stress. Some authorities prefer to derive the working stress from the 'yield-point' stress.

$$\text{Working stress} = \frac{\text{Ultimate stress}}{\text{Factor of safety}}.$$

* Reproduced by courtesy of Engineering Laboratory Equipment Ltd.

EXAMPLES:

(i) *In a tensile test to destruction on a mild-steel specimen the maximum load indicated by the testing machine was* 173·6 *kN. The sectional area of the specimen was* 320 *mm². Calculate the working tensile stress, for steel of this quality, assuming a factor of safety of* 3·5.

$$\text{Ultimate stress} = \frac{\text{Maximum load during test}}{\text{Sectional area of specimen}}$$

$$= \frac{173·6 \times 1\,000}{320} = 542·5 \text{ N/mm}^2.$$

$$\text{Working stress} = 542·5/3·5 \text{ N/mm}^2$$
$$= 155 \text{ N/mm}^2.$$

(ii) *A* 150 *mm cube of concrete crushed, in a compression test, at a load of* 500 *kN. Assuming a factor of safety of* 6, *determine suitable dimensions for the section of a short column, of the same quality concrete, which has to carry an axial load of* 230 *kN.*

$$\text{Ultimate compressive stress} = \frac{\text{Max. load during test}}{\text{Original sectional area}}$$

$$= \frac{500 \times 1\,000}{150 \times 150} \text{ N/mm}^2 = 22·2 \text{ N/mm}^2$$

$$\text{Working stress} = \frac{\text{Ultimate stress}}{\text{Factor of safety}} = \frac{22·2}{6} \text{ N/mm}^2$$

$$= 3·7 \text{ N/mm}^2$$

$$\text{Stress} = \frac{\text{Load}}{\text{Area}} \quad \therefore \quad \text{Area} = \frac{\text{Load}}{\text{Stress}}$$

$$= \frac{230 \times 1\,000}{3·7} = 62\,160 \text{ mm}^2.$$

Assuming a square section, say 250 mm + 250 mm.

(iii) *In a shear test on rivet steel the ultimate load was* 216 *kN, the total area under shear being* 480 *mm². In a joint (as in Fig.* 232), 4 *No.* 16 *mm dia. rivets (in 18 mm dia. holes) of this quality steel supported a shear load of 101·8 kN. What factor of safety does this represent?*

$$\text{Ultimate shear stress} = \frac{\text{Ultimate load}}{\text{Shear area provided}}$$

$$= \frac{216 \times 1\,000\text{ N}}{480\text{ mm}^2} = 450\text{ N/mm}^2.$$

$$\text{Actual working stress of rivets in joint} = \frac{\text{Total load}}{\text{Total area}}$$

$$\frac{101\cdot8 \times 1\,000}{4 \times 254\cdot5}\text{ N/mm}^2 = 100\text{ N/mm}^2$$

(Gross area of one 16 mm dia. rivet = 254·5 mm^2)

$$\text{Factor of safety} = \frac{\text{Ultimate stress}}{\text{Working stress}} = \frac{450\text{ N/mm}^2}{100\text{ N/mm}^2} = 4\cdot5.$$

The reader is referred to regulations issued by the Greater London Council (G.L.C. By-laws for steelwork, reinforced concrete and timber design, etc.), and the British Standards Institution (B.S. 449, etc.), and to current codes of practice, for detailed lists of working stresses. From time to time revision of these working stresses is made necessary by improved methods of manufacture and the application of the results of research.

Strain

When a member is loaded and its fibres are put into a state of stress, some alteration is bound to take place in the dimensions or shape of the member. A member subjected to such dimensional change is said to be in a state of *strain*. The effect of load is therefore to develop in the fibres of the member both *stress* and *strain* simultaneously.

Tensile Strain.—This is the type of strain which the fibres of a tie member experience. Tensile strain is associated with the elongation of members.

The numerical value of the '*strain*' is not the extension itself. The tensile strain in the member shown in Fig. 239 is obtained by dividing the *extension* by the *original length* of the member.

$$\text{Tensile strain} = \frac{\text{Extension}}{\text{Original length}} = \frac{x \text{ mm}}{l \text{ mm}} = x/l.$$

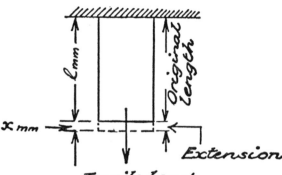

FIG. 239.—TENSILE STRAIN.

EXAMPLE.—*Calculate the tensile strain in a steel rod* 2 *m long which undergoes an extension of* 1 *mm when the load is applied.*

$$\text{Tensile strain} = \frac{\text{Extension}}{\text{Original length}} = \frac{1 \text{ mm}}{(2 \times 1\,000) \text{ mm}}$$

$$= \frac{1}{2\,000} = 0.0005 \text{ (expressed as a '}\textit{number}\text{')}.$$

Compressive Strain.—When a member shortens under the action of a thrust, the material of the member is subjected to *compressive strain*. This is the type of strain which occurs in columns (Fig. 240).

$$\text{Compressive strain} = \frac{\text{Contraction}}{\text{Original length}} = \frac{x \text{ mm}}{l \text{ mm}} = x/l .$$

EXAMPLE (i).—*A short reinforced concrete column,* 1·5 *m high, shortens by* 0·45 *mm under the applied axial load. Calculate the strain in the concrete and in the steel reinforcement.*

As the concrete and the steel contract in length equally, they are both subjected to equal '*strain.*'

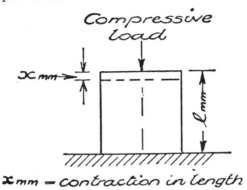

FIG. 240.—COMPRESSIVE STRAIN.

Compressive strain in both materials

$$= \frac{0 \cdot 45 \text{ mm}}{1\ 500 \text{ mm}} = 0 \cdot 0003.$$

EXAMPLE (ii).—*A 150 mm timber cube in a test is loaded so that the compressive strain is* 0·0004. *Calculate the contraction in length of the specimen.*

$$\text{Compressive strain} = \frac{\text{Contraction in length}}{\text{Original length}}.$$

Let x mm = Contraction

$$0 \cdot 0004 = \frac{x \text{ mm}}{150 \text{ mm}}$$

$$\therefore \quad x = (150 \times 0 \cdot 0004) \text{ mm} = 0 \cdot 06 \text{ mm.}$$

Shear Strain.—The piece of material illustrated in Fig. 241 is distorted by the action of the shear load shown.

The value of the shear strain is obtained by the ratio x mm/l mm, i.e. x/l.

When a bolt is turned by a spanner the material of the bolt is in a state of shear strain. The name '*torsional strain*' is given to this particular type of shear strain.

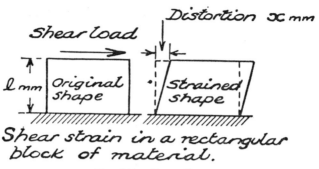

FIG. 241.—SHEAR STRAIN.

Relationship between Stress and Strain

Hooke's Law.—The two physical states '*stress*' and '*strain*'—which we have seen are always co-existent—are related by a law known as '*Hooke's Law.*' It is a law concerned with bodies in an 'elastic state.'

An '*elastic*' body is one which, having been deformed by an applied force, will regain its original size and shape when the deforming force is removed.

As working stresses in structural work are so chosen as to ensure that the property of elasticity is maintained, it is clear that Hooke's law of elasticity is of great importance.

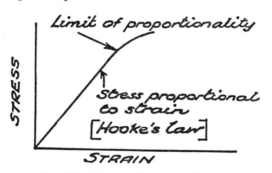

FIG. 242.—ELASTIC STRESS AND STRAIN.

The law states that, provided a member remain perfectly elastic, the stress induced in it will always be directly proportional to the accompanying strain. This means that should the load the member is carrying alter in magnitude (and thereby cause an increase or a decrease in the stress value), the strain value would increase or decrease proportionately with the variation in stress. Thus if the stress increases by 50%, the strain would increase by 50%. The *stress-strain* graph for an elastic member is therefore a straight line.

Young's Modulus of Elasticity.—Let us suppose that we have conducted a tensile or compressive test on a suitable elastic specimen, and have tabulated corresponding 'stress' and 'strain' values as in Fig. 243. The figures shown in the table represent a perfect set of results in that they comply exactly with the terms of '*Hooke's Law*.' In the last column the values obtained by dividing a 'stress result' by the corresponding 'strain result' are shown. The values in each case is the same.

Material.	Stress N/mm²	Strain.	Ratio of stress to strain $= \dfrac{\text{Stress}}{\text{Strain}}$
1:2:4 Concrete tested in compression	2·1	0·0001	$\dfrac{2 \cdot 1}{0 \cdot 0001} = 21 \text{ kN/mm}^2$
	4·2	0·0002	$\dfrac{4 \cdot 2}{0 \cdot 0002} = 21 \text{ kN/mm}^2$
	6·3	0·0003	$\dfrac{6 \cdot 3}{0 \cdot 0003} = 21 \text{ kN/mm}^2$
	8·4	0·0004	$\dfrac{8 \cdot 4}{0 \cdot 0004} = 21 \text{ kN/mm}^2$

FIG. 243.

Thus if stress varies as strain we may say that the ratio $\dfrac{\text{Stress}}{\text{Strain}}$ is a constant value. If we repeated the test, using a specimen of different dimensions and form of cross-section but of the same material, we would get the same value for the 'stress-strain' ratio. Such a constant, which depends simply upon the nature of the material and not upon the dimensions of the tested specimen, is termed a '*physical constant*.' To the physical constant represented by the ratio $\dfrac{\text{Stress}}{\text{Strain}}$ the name 'Young's modulus' is given. The letter 'E' is used to denote 'Young's modulus':

$$E = \frac{\text{Stress}}{\text{Strain}}.$$

Just as '*density*' (another 'physical constant') tells us how much weight is associated with a certain volume of a particular substance, so Young's modulus tells us how much stress accompanies a given strain in the material of a given structural member.

Units for 'E.'—As 'strain' is merely a number, the units in which Young's modulus values are expressed are those of 'stress,' e.g. 'N/mm^2' or 'kN/mm^2.'

Typical values of 'E':

Structural or mild steel: 210 kN/mm²
Concrete: 14–28 kN/mm²
Timber (average):   8·4 kN/mm²
Rubber, say: 0·7 N/mm².

The value for rubber has been inserted to indicate that materials which, under similar test conditions, stretch considerably more than others have much lower 'E' values.

Practical Determination of Young's Modulus

In a practical test on a material like steel or concrete, the alteration in length of a specimen will be very small if we keep the stress within the elastic range. Such alterations in length are not visible to the naked eye.

To effect accurate readings of extensions or contractions in these circumstances, special instruments known as *'extensometers'* are employed. The underlying principles, upon which the operation of extensometers depends, vary considerably. A number employ the principles of optics.

Fig. 244 illustrates the employment of an extensometer in the derivation of 'E' for a steel bar, tested in tension.

EXAMPLES:

(i) *A bar of sectional area* 1 250 mm² *and* 2 m *in length extended* 0·4 *mm when an axial load of* 52·5 kN *was applied. Calculate Young's modulus for the material of the bar.*

$$E = \frac{\text{Stress}}{\text{Strain}}.$$

$$\text{Stress} = \frac{\text{Load}}{\text{Area}} = \frac{52 \cdot 5 \times 1\,000 \text{ N}}{1\,250 \text{ mm}^2} = 42 \text{ N/mm}^2.$$

$$\text{Strain} = \frac{\text{Extension}}{\text{Original length}} = \frac{0 \cdot 4 \text{ mm}}{2\,000 \text{ mm}} = 0 \cdot 0002.$$

$$E = \frac{42}{0 \cdot 0002} \text{ N/mm}^2 = 210\,000 \text{ N/mm}^2.$$

FIG. 244.—EXTENSOMETER.

'Young's modulus' is one of the physical constants by which we can identify a given material. The result of the calculations in this problem indicates that the bar was composed of mild steel.

(ii) *Calculate the contraction in length of a short concrete column, 300 mm × 300 mm square section, when carrying an axial load of 360 kN. The original unloaded length of the column was 3 m. Assume 'E' for concrete to be 14 000 N/mm².*

Let x mm = contraction in length.

$$\text{Compressive stress in concrete} = \frac{360\ 000}{300 \times 300}\ \text{N/mm}^2$$

$$= \frac{360\ 000}{90\ 000}\ \text{N/mm}^2 = 4\ \text{N/mm}^2.$$

$$E = \frac{\text{Stress}}{\text{Strain}} = \frac{4}{x/3\ 000} = \frac{4 \times 3\ 000}{x}$$

$$\therefore \quad 14\ 000 = \frac{4 \times 3\ 000}{x}$$

$$x = \frac{4 \times 3\ 000}{14\ 000}\ \text{mm} = 0\cdot86\ \text{mm}.$$

(iii) *A short timber post of rectangular section has one side of its section twice the other. When the post is loaded axially with 9·8 kN it contracts 0·119 mm per metre of length. If 'E' for this timber = 8·4 kN/mm², calculate the sectional dimensions of the post.*

Let x mm = smaller side, $\therefore$ $2x$ mm = larger side.

$$\text{Area of section} = 2x^2\ \text{mm}^2$$

$$\text{Stress} = \frac{\text{Load}}{\text{Area}} = \frac{9\ 800}{2x^2}\ \text{N/mm}^2 = \frac{4\ 900}{x^2}\ \text{N/mm}^2.$$

$$\text{Strain} = \frac{0\cdot119\ \text{mm}}{1\ 000\ \text{mm}} = 0\cdot000119.$$

$$E = \frac{\text{Stress}}{\text{Strain}}.$$

$$8\,400 = \frac{4\,900/x^2}{0 \cdot 000119}$$

$$\therefore \quad \frac{4\,900}{x^2} = 1, \qquad \therefore \quad x^2 = \frac{4\,900}{1} = 4\,900$$

$$\therefore \qquad x = 70 \text{ mm.}$$

∴ Sectional dimensions are 70 mm × 140 mm.

(iv) *Fig. 245 shows the section of a reinforced concrete short column. Calculate the stress in the concrete and the stress in the steel if an axial load of 753·5 kN be applied to the column.*
'*E' for steel* = 210 *kN/mm²*, '*E' for concrete* = 14 *kN/mm²*.

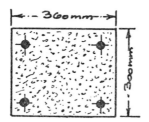

FIG. 245.—REINFORCED CONCRETE COLUMN SECTION.

In such problems (as in the practical design of reinforced concrete members) it is assumed that there is no slipping of the steel in the concrete, i.e. that there is perfect '*bond*' between the two materials. The steel and the concrete will therefore be subjected to equal strain as the column shortens under the load.

Let '*c*' N/mm² = stress in concrete and '*t*' N/mm² be the stress in the steel reinforcement.

Area of steel = (4 × 314) mm² = 1 256 mm².
∴ Load carried by steel bars = (1 256 × *t*) N.

Area of concrete = [(360 mm × 300 mm) − 1 256] mm²
= 106 744 mm².

∴ Load carried by concrete = (106 744 × *c*) N.
But load carried by steel plus load carried by concrete = total load on column,

$$\therefore \quad 1\,256t + 106\,744c = 753\,500 \qquad \text{(i)}$$

We get a second equation connecting 't' and 'c' from the equal-strain property.

$$210\,000 = \frac{t}{\text{Strain in steel}}$$

$$\therefore \quad \text{Strain in steel} = \frac{t}{210\,000}$$

$$\text{Also } 14\,000 = \frac{c}{\text{Strain in concrete}}$$

$$\therefore \quad \text{Strain in concrete} = \frac{c}{14\,000}$$

$$\therefore \quad \text{As strains are equal, } \frac{t}{210\,000} = \frac{c}{14\,000}$$

$$\therefore \quad t = c \times \frac{210\,000}{14\,000} = 15c \qquad \text{(ii)}$$

This is a well-known and important result much used in reinforced concrete calculations. Substituting in (i):

$$(1\,256 \times 15c) + 106\,744c = 753\,500$$
$$125\,584c = 753\,500$$
$$c = 6 \text{ N/mm}^2$$
$$\text{and } t = 15 \times c = 90 \text{ N/mm}^2.$$

Definitions

Ductility.—A specimen of mild steel when tested in tension will not show any *visible* signs of dimensional alteration during the early stages of the test, i.e. during the 'elastic range of stress.' Shortly after the maximum elastic stress has been reached the specimen stretches visibly and extensions can be measured by means of dividers. The cross-sectional area becomes much smaller and the area at fracture may be less than half the original sectional area. This property of marked dimensional change with increasing applied stress is termed '*ductility.*' Steel to be used in a steel-framed building must exhibit the property of ductility when representative specimens are tested.

Brittleness.—'*Brittleness*' denotes absence of ductility. Cast iron is brittle and specimens of cast iron tested to destruction undergo very little dimensional change. Fig. 246 shows a typical stress-strain graph for a brittle specimen tested in tension.

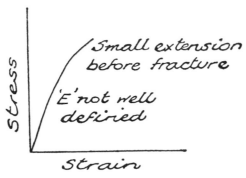

FIG. 246.—BRITTLE SPECIMEN IN TENSILE TEST.

Elastic Limit.—The elastic limit stress is the stress intensity up to which it is perfectly safe to go without causing the material to lose its property of perfect elasticity. If we stress a piece of material beyond its elastic limit stress, it will not regain its original size or shape when the deforming force producing the stress is removed.

Permanent Set.—This is the dimensional change which persists after all load is removed. It is due to stressing the material beyond its elastic limit.

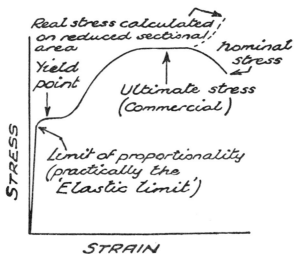

FIG. 247.—TENSILE TEST. MILD STEEL.

Limit of Proportionality.—As the term implies, this is the stress value at which the proportional law of Hooke breaks down. In most cases little practical distinction need be made between this stress and the 'elastic limit stress.'

Yield Point.—As will be seen in Fig. 247, at a point just above the 'limit of proportionality' a considerable increase in strain takes place in ductile materials with little increases of stress. The stress value at which this big increase takes place is termed the '*yield point*' of the material. As the '*working stress*' in a member must be kept well below the yield point, in order to be able to take advantage of the high ultimate stresses of certain modern steels, it is necessary to show by test specimens that the 'yield point' has also been suitably raised.

, The reader is referred to books on '*Strength of Materials*' for a more detailed account of the behaviour of various structural materials under test.

Examples of Bolted and Riveted Joints in Structural Steelwork

The general principles involved in the calculation of the strength of riveted (or bolted) joints will be explained by means of a few typical examples.

Value of One Rivet (or Bolt)

In the usual type of structural connection—in which steel plates are lapped or butted together—a rivet has two types of strength: (i) *a shear strength*, (ii) *a bearing strength*.

Shear Strength of One Rivet

A rivet may be (i) in *single* shear or (ii) in *double* shear, according to the type of joint.

Single Shear.—When the shearing tendency is across one sectional plane only, the rivet is said to be in '*single shear*.'

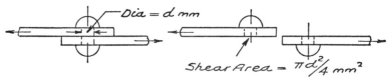

FIG. 248.—RIVET IN SINGLE SHEAR.

Double Shear.—If the rivet is subjected to shear across two sections, as in Fig. 249, the rivet is in '*double shear*.'

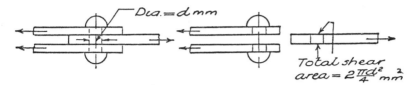

FIG. 249.—RIVET IN DOUBLE SHEAR.

If 'f_s' N/mm² be the safe working stress in shear in the rivet material and 'd' mm be the rivet diameter, the safe shear load, in N per rivet, will be as follows:

(i) Single shear: Safe load = Working stress × sectional area
$$= (f_s \times \pi d^2/4) \text{ N.}$$
(ii) Double shear: Safe load $= (f_s \times 2\pi d^2/4)$ N.

∴ *Single shear strength of one rivet (or bolt)* $= \dfrac{\pi d^2}{4} f_s$ N.

Double shear strength of one rivet (or bolt) $= \dfrac{2\pi d^2}{4} f_s$ N.

It should be noted that it is usual, in building work, for gross area to be used in the case of rivets. The gross area is the cross-sectional area of the rivet hole which is usually 2 mm greater than the nominal diameter of the rivet, so as to allow easy entry of the rivet when hot. For bolts the gross area of the bolt is the gross area calculated on the nominal diameter of the bolt.

Bearing Strength of One Rivet

A thin plate pressing against the shank of a rivet may do greater damage by imposing a high intensity of '*bearing*' or contact stress than by tending to cause shear in the manner already explained.

It is necessary therefore to limit the intensity of the bearing stress on the rivet.

If 'f_b' N/mm² be the maximum safe bearing stress, and the area resisting bearing be taken as the projected area normal to the direction of pressure (see Fig. 250), the safe bearing strength of one rivet (or bolt)

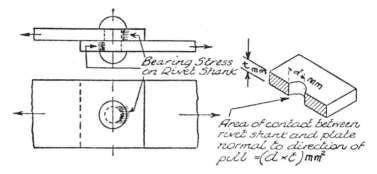

FIG. 250.—BEARING STRESS IN RIVETS.

will be given by the expression $d \times t \times f_b$ N, where t mm = plate thickness. Note that, in simple bearing, the permissible stress is 80% of 'f_b.'

Summary

S.S. strength of one rivet (or bolt) $= \dfrac{\pi d^2}{4} f_s$ N.

D.S. strength of one rivet (or bolt) $= \dfrac{2\pi d^2}{4} f_s$ N.

Bearing strength of one rivet (or bolt) $= dt f_b$ N.

The actual safe load for a rivet (or bolt) in any given case will be the *lesser* of its 'shear' and 'bearing' values.

Working Stresses for Rivets and Bolts

The tables given on pages 183 and 184 are based on the working stresses specified at the head of the respective tables. The reader will have no difficulty in amending the tabular values for the case of a working stress not included in either of the two tables given. The safe value for a bolt or a rivet is directly proportional to the appropriate working stress selected.

EXAMPLE (i).—*Calculate the safe load in (i) double shear, (ii) bearing for a 18 mm dia. rivet in a 20 mm dia. hole, assuming a plate thickness of 15 mm and the following working stresses: $f_s = 100$ N/mm², $f_b = 290$ N/mm².*

SHEARING AND BEARING VALUES

IN KILONEWTONS FOR POWER-DRIVEN SHOP RIVETS OF STEEL HAVING A YIELD STRESS OF 250 N/mm²

BASED ON
BS 449
1969

Gross Dia. of Rivet after driving in mm	Area in cm²	Shearing Value @ 110 N/mm²		Simple Bearing Value @ 80% of 315 N/mm² and Enclosed Bearing Value @ 315 N/mm² (see footnote) Thickness in mm of plate passed through or of enclosed plate											
		Single Shear	Double Shear	5	6	7	8	9	10	12	15	18	20	22	25
10	0.79	8.64	17.3	12.6	15.1	17.6	20.2	22.7							
				15.8	18.9	22.0	25.2								
12	1.13	12.4	24.9	15.1	18.1	21.2	24.2	27.2	30.2	36.3					
				18.9	22.7	26.5	30.2	34.0							
14	1.54	16.9	33.9	17.6	21.2	24.7	28.2	31.8	35.3	42.3	52.9				
				22.0	26.5	30.9	35.3	39.7	44.1						
16	2.01	22.1	44.2	20.2	24.2	28.2	32.3	36.3	40.3	48.4	60.5	72.6			
				25.2	30.2	35.3	40.3	45.4	50.4	60.5					
18	2.54	28.0	56.0	22.7	27.2	31.8	36.3	40.8	45.4	54.4	68.0	81.6	90.7		
				28.3	34.0	39.7	45.4	51.0	56.7	68.0	85.0				
20	3.14	34.6	69.1	25.2	30.2	35.3	40.3	45.4	50.4	60.5	75.6	90.7	101		
				31.5	37.8	44.1	50.4	56.7	63.0	75.6	94.5	113			
22	3.80	41.8	83.6	27.7	33.3	38.8	44.4	49.9	55.4	66.5	83.2	99.8	111	122	
				34.6	41.6	48.5	55.4	62.4	69.3	83.2	104	125	139		
24	4.52	49.8	99.5	30.2	36.3	42.3	48.4	54.4	60.5	72.6	90.7	109	121	133	
				37.8	45.4	52.9	60.5	68.0	75.6	90.7	113	136	151		
27	5.73	63.0	126	34.0	40.8	47.6	54.4	61.2	68.0	81.6	102	122	136	150	170
				42.5	51.0	59.5	68.0	76.5	85.0	102	128	153	170		

Upper line Bearing Values for each diameter of rivet are Simple Bearing Values.
Lower line Bearing Values for each diameter of rivet are Enclosed Bearing Values.
For areas to be deducted from a bar for one hole, see table on page 80.
For explanation of table, see Notes on page 131.
1 kilonewton may be taken as 0.102 metric tonne (megagramme) force, but see page 102.

SHEARING AND BEARING VALUES

BASED ON BS 449 1969

IN KILONEWTONS FOR CLOSE TOLERANCE AND TURNED BOLTS OF STEEL OF STRENGTH GRADE DESIGNATION 4.6

Dia. of Bolt Shank in mm	Area in cm²	Shearing Value @ 95 N/mm²		Simple Bearing Value @ 80% of 300 N/mm² and Enclosed Bearing Value @ 300 N/mm² (see footnote) Thickness in mm of plate passed through or of enclosed plate											
		Single Shear	Double Shear	5	6	7	8	9	10	12	15	18	20	22	25
10	0.79	7.46	14.9	12.0	14.4	16.8	19.2	21.6							
				15.0	18.0	21.0	24.0								
12	1.13	10.7	21.5	14.4	17.3	20.2	23.0	25.9	28.8						
				18.0	21.6	25.2	28.8								
14	1.54	14.6	29.2	16.8	20.2	23.5	26.9	30.2	33.6	40.3					
				21.0	25.2	29.4	33.6	37.8							
16	2.01	19.1	38.2	19.2	23.0	26.9	30.7	34.6	38.4	46.1	57.6				
				24.0	28.8	33.6	38.4	43.2	48.0						
18	2.54	24.2	48.3	21.6	25.9	30.2	34.6	38.9	43.2	51.8	64.8	77.8			
				27.0	32.4	37.8	43.2	48.6	54.0	64.8					
20	3.14	29.8	59.7	24.0	28.8	33.6	38.4	43.2	48.0	57.6	72.0	86.4	96.0		
				30.0	36.0	42.0	48.0	54.0	60.0	72.0	90.0				
22	3.80	36.1	72.2	26.4	31.7	37.0	42.2	47.5	52.8	63.4	79.2	95.0	106		
				33.0	39.6	46.2	52.8	59.4	66.0	79.2	99.0	119			
24	4.52	43.0	86.0	28.8	34.6	40.3	46.1	51.8	57.6	69.1	86.4	104	115		
				36.0	43.2	50.4	57.6	64.8	72.0	86.4	108	130			
27	5.73	54.4	109	32.4	38.9	45.4	51.8	58.3	64.8	77.8	97.2	117	130	143	
				40.5	48.6	56.7	64.8	72.9	81.0	97.2	122	146	162		

Upper line Bearing Values for each diameter of bolt are Simple Bearing Values.
Lower line Bearing Values for each diameter of bolt are Enclosed Bearing Values.
For areas to be deducted from a bar for one hole, see table on page 80.
For explanation of table, see Notes on page 131.
1 kilonewton may be taken as 0.102 metric tonne (megagramme) force, but see page 102.

(a) *By use of formulae*:

$$\text{D.S. value} = \frac{2\pi d^2}{4} f_s \text{ N} = 2 \times \frac{\pi \times 20^2}{4} \times 100 \text{ N}$$

$$= 62\ 800 \text{ N} = 62 \cdot 8 \text{ kN}.$$

$$\text{Bearing value} = dtf_b \text{ N} = 20 \times 15 \times 290 \text{ N}$$

$$= 87000 \text{ N} = 87 \text{ kN}.$$

(b) *By use of tables:*

Using the first table, evaluated for $f_s = 110 \text{ N/mm}^2$ and $f_b = 315 \text{ N/mm}^2$, we have, for the given stresses:

$$\text{D.S. value} = \left(69 \cdot 1 \times \frac{100}{110} \right) \text{kN} = 62 \cdot 8 \text{ kN}.$$

$$\text{Bearing value} = \left(94 \cdot 5 \times \frac{290}{315} \right) \text{kN} = 87 \text{ kN}.$$

The safe load for a 18 mm dia. rivet in the given circumstances will be 62·8 kN.

EXAMPLE (ii).—*Calculate the safe load for the joint shown in Fig. 251, from the point of view of the rivets in the connection.*

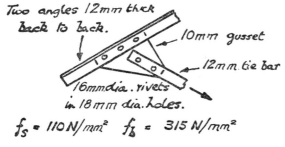

Fig. 251.—RIVETED JOINT.

In this example the tie-member laps on to the gusset plate so that the two connecting rivets are in single shear. These rivets are also under bearing stress in plate thicknesses, respectively of 10 mm and 12 mm. The 10 mm thickness would clearly produce the higher stress. In the case of a lap joint with plates of different thicknesses, take the *smaller thickness* for bearing-strength calculations.

Single shear strength of one 16 mm dia. rivet at 110 N/mm²

$$= \frac{\pi d^2}{4} f_s = \frac{\pi \times 18^2}{4} \times \frac{110}{1000} = 28 \text{ kN}.$$

Bearing strength of one 16 mm dia. rivet in 10 mm plate thickness (simple bearing) $= dtf_b = 18 \times 10 \times 0.8 \times \dfrac{315}{1000} = 45.4$ kN.

The *lesser* of these two strengths $= 28$ kN, therefore one rivet is worth 28 kN in the joint.

∴ Safe total load $= 2 \times 28$ kN $= 56$ kN.

The connection between the gusset plate and the angles is clearly stronger because we have here 3 rivets in *double shear* or bearing in 10 mm thickness.

EXAMPLE (iii).—*Fig. 252 shows a plate, 150 mm wide × 20 mm thick, connected to two 12 mm plates by means of 4 No. 24 mm dia. turned and fitted bolts. Calculate the safe load the connection can transmit.* $f_s = 95$ N/mm², $f_b = 300$ N/mm², $f_t = 155$ N/mm².

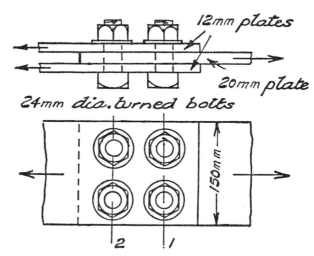

FIG. 252.—BOLTED JOINT.

In this case three distinct strengths will have to be considered:

(i) Bolt strength, (ii) 20 mm plate strength, (iii) the combined strength of the two 12 mm plates.

Bolt strength:

The bolts are in double shear and bearing in 20 mm plate thickness. The 12 mm plate thickness is not taken for bearing-strength calculation, although less than 20 mm, because the two 12 mm plates act in conjunction and are equivalent to a thickness of 12 mm + 12 mm = 24 mm.

Double shear strength of one 24 mm dia. bolt $= 2 \times \dfrac{\pi d^2}{4} f_s$ N

$$= 2 \times \frac{\pi \times 24^2}{4} \times \frac{95}{1000} \text{ kN} = 86 \text{ kN}.$$

Bearing strength of one 24 mm dia. bolt in 20 mm plate thickness =
$dft_b = \left(24 \times 20 \times \dfrac{300}{1000} \right) \text{kN} = 144 \text{ kN}.$

∴ Actual value of one bolt = 86 kN
∴ Bolt strength of connection = 4 × 86 kN = 344 kN.

20 mm plate strength:

This plate is in tension and will tend to fail at a section weakened by bolt holes. When there are several sections weakened by reduction of sectional area, each section should be separately considered.

Section 1.—The *'effective'* or solid plate width at this section = 150 mm − (2 × 24) mm = 102 mm.

∴ Effective or net area = 102 mm × 20 mm = 2 040 mm².

Safe tensile load = net area × working tensile stress = $\left(2\,040 \times \dfrac{155}{1000} \right) \text{kN} = 316 \cdot 2 \text{ kN}.$

Section 2.—If the plate failed at this section it could not come out from between the two 12 mm plates owing to the two bolts at section 1 acting as stop pegs. Therefore the strength of these two bolts should be added on to the 20 mm plate strength at section 2, in order to compute the total strength there (see also Example (iv)). The plate is therefore stronger at section 2.

Combined 12 mm *plates:*—(no 'stop peg' allowance this time).

Tensile strength = Effective area × 155 N/mm²

$$= [(150 - 48) \times (2 \times 12)] \times \frac{155}{1000} \text{ kN} = 379 \text{ kN}.$$

Summary: Bolt strength = 344 kN.
      20 mm pl. strength = 316·2 kN.
      2/12 mm pl. strength = 379 kN.

The safe load for the connection is the smallest of all these values, i.e. 316·2 kN.

EXAMPLE (iv).—*Calculate the safe load for the double-covered butt joint given in Fig. 253. Use the following working stresses:*

$$f_s = 110 \ N/mm^2, f_b = 315 \ N/mm^2, f_t = 155 \ N/mm^2.$$

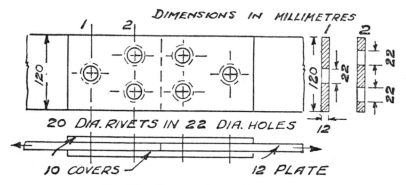

FIG. 253.—DOUBLE-COVERED BUTT JOINT.

The rivet arrangement, giving a single or '*leading*' rivet at section 1, results in an efficient or economical joint, as only one rivet hole is deducted from the solid plate section without some compensating rivet strength (see below). The reader will note the usual allowance of 2 mm clearance on the *nominal* rivet diameter.

$$\text{D.S. value of one rivet} = 2 \times \frac{\pi d^2}{4} \times f_s = 2 \times \frac{\pi \times 22^2}{4} \times \frac{110}{1000}$$

$$= 83 \cdot 6 \text{ kN.}$$

$$\text{Bearing value of one rivet} = dtf_b = 22 \text{ mm} \times 12 \text{ mm} \times \frac{315}{1000}$$

$$= 83 \cdot 2 \text{ kN}$$

∴ Strength of one rivet $= 83 \cdot 2$ kN

∴ Total rivet strength $= 3 \ (not \ 6) \times 83 \cdot 2 \text{ kN} = 249 \cdot 6 \text{ kN.}$

12 mm plate strength.

Section 1.—Tensile strength $= (120 - 22) \times 12 \times \dfrac{155}{1000}$

$$= 182 \text{ kN.}$$

Section 2.—Total strength $= (120 - 2 \times 22) \times 12 \times \dfrac{155}{1000}$

plus the strength of one rivet $= 141 \cdot 4 + 83 \cdot 2 = 224 \cdot 6 \text{ kN.}$

∴ Actual 12 mm plate strength $= 182$ kN.

Cover-plate strength:

The weakest section for the covers is 'section 2.' There is no rivet strength to be added on. The 'leading rivet' at section 1 would simply come away with the fractured joint.

$$\text{Tensile strength} = (120 - 2 \times 22) \times 2 \times 10 \times \frac{155}{1000} \text{ kN}$$

$$= 235 \cdot 6 \text{ kN}.$$

Summary: Rivet strength = 249·6 kN.
               $\frac{1}{2}$ in plate strength = 182 kN.
               Cover-plate strength = 235·6 kN.
∴ Actual safe load for joint = 182 kN.

EXAMPLE (v).—*Calculate the safe uniformly distributed load 'W'* (Fig. 254) *from the point of view of the end connections of the steel beam.*

$$f_s = 110 \ N/mm^2, f_b = 315 \ N/mm^2.$$

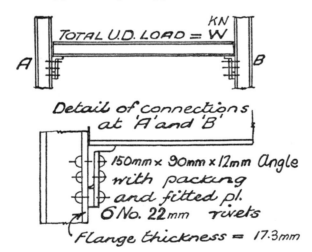

FIG. 254.—BRACKET CONNECTION.

Use gross dia.

In practical connections of the type illustrated, the single shear strength of the rivets is the criterion. Thus in this case:

$$\text{S.S. strength of one rivet} = \frac{\pi \times 24^2}{4} \times \frac{110}{1000} = 49 \cdot 8 \text{ kN}.$$

Bearing strength of one rivet $= dtf_b = 24 \times 12 \times \dfrac{315}{1000}$

$$= 90 \cdot 7 \text{ kN.}$$
$$\therefore \text{ Actual strength} = 49 \cdot 8 \text{ kN.}$$

[Sometimes a slight reduction is made in the rivet strength to allow for a tendency for the bracket to bend.]

$\therefore$ Total rivet strength $= 6 \times 49 \cdot 8 = 298 \cdot 8$ kN.

This value is the maximum safe reaction for the beam, therefore the maximum safe value of 'W' $= 2 \times 298 \cdot 8 = 597 \cdot 6$ kN.

EXERCISES 9

(1) A tie-bar has a rectangular section, 100 mm × 16 mm. It is subjected to an axial pull of 240 kN. Find the stress in the material of the tie-bar. Replace the 16 mm thickness given by another thickness which would correspond to a stress of 100 N/mm².

(2) A mild steel tie-rod has to carry an axial load of 53·2 kN. The maximum permissible stress = 140 N/mm². Calculate a suitable diameter for the rod.

(3) A reinforced concrete column transmits a total axial load of 1 315 kN to a footing slab of 60 kN estimated weight. Assuming the ground to have a safe bearing pressure of 220 kN/m², calculate suitable dimensions, in plan, for the slab.

(4) A short square pier is built of blue brickwork in cement mortar. Calculate the necessary side of the square section if the pier is to carry a concentric load of 120 kN. Assume 1 100 kN/m² to be the safe load for the brickwork. (Make no allowance for self-weight of pier.)

(5) How many 18 mm dia. rivets in single shear in 20 mm dia. holes would be required to transmit a shear load of 346 kN, if the maximum permissible stress were 110 N/mm²?

(6) Calculate Young's modulus in compression for a specimen of timber, 100 mm × 100 mm in section, which shortened by 0·04 mm on a gauge length of 200 mm when the test load (axially applied) was 16·8 kN.

(7) A mild steel tie-rod 3 m long, having a sectional area of 200 mm², is subjected to an axial pull of 94·5 kN. Assuming 'E' to be 210 kN/mm², calculate the extension in the rod.

(8) A short compression specimen is composed of concrete and steel. The total sectional area = 15 000 mm², comprised of 2 500 mm² of

steel surrounded by 12 500 mm² of concrete. The specimen is subjected
to an axial test load which produces a stress of 6 N/mm² in the concrete.
Calculate (i) the stress in the steel, (ii) the test load referred to.

$$E_{steel} = 210 \text{ kN/mm}^2, \qquad E_{concrete} = 14 \text{ kN/mm}^2.$$

(9) The following results were obtained in a laboratory test on a
steel specimen:

Sectional area of specimen = 400 mm².
Gauge length = 200 mm.
Applied load = 21 kN.
Corresponding extension on gauge length = 0·05 mm.
Maximum load in test = 200 kN.

Calculate (i) Young's modulus, (ii) ultimate stress, (iii) necessary
thickness for a tie-bar of this quality of steel, 100 mm wide, to safely
carry an axial load of 156·25 kN, using a factor of safety of 4.

(10) A specimen of mild steel, 100 mm × 20 mm, broke in a tensile
test at a maximum load of 950 kN. Calculate the necessary diameter of
a circular bar of the same quality steel which has to carry an axial pull
of 19·1 kN. Use a factor of safety of 5.

(11) Briefly explain the meaning of the following: 'ultimate stress,'
'factor of safety' and 'Young's modulus.' Draw a fully referenced
diagram indicating the stress-strain relationship for a mild steel specimen tested to destruction.

A short compression specimen of concrete, 100 mm × 100 mm in
section, was tested between the same compression plates as a timber
specimen, 150 mm × 150 mm in section.

Calculate the load carried by the concrete and the load carried by
the timber if the testing machine registered a total load of 23·5 kN.

$$E_{concrete} = 14 \text{ kN/mm}^2, \qquad E_{timber} = 8·4 \text{ kN/mm}^2.$$

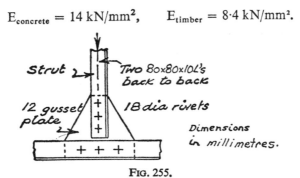

FIG. 255.

S.M.(M)—7*

(12) Fig. 255 shows the end connection of a strut member. Calculate the safe load for the strut from the point of view of the rivets at its end, assuming the working stresses in shear and bearing to be 110 N/mm² and 315 N/mm² respectively. Use gross dia.

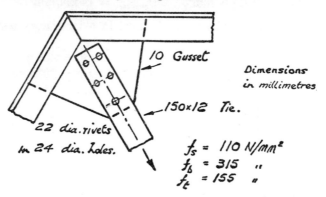

FIG. 256.—GUSSET CONNECTION.

(13) Calculate the safe load for the tie-bar joint given in Fig. 256. The tensile stress has to be limited to 155 N/mm².

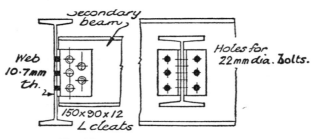

FIG. 257.—END CONNECTION FOR BEAM.

(14) Obtain the safe end reaction for the beam shown in Fig. 257 from the point of view of the strength of the bolts in the web cleats.

$$f_s = 95 \text{ N/mm}^2, f_b = 0.8 \times 300 \text{ N/mm}^2$$

CHAPTER X

SIMPLE BEAMS. BENDING MOMENT AND SHEAR FORCE

THE theory underlying the calculation of the load-carrying capacity of a simple beam will be developed by means of an experimental model beam. The more mathematical treatment of the theory of bending will be found in other text-books on this subject.

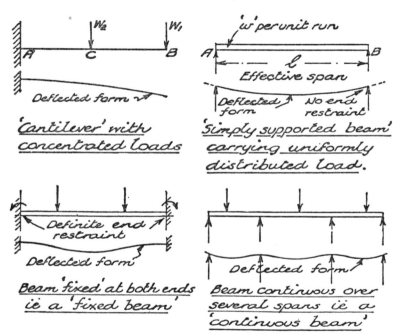

FIG. 258.—TYPES OF BEAMS.

Fig. 258 illustrates various types of beams. The type most commonly occurring in beam calculations is the simply supported beam of one span. The end connections of such a beam are not assumed to be able to develop any appreciable degree of '*fixity*' in the beam.

Failure Tendency at a Beam Section

Consider section 'XX' of the cantilever shown in Fig. 259. If we saw-cut through the cantilever at this section, collapse of the portion to the right of the section plane would take place.

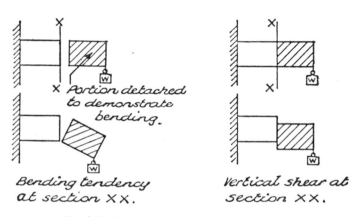

FIG. 259.—BENDING MOMENT AND SHEAR FORCE.

The failure may be attributed to two distinct reasons: (i) the load 'W' produces a moment at section 'XX' which the cantilever, when cut, is unable to balance; (ii) the load 'W' is able to cause vertical slide at the section when the '*shear*' resistance there is destroyed by the saw-cut.

It is important to realise that to prevent failure of a beam at a section such as 'XX' we must *independently* counterbalance the two failure tendencies indicated above. This is borne out by trial with the beam model shown in Fig. 260.

(i) If the pull in the vertical string be reduced the cantilever fails. The tension in the chains and the thrust in the horizontal bars do not assist vertical equilibrium.

(ii) If either the chains or bars be removed collapse takes place. The pull in the vertical string is not able to maintain equilibrium.

A further experimental result should be carefully noted. If an additional weight, say 2 N, be placed on the detached portion of the cantilever, 2 N extra weight will have to be placed in the scale pan to preserve equilibrium. But the exact *position* in which the 2 N weight is placed on the cantilever is quite immaterial provided it is placed to the right of the vertical string (see the definition of *'shear force'* later).

To demonstrate the failure tendencies with the beam model, it is necessary to have a gap in the cantilever as shown in Fig. 260. In the

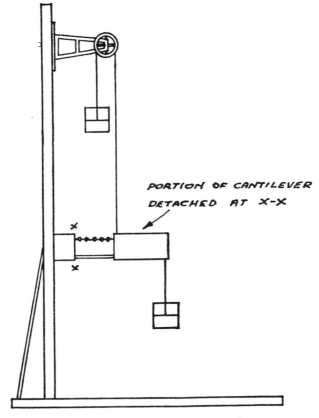

PORTION OF CANTILEVER
DETACHED AT X-X

FIG. 260.—FORCES ACTING AT A BEAM SECTION.

following development of the theory this gap is assumed to be extremely small, so that section 'XX' may be taken to be at the left end of the detached portion of the cantilever.

Bending Moment at Section 'XX' (Fig. 261)

The load 'W' and the string tension 'S,' being equal and unlike parallel forces acting out of line, form a couple. It is the moment of this couple which causes the *'bending-off'* tendency of failure at section 'XX.' This moment, the magnitude of which is 'W $\times$ *l*,' is termed the *'bending moment'* at section 'XX.'

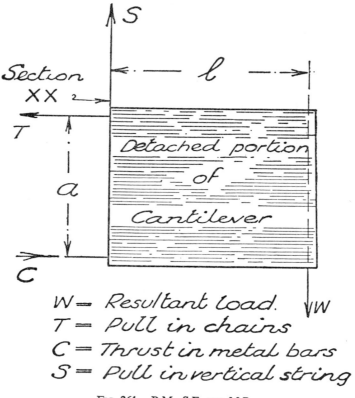

W = *Resultant load.*
T = *Pull in chains*
C = *Thrust in metal bars*
S = *Pull in vertical string*

Fig. 261.—B.M., S.F. and M.R.

Shear Force at Section 'XX'

The load 'W,' the resultant force tending to cause vertical shearing at 'XX,' is termed the *'shear force'* at the section.

The first portion of the work on beams will be devoted to the investigation of the magnitudes of the two quantities *'bending moment'* and *'shear force'* respectively, in certain beam examples. Later we will have to consider how the internal fibre forces at a beam section, such as 'XX,' build up a *'couple,'* able to perform the task of the chains and the metal bars in the model beam apparatus.

Definitions

Bending Moment (B.M.).—*The bending moment at any section of a beam is the resultant moment about that section of all the forces acting to one side of the section.*

Shear Force (S.F.).—*The shear force at any section of a beam is the resultant vertical force of all the forces acting to one side of the section.*

Note.—It is necessary, first of all, to decide which side of the given beam section is going to be taken. Both in 'B.M.' and 'S.F.' calculations, whichever side be selected, the respective answers will be the same. In cantilevers it is easier to consider the forces which act to the 'free end' side of the section considered. Shear force calculations, as indicated in the definition of S.F. above, merely involve the algebraic addition of vertical loads. The actual positions of the loads, therefore, do not affect the S.F. value, provided they lie to one side of the section (cf. experimental result on page 194).

EXAMPLE.—*Calculate the B.M. and S.F. at section 'XX' of the given cantilever (Fig. 262).*

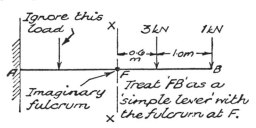

FIG. 262.—BENDING MOMENT AND SHEAR FORCE.

We consider the forces to the *right* of section 'XX,' completely ignoring those to the left. We then imagine a hinge at the point 'F' and treat the portion 'FB' as a simple lever with the fulcrum at 'F.'

Resultant moment about 'F' = $(1 \text{ kN} \times 1\cdot6 \text{ m}) + (3 \text{ kN} \times 0\cdot6 \text{ m})$
$$= (1\cdot6 + 1\cdot8) \text{ kN m}$$
$$= 3\cdot4 \text{ kN m}$$
∴   B.M. at section 'XX' = 3·4 kN m.

Bending moment values are expressed in the usual units for '*moments.*'

The resultant vertical force, considering all the forces which lie to the right of section 'XX,' = 3 kN + 1 kN = 4 kN.

∴   S.F. at section 'XX' = 4 kN.

Shear force values are, of course, expressed in *force* units.

Convention of Signs for B.M. and S.F.

Fig. 263 illustrates the two possible types of curvature which a given bending moment may impose on a beam. These are distinguished by employing *'plus'* and *'minus'* signs. The convention adopted throughout the book is shown in the figure.

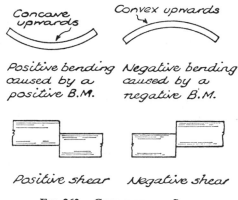

FIG. 263.—CONVENTION OF SIGNS.

Shear force may tend to cause shear in the two ways indicated. The convention adopted in this case is also shown in Fig. 263.

B.M. and S.F. Diagrams

It is sometimes necessary to draw a graph showing the variation of bending moment along the span of a beam. Such a graph is known as a *'bending moment diagram.'* A 'B.M. diagram' has two independent scales: (i) a linear scale for the span (e.g. 10 mm = 1 m), and (ii) a B.M. scale for the vertical ordinates (e.g. 5 mm = 1 kN m).

In the same way 'S.F. diagrams' are constructed to *'linear'* and *'force'* scales. The general form of B.M. and S.F. diagrams in certain standard cases will now be considered.

Cantilever with Single Point Load at the Free End

Fig. 264 shows a cantilever 'AB' with a point load of 2 kN at 'B.' B.M. values:

At 'B,' the free end: B.M. = 2 kN × 0 m   = 0 kN m
At 0·3 m from 'B': B.M. = 2 kN × 0·3 m = 0·6 kN m

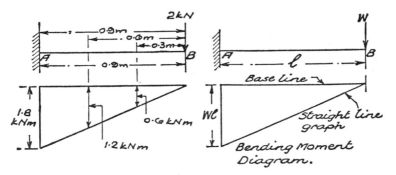

FIG. 264.—BENDING MOMENT DIAGRAM.

At 0·6 m from 'B': B.M. = 2 kN × 0·6 m = 1·2 kN m
At 0·9 m from 'B': B.M. = 2 kN × 0·9 m = 1·8 kN m

The form of the graph is now clear, as each ordinate representing a B.M. value is proportional to the distance of the particular section from 'B.'

The maximum bending moment will obviously be at the support in this case.

$$\text{B.M.}_{max.} = 2 \text{ kN} \times 0·9 \text{ m} = 1·8 \text{ kN m (negative)}.$$

If 'W' = the point load and 'l' = the length of the cantilever, the max. B.M. will be given by the expression 'Wl.' The B.M. diagram will be bounded by a straight line as indicated in Fig. 264.

S.F. Values

In cases in which the length of bearing of a load on a beam is small compared with the beam length, the load may be regarded as acting at a *'point'* for purposes of B.M. and S.F. diagram construction. This accounts for the sudden vertical jumps which are common in S.F. diagrams. In actual fact, of course, there must be a transition from one 'shear' value to another, however rapid, and no point in a beam can simultaneously have two S.F. values.

Just inside point 'B' (Fig. 265) the S.F. = 2 kN, i.e. the load to the right of the section.

At 0·3 m from 'B,' S.F. = load to right of section = 2 kN
At 0·6 m from 'B,' S.F. = load to right of section = 2 kN

The shear force is 2 kN for every section in the cantilever. Expressed in symbols, S.F. = constant = 'W.' The S.F. graph will be a straight line parallel to base (Fig. 265). The B.M. diagram is drawn below the base line and the S.F. diagram above the base line in agreement with the convention of signs adopted.

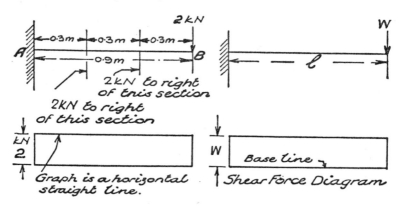

FIG. 265.—SHEAR FORCE DIAGRAM.

Addition of B.M. (or S.F.) Values and Diagrams

When a beam carries several loads, or more than one type of load system, it is sometimes convenient to deal separately with suitable component portions of the total load. The B.M. (or S.F.) values for a given beam section, contributed by the various component loads, are then algebraically added. Thus if a total load system be divided up into three component systems which taken *quite independently* produce, respectively, B.M. values of 5 N m, 3 N m and −2 N m at a given section 'C' in a beam, the actual B.M. at 'C' = (5 + 3 − 2) = 6 N m.

In like manner B.M. (or S.F.) diagrams may be algebraically added by geometrical means.

Cantilever with Several Point Loads

Fig. 266 shows the combination of two B.M. diagrams in order to arrive at the actual final B.M. diagram. Also the figure illustrates the building up of the net S.F. diagram. The component-diagram method of treatment is not commonly used in this type of example. It has been used here to demonstrate the nature of the complete diagrams.

As the diagrams are both bounded by straight lines, they are more quickly constructed by direct calculation of the values at the points where the loads occur (see following example).

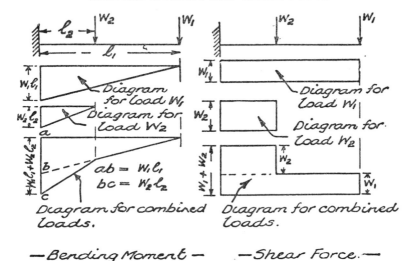

FIG. 266.—ADDITION OF COMPONENT DIAGRAMS.

EXAMPLE.—*Construct the B.M. and S.F. diagrams for the cantilever given in Fig. 267. Calculate the B.M. and S.F. for a section 0·4 m from the support. Write down the values of B.M.$_{max}$ and S.F.$_{max}$ respectively.*

B.M. at 'B' = 0. The bending moment is zero at the free end of a cantilever.

B.M. at 'D.' Regard 'D' as a fulcrum and 'DB' as a simple lever.
$$\text{B.M.}_D = 200 \text{ N} \times 0·4 \text{ m} = 80 \text{ N m.}$$

B.M. at 'C.' Regard 'C' as a fulcrum and 'CB' as a simple lever.

$$\text{B.M.}_C = [(200 \times 0·8) + (400 \times 0·4)] \text{ N m}$$
$$= (160 + 160) \text{ N m} = 320 \text{ N m.}$$

B.M. at the support 'A.' Regard 'A' as a fulcrum.

$$\text{B.M.}_A = [(200 \times 1·0) + (400 \times 0·6) + (300 \times 0·2)] \text{ N m}$$
$$= (200 + 240 + 60) \text{ N m}$$
$$= 500 \text{ N m}$$

(All these bending movements are negative, according to the convention of signs we have adopted.)

Special section 0·4 m from support (point 'E'):
Regard 'E' as the fulcrum of a simple lever 'EB.'

$$\text{B.M.}_E = [(200 \times 0·6) + (400 \times 0·2)] \text{ N m}$$
$$= (120 + 80) \text{ N m} = 200 \text{ N m.}$$

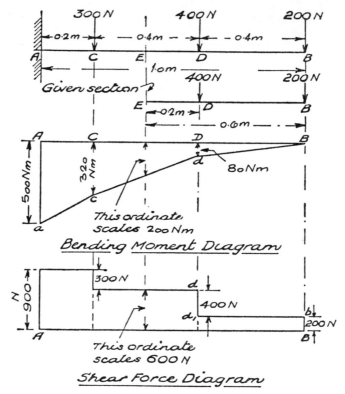

FIG. 267.—CANTILEVER WITH POINT LOADS.

Shear Force.—Between points 'B' and 'D' the shear force is 200 N. At any point between 'D' and 'C' there is a total load of (200 + 400) N to the right, hence the S.F. between 'D' and 'C' = 600 N.

S.F. at 'E' = 600 N.
S.F. between 'C' and 'A' = (200 + 400 + 300) N = 900 N.

$$\text{B.M.}_{max} = 500 \text{ N m.}$$
$$\text{S.F.}_{max} = 900 \text{ N.}$$

Construction of Diagrams

Bending Moment.—As B.M.s are all negative, the ordinates are drawn downwards from the base line. At 'D' draw 'D*d*' to represent 80 N m. At 'C' draw 'C*c*' to represent 320 N m, and at the support draw 'A*a*' to represent 500 N m. Join up B*dca*.

Shear Force.—At 'B' draw 'Bb' to represent 200 N and draw 'bd_1' parallel to base. 'd_1d' must then be drawn to scale to represent 400 N. At each load point the S.F. diagram jumps vertically upwards by an amount which represents the corresponding load, to the force scale of the diagram.

The diagrams should be constructed to the following scales:

B.M. diagram: 100 mm = 1 m and 10 mm = 50 N m.
S.F. diagram: 100 mm = 1 m and 10 mm = 100 N.

The values calculated for the special section may be checked by means of the two diagrams.

Cantilever with Uniformly Distributed Load

The cantilever given in Fig. 268 carries a load of '2 kN per metre run.' To find the B.M. at any particular section of the cantilever we imagine, as in the last case, a fulcrum at the section and the portion to the right to be a simple lever.

For purposes of taking *simple moments* a uniformly distributed load may be regarded as being concentrated at the centre of its length (i.e. its centre of gravity).

Let 'x' = distance from 'B' of a typical section of the beam. When $x = 1$ m (see Fig. 268), (2×1) kN lie to the right of the section. The centre of this load is $\frac{1}{2}$ m from the section, hence the B.M. at the section = 2 kN × $\frac{1}{2}$ m = 1 kN m.

When $x = 2$ m,   B.M. = $[(2 \times 2)\,\text{kN} \times 1\,\text{m}] = 4$ kN m.
When $x = 3$ m,   B.M. = $[(2 \times 3)\,\text{kN} \times \frac{3}{2}\,\text{m}] = 9$ kN m.
When $x = 4$ m,   B.M. = $[(2 \times 4)\,\text{kN} \times 2\,\text{m}] = 16$ kN m.
The 'x' values are in the ratio 1:2:3:4.
The B.M. values are in the ratio 1:4:9:16,   i.e. $1^2:2^2:3^2:4^2$.

This means that the B.M. value varies as the square of the distance from 'B.' The B.M. graph is therefore a *parabola*—a curve satisfying the given conditions.

If 'w' = load per unit run and 'l' = length of the cantilever, the total load carried by the cantilever = wl = W.

B.M.$_{\text{max}}$ will occur at the support and will equal W × $l/2 = \dfrac{Wl}{2}$.

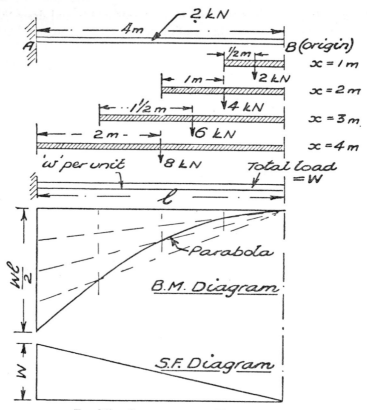

FIG. 268.—CANTILEVER WITH UNIFORM LOAD.

Geometrical Construction of a Parabola

The base line representing the length of the cantilever is divided into a convenient number of equal parts and the ordinate which represents the maximum bending moment is divided into the *same number* of equal parts. Verticals and radials are drawn in, as shown in Fig. 268. The intersections of corresponding verticals and radials, in the manner indicated, give points on the parabola.

Shear Force—Referring to Fig. 268, when $x = 1$ m the load to right of the section $= 2$ kN. When $x = 2$ m the load $= 4$ kN. In this case these loads, which represent S.F. values, increase in direct proportion with the distance from the free end of the cantilever. The shear force graph is therefore an inclined straight line. The max. S.F. will occur at the support and will equal 'W,' the total load on the cantilever.

Note.—The max. S.F. for a cantilever with *any* load system equals the total load carried by the cantilever.

EXAMPLE.—*Draw the B.M. and S.F. diagrams for a cantilever 2 m long, which carries a uniformly distributed load of total value 16 kN. Calculate the B.M. and S.F. values for the centre point of the cantilever.*

$$\text{B.M.}_{max} = \frac{Wl}{2} = \frac{16 \times 2}{2} \text{ kN m} = 16 \text{ kN m (negative)}.$$

S.F.$_{max}$ = total load = W = 16 kN (positive).

The diagrams are shown in Fig. 269.

B.M. at centre of cantilever = (8×0.5) kN m = 4 kN m.

S.F. at centre of cantilever = 8 kN.

Suggested scales for diagrams:

50 mm = 1 m,   50 mm = 1 kN m (B.M.),   20 mm = 1 kN (S.F.).

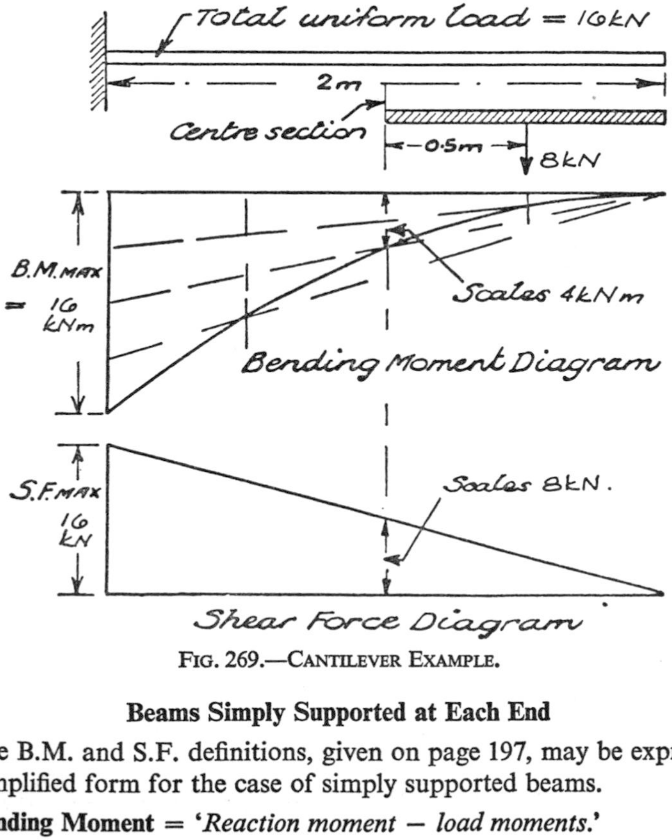

FIG. 269.—CANTILEVER EXAMPLE.

Beams Simply Supported at Each End

The B.M. and S.F. definitions, given on page 197, may be expressed in simplified form for the case of simply supported beams.

Bending Moment = '*Reaction moment − load moments.*'

The 'reaction' and 'loads' must be taken to one side of the section and 'moments' taken about the section.

Shear Force = '*Left-end reaction* − *sum of loads up to section.*'

If the left side of section be taken always for S.F. computation, the correct *sign* will be obtained for the S.F. value.

Simply Supported Beam with Single Non-central Point Load

The usual first step in all beam problems is to calculate the support reactions.

Left-end Reaction.—Moments about 'B':

$$R_A \times l = W \times b. \qquad \therefore \quad R_A = \frac{Wb}{l}.$$

Right-end Reaction.—Moments about 'A':

$$R_B \times l = W \times a. \qquad \therefore \quad R_B = \frac{Wa}{l}.$$

Consider portion 'AC' of the beam.

$$\text{At 1 m from end A, B.M.} = \frac{Wb}{l} \times 1 = \frac{Wb}{l}.$$

$$\text{At 2 m from end A, B.M.} = \frac{Wb}{l} \times 2 = \frac{2Wb}{l}.$$

$$\text{At 3 m from end A, B.M.} = \frac{Wb}{l} \times 3 = \frac{3Wb}{l}.$$

The B.M. value increases in *direct proportion* with the distance from

end 'A' and at the point 'C' the value is $\dfrac{Wb}{l} \times a = \dfrac{Wab}{l}.$

Similarly, regarded from end 'B', the B.M. at 'C' $= \dfrac{Wa}{l} \times b = \dfrac{Wab}{l}.$

The B.M. diagram will be triangular in form.

Between points 'A' and 'C' the S.F. $= R_A - 0 = R_A = \dfrac{Wb}{l}.$

Between 'C' and 'B' the S.F. $= R_A - W = \dfrac{Wb}{l} - W$

$$= \frac{Wb}{a+b} - W = \frac{Wb - Wa - Wb}{a+b} = -\frac{Wa}{l} = -R_B.$$

The S.F. diagram will therefore be as indicated in Fig. 270.

The B.M. and S.F. diagrams are drawn to agree with the convention of signs laid down.

EXAMPLE.—*A simply supported beam of 9 m effective span carries a concentrated load of 6 kN at 3 m from the left support. Construct the B.M. and S.F. diagrams for the beam (See Fig. 270.)*

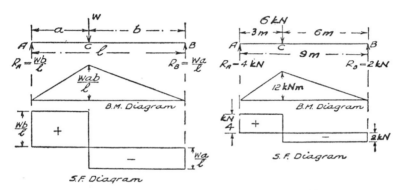

FIG. 270.—SIMPLE BEAM WITH NON-CENTRAL POINT LOAD.

Note.—The '*effective span*' of a beam is the span '*centre to centre*' of the bearings. This span is used in diagram construction and for calculation of beam loading, etc.

$$R_A \times 9 = 6 \times 6 = 36$$
$$R_A = 36/9 \quad = 4 \text{ kN.}$$
$$R_B \times 9 = 6 \times 3 = 18$$
$$R_B = 18/9 \quad = 2 \text{ kN.}$$
$$\text{B.M.}_C = R_A \times 3 \quad \text{or} \quad R_B \times 6$$
$$= (4 \times 3) \quad \text{or} \quad (2 \times 6) \text{ kN m}$$
$$= 12 \text{ kN m.}$$

By formula: $\text{B.M.}_C = \dfrac{Wab}{l} = \dfrac{6 \times 3 \times 6}{9} = 12$ kN m.

The B.M. diagram is a triangle with a maximum height of 12 kN m (at point 'C'). To construct the S.F. diagram, erect at 'A' an ordinate to represent 'R_A,' i.e. '4 kN.' The completion of the diagram is shown in Fig. 270. General rules for the construction of shear force diagrams are given on page 220.

Simply Supported Beam with Single Central-point Load

The max. B.M. for this case may be derived from the previous case by putting $a = b = \dfrac{l}{2}$. (Fig. 271.)

$$\text{B.M.}_C = \frac{Wab}{l} = \frac{W \times \frac{l}{2} \times \frac{l}{2}}{l} = \frac{Wl}{4}.$$

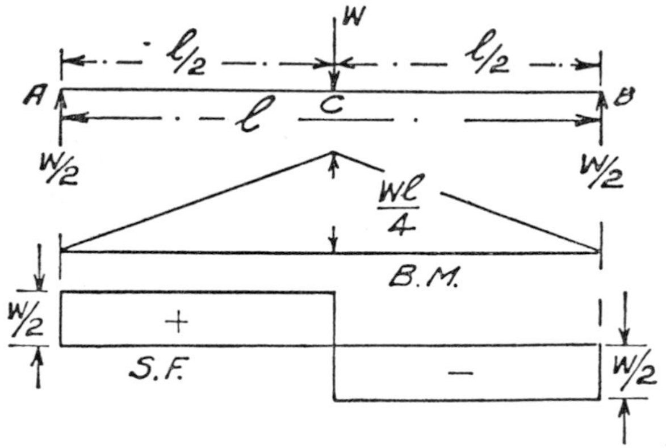

FIG. 271.—SIMPLE BEAM WITH CENTRAL POINT LOAD.

Alternatively, by direct method, B.M.$_C$ = Reaction moment

$$= \frac{W}{2} \times \frac{l}{2} = \frac{Wl}{4}.$$

For the shear force diagram we have: $R_A = R_B = \dfrac{W}{2}$.

Simply Supported Beam with Several Point (or Concentrated) Loads

The nature of the complete B.M. and S.F. diagrams in this case may be demonstrated by using the principle of the *algebraic addition of component diagrams*.

Although this method of 'diagram addition' is not normally employed, it is a very useful method in dealing with beams with complicated load systems or systems of an unusual character. In this particular case it will be noted that the final B.M. diagram is bounded by straight lines which change their slopes at the load points. If we calculate, and plot to scale the net B.M. values at the respective load points, the B.M. diagram may be completed by merely joining up the tops of the ordinates.

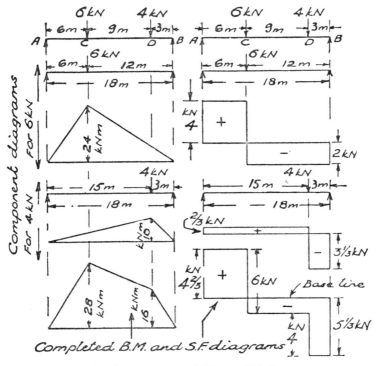

FIG. 272.—COMPOSITION OF B.M. AND S.F. DIAGRAMS.

The S.F. diagram is a stepped diagram, the vertical drops taking place at the load points and representing the corresponding loads to scale. The reader should check the values given on the component diagrams.

EXAMPLE.—*Fig. 273 shows a simply supported beam carrying a concentrated load system. Calculate the B.M. and S.F. values for a section 6 m from the left end. Check the values obtained by constructing the B.M. and S.F. diagrams for the beam.*

The B.M. and S.F. diagrams for this example will be developed in the straightforward way usually adopted.

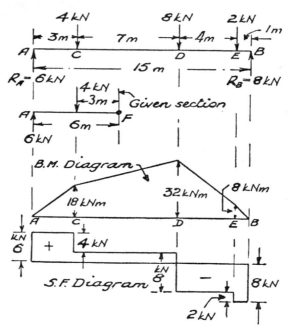

FIG. 273.—SIMPLE BEAM WITH SEVERAL POINT LOADS.

$$R_A \times 15 = (4 \times 12) + (8 \times 5) + (2 \times 1)$$
$$= 48 + 40 + 2 = 90$$

$$R_A = \frac{90}{15} \text{ kN} = 6 \text{ kN.}$$

$$R_B \times 15 = (2 \times 14) + (8 \times 10) + (4 \times 3)$$
$$= 28 + 80 + 12 = 120$$

$$R_B = \frac{120}{15} \text{ kN} = 8 \text{ kN.}$$

B.M._C = Reaction moment − Load moment
$$= [(6 \times 3) - 0] = 18 \text{ kN m.}$$
B.M._D = Reaction moment − Load moment
$$= [(6 \times 10) - (4 \times 7)] \text{ kN m}$$
$$= (60 - 28) \text{ kN m} = 32 \text{ kN m.}$$

Alternatively, taking forces to right of section 'D.'

$$\text{B.M.}_D = \text{Reaction moment} - \text{Load moment}$$
$$= [(8 \times 5) - (2 \times 4)] \text{ kN m}$$
$$= (40 - 8) \text{ kN m} = 32 \text{ kN m.}$$
$$\text{B.M.}_E = \text{Right-end reaction moment}$$
$$= (8 \times 1) \text{ kN m} = 8 \text{ kN m.}$$

Construction of B.M. Diagram.—At 'C' an ordinate is erected to scale 18 kN m, at 'D' to scale 32 kN m and at 'E' to scale 8 kN m. The complete diagram is given in Fig. 273.

To obtain the S.F. diagram an ordinate must be erected at 'A' to represent 6 kN. At 'C' the diagram must drop vertically a distance representing '4 kN' and so on. The diagram is checked by the fact that it should give '$-R_B$' as the S.F. value at the right end.

Values at the special section at 'F':

The forces are taken to the left or the right of section as most convenient.

Taking forces to the left:

$$\text{B.M.}_F = \text{Reaction moment} - \text{Load moment}$$
$$= [(6 \times 6) - (4 \times 3)] \text{ kN m}$$
$$= (36 - 12) \text{ kN m} = 24 \text{ kN m.}$$

[The reader should check this value by taking the forces to the right of 'F.']

$$\text{S.F.}_F = \text{Left-end reaction} - \text{Loads up to section}$$
$$= (6 - 4) \text{ kN} = 2 \text{ kN (positive).}$$

Suggested scales for diagrams:

10 mm = 1 m, 10 mm = 4 kN m (B.M.), 10 mm = 2 kN (S.F.).

Simply Supported Beam with Uniformly Distributed Load

A convenient method of dealing with this case is to investigate the value of the B.M. (or S.F.) at a typical section of the beam. The value is expressed in terms of 'x,' the distance of the section from 'A.' The form of the expression obtained gives the clue to the nature of the B.M. (or S.F.) diagram.

$$\text{B.M.}_x = \textit{Reaction moment} - \textit{Load moment}$$
$$= \left(\frac{wl}{2} \times x\right) - \left(wx \times \frac{x}{2}\right) = \frac{wlx}{2} - \tfrac{1}{2}wx^2.$$

If we plotted this expression for values of 'x' from '0' (i.e. at 'A') to 'l' (i.e. at ('B'), we would obtain a graph in the form of a *parabola*. The maximum ordinate of the parabola will occur for $x = l/2$, i.e. at mid-span.

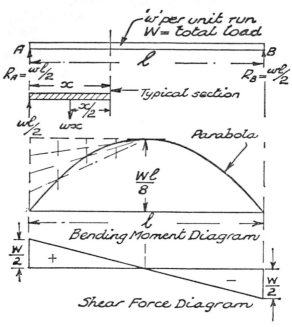

FIG. 274.—SIMPLE BEAM WITH UNIFORM LOAD.

$$\therefore \quad \text{B.M.}_{\text{max}} = \left(\frac{wl}{2} \times \frac{l}{2}\right) - \tfrac{1}{2}w \times \left(\frac{l}{2}\right)^2 = \frac{wl^2}{4} - \frac{wl^2}{8} = \frac{wl^2}{8}.$$

Replacing 'wl' by 'W,' the total load on the beam,

$$\text{B.M.}_{\text{max.}} = \frac{Wl}{8}.$$

The shear force at the typical section = *Left-end reaction — Load up*

to section $= \dfrac{wl}{2} - wx.$

The expression indicates that the shear force graph will be of a linear character.

When $x = 0$ (i.e. at A), S.F. $= \dfrac{w}{2} - 0 = \dfrac{w}{2} = \dfrac{W}{2}$.

When $x = l$ (i.e. at B), S.F. $= \dfrac{wl}{2} - wl = -\dfrac{wl}{2} = -\dfrac{W}{2}$.

In all simple beams of the type now being considered, the shear force at the left end of the beam is the 'left-end reaction,' and at the right end of the beam it is the 'right-end reaction.'

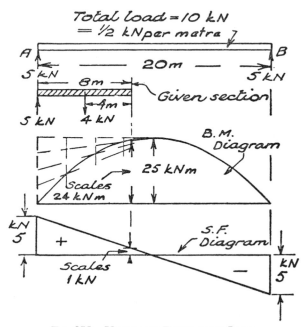

FIG. 275.—UNIFORMLY DISTRIBUTED LOAD.

EXAMPLE.—*A simply supported beam of* 20 m *effective span carries a uniformly distributed load of total value* 10 kN. *Construct the B.M. and S.F. diagrams for the beam. Calculate the values of B.M. and S.F. respectively for a section* 8 m *from the left support.*

$$\text{B.M.}_{\text{max}} = \frac{Wl}{8} = \frac{10 \times 20}{8} \text{ kN m} = 25 \text{ kN m.}$$

$$\text{S.F.}_{\text{max}} = \pm \frac{W}{2} = 5 \text{ kN at 'A' and } -5 \text{ kN at 'B.'}$$

B.M. at 8 m from left support:

> B.M. = Reaction moment − Load moment
> $$= (5 \times 8) - (4 \times 4)$$
> $$= (40 - 16) \text{ kN m} = 24 \text{ kN m}.$$

S.F. at 8 m from left support:

> S.F. = Left-end reaction − Load up to section
> $$= 5 \text{ kN} - 4 \text{ kN} = 1 \text{ kN (positive)}.$$

Suggested scales for diagrams:
20 mm = 1 m, 10 mm = 5 kN m (B.M.), 10 mm = 1 kN (S.F.).

Example on Timber and Steel Floor

The diagram (Fig. 276) shows a floor in which timber beams are supported by steel joists. Assuming the total floor load to be 6 000 N per sq. metre (inclusive of the self-weight of the floor), draw the B.M. and S.F. diagrams for one of the timber beams. Calculate the maximum bending moment in each of the steel beams 'AB' and 'CD' respectively.

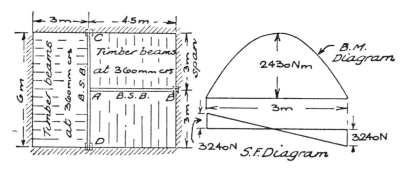

FIG. 276.—TIMBER AND STEEL FLOOR.

Timber Beams.—Area of floor supported by one beam

$$= 3 \text{ m} \times 0.36 \text{ m} = 1.08 \text{ m}^2.$$

U.D. load carried by one beam = $(1.08 \times 6\,000) \text{ N} = 6\,480 \text{ N}.$

Max. B.M. $= \dfrac{Wl}{8} = \dfrac{6\,480 \times 3}{8} \text{ N m} = 2\,430 \text{ N m}.$

Max. S.F. $= \pm \dfrac{W}{2} = \pm 3\,240 \text{ N}.$

Steel Beam 'AB.'—This beam carries the reaction loads from the timber beams in the spans on either side. It is usual to assume that such a load system may be taken as uniformly distributed for the supporting beam. Beam 'AB' will have to carry a 'uniform load' equal to that carried on a floor area $(1\cdot5 + 1\cdot5)$ m $\times$ $4\cdot5$ m $= 13\cdot5$ m^2.

$$\text{Total load} = (13\cdot5 \times 6\,000)\ \text{N} = 81\ \text{kN}.$$

$$\text{Max. B.M. in beam 'AB'} = \frac{Wl}{8} = \frac{81 \times 4\cdot5}{8}\ \text{kN m} = 45\cdot56\ \text{kN m}.$$

Steel Beam 'CD'—Beam 'AB' is supported at end 'A' by beam 'CD.' Beam 'CD' therefore carries a reaction load, at its centre, equal to half the load carried by beam 'AB.'

$$\therefore \quad \text{Load carried by beam CD at its centre} = \frac{81}{2}\ \text{kN} = 40\cdot5\ \text{kN}.$$

B.M. at centre of beam 'CD' due to this load alone

$$= \frac{Wl}{4} = \frac{40\cdot5 \times 6}{4}\ \text{kN m} = 60\cdot75\ \text{kN m}.$$

Beam 'CD' carries in addition the reaction loads from the timber beams in the left-end bay. As in the case of beam 'AB,' it is usual to consider such load as being uniformly distributed along 'CD.'

$\therefore$ Total 'U.D.' load $= (1\cdot5 \times 6 \times 6\,000)\ \text{N} = 54\ \text{kN}$, i.e. half the total load on the left-end bay.

$$\text{B.M.}_\text{max}\ \text{in CD due to this load} = \frac{Wl}{8} = \frac{54 \times 6}{8}\ \text{kN m} = 40\cdot5\ \text{kN m}.$$

As both the maximum bending moments calculated *occur at the same beam section*, i.e. at the centre of span, we may add them together.

Total max. B.M. in beam CD $= (60\cdot75 + 40\cdot5)\ \text{kN m} = 101\cdot25\ \text{kN m}.$

Exercises 10

(Self-weight of beams may be neglected.)

(1) A cantilever projects $1\cdot2$ m horizontally from its support. A rope tackle fixed at the free end is capable of exerting a total vertical pull of 2 kN. Calculate the values of B.M.$_\text{max}$ and S.F.$_\text{max}$ respectively. Construct the B.M. and S.F. diagrams for the cantilever.

(2) Fig. 277 shows a cantilever supporting a number of point loads. Obtain the following values: (i) maximum bending moment, (ii) maximum shear force, (iii) bending moment at mid-point of cantilever, (iv) shear force at 1 m from the support. Construct the bending moment and shear force diagrams for the cantilever.

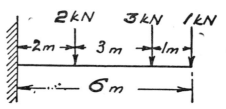

FIG. 277.

(3) A solid concrete balcony floor is supported by embedded steel cantilevers at 0·6 m centres, the cantilevers and floor projecting 1·5 m from a wall (Fig. 278). Calculate the total safe load per m² of balcony floor (inclusive of self-weight of floor) if the maximum bending moment in one cantilever is not to exceed 6 750 N m. Draw the B.M. and S.F.

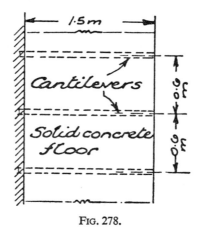

FIG. 278.

diagrams for one of the cantilevers assuming the given maximum bending moment.

(4) A simply supported beam of 2 m effective span carries a central-point load of 100 kN. Calculate the value of the maximum bending moment (i) by formula, (ii) by first principles (i.e. by first finding the support reactions).

Sketch the B.M. and S.F. diagrams for the beam. [All important values should be indicated on a 'sketch' diagram.]

(5) A given steel beam is able to resist safely a maximum bending moment of 13·5 kN m. The beam is used to carry the loads shown in Fig. 279.

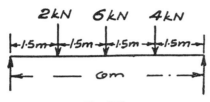

FIG. 279.

(i) Check the safety of the beam from the bending moment point of view.

(ii) Draw the B.M. and S.F. diagrams for the beam.

(6) Construct the B.M. and S.F. diagrams for the simply supported beam shown in Fig. 280. Calculate the B.M. and S.F. values respectively for a section 2·7 m from the right end. Verify your values by means of the diagrams you have constructed.

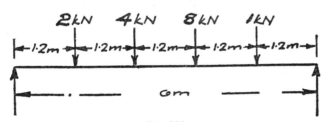

FIG. 280.

(7) A simple beam of 12 m effective span carries a uniform load of total value 24 kN. Find the maximum bending moment and the maximum shear force. Draw the B.M. and S.F. diagrams for the beam. Calculate the B.M. and S.F. values for a section 4 m from the left end.

(8) A simple beam of 10 m effective span carries a uniform load which produces a bending moment of 24 kN m at a section 4 m from the left support. Calculate the value of the load in kN per metre run. Also obtain the maximum bending moment in the beam.

(9) Fig. 281 shows a floor with timber beams of 2·4 m effective span supported by a steel beam. Find the maximum bending moment and

maximum shear force respectively for one of the timber beams and also for the steel beam. Construct (i) the B.M. diagram for a timber beam and (ii) the S.F. diagram for the steel beam. The total inclusive floor load may be taken as 5 kN/m².

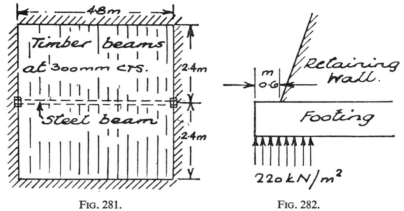

FIG. 281.   FIG. 282.

(10) The resultant upward pressure on the footing in the example given in Fig. 282 is 220 kN/m². Calculate the maximum bending moment the footing will have to resist per metre run of wall, assuming the maximum bending moment to occur at the face of the wall.

(11) A simply supported beam of 6 m effective span carries a point load of 4 kN at the centre. Calculate the load which would produce the same maximum bending moment if (i) uniformly distributed, (ii) concentrated at 2 m from the left-end support.

(12) A lintel beam of 3 m effective span carries a brick wall, 225 mm thick and of 1·5 m uniform height. Assuming the average density of the brickwork to be 19·2 kN/m³, obtain the B.M. and S.F. values for a section 1·2 m from the left end of beam. Construct the B.M. and S.F. diagrams for the beam.

(13) A steel chimney stack of circular section, 1 m external diameter, is 20 m high. It is subjected to a horizontal wind pressure of uniform intensity, 1·5 kN/m². Assuming the area acted upon by the wind to be 0·6 of the area 'height × diameter' (a common practical allowance for circular chimneys), calculate the B.M. and S.F. values for a section at mid-height of the chimney and also for the base.

(14) A crane girder carries two wheel loads of 2 kN each at a fixed distance of 1·2 m apart. The effective span of the girder is 4·8 m. Calculate the two distances the left-hand wheel load may be from the

left support so that the bending moment in the girder at the position of the other wheel is 3·6 kN m.

(15) Calculate the total uniformly distributed load the steel beam shown in Fig. 283 could carry, in addition to the point loads shown, if the beam were capable of resisting safely a bending moment of 22 kN m.

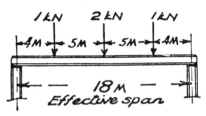

FIG. 283.

Find the 'equivalent uniform load' which would produce (i) the same maximum bending moment, (ii) the same maximum shear force as the *point load* system shown in Fig. 283.

CHAPTER XI

BENDING MOMENT AND SHEAR FORCE. FURTHER EXAMPLES

THE examples of the last chapter have illustrated certain facts which will now be summarised.

(i) **A concentrated load system will lead to a B.M. diagram bounded by straight lines, which change their slope at the load points.**

(ii) **A distributed load will have some form of curve for the B.M. diagram. If the load be uniformly distributed a parabola will be involved in the construction of the diagram.**

(iii) **For a concentrated load system the S.F. diagram will be stepped. A step will take place at every load point, the vertical drop representing the load to scale.**

(iv) **The S.F. diagram for a uniformly distributed load will be of a uniformly sloping character. For our convention of signs the slope will be downwards towards the right.**

The foregoing rules will be exemplified in the following cases.

Overhanging Beam with Concentrated Load System

EXAMPLE.—*Construct the B.M. and S.F. diagrams for the overhanging beam given in Fig. 284.*

Moments about 'B':

$$(R_A \times 3) + (2 \times 1 \cdot 5) = (8 \times 0 \cdot 6) + (6 \times 1 \cdot 8) + (2 \times 4 \cdot 2)$$
$$= 4 \cdot 8 + 10 \cdot 8 + 8 \cdot 4 = 24$$
$$3R_A + 3 = 24$$
$$3R_A = 24 - 3 = 21$$
$$\therefore \quad R_A = 7 \text{ kN}.$$

Moments about 'A':

$$(R_B \times 3) + (2 \times 1 \cdot 2) = (6 \times 1 \cdot 2) + (8 \times 2 \cdot 4) + (2 \times 4 \cdot 5)$$
$$3R_B + 2 \cdot 4 = 7 \cdot 2 + 19 \cdot 2 + 9 \cdot 0 = 35 \cdot 4$$
$$3R_B = 35 \cdot 4 - 2 \cdot 4 = 33$$
$$R_B = 11 \text{ kN}.$$

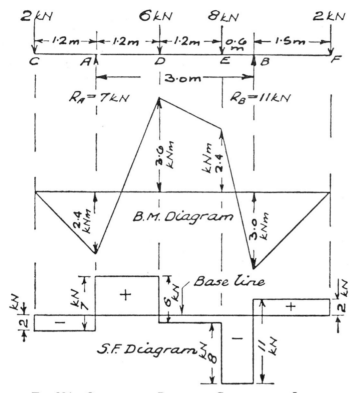

Fig. 284.—Overhanging Beam with Concentrated Loads.

B.M.$_C$ = 0 (free end of beam).

B.M.$_A$ = Moment of load to left of section

  = $(2 \times 1 \cdot 2)$ kN m = $2 \cdot 4$ kN m (*negative*).

B.M.$_D$ = Reaction moment − Load moment

  = $(7 \times 1 \cdot 2) - (2 \times 2 \cdot 4)$ kN m

  = $(8 \cdot 4 - 4 \cdot 8)$ kN m = $3 \cdot 6$ kN m

B.M.$_E$ = Reaction moment − Load moment

  = $(11 \times 0 \cdot 6) - (2 \times 2 \cdot 1)$ kN m (*taking forces to right of section*)

  = $(6 \cdot 6 - 4 \cdot 2)$ kN m = $2 \cdot 4$ kN m

B.M.$_B$ = $(2 \times 1 \cdot 5)$ kN m = 3 kN m (*negative*).

The shear force diagram is constructed without any further detailed calculation.

Suggested scales:

  25 mm = 1 m, 10 mm = 1 kN m (B.M.), 10 mm = 1 kN (S.F.).

Overhanging Beam with Uniformly Distributed Load

EXAMPLE.—*The beam given in Fig. 285 carries a uniform load of 5 kN per metre run. Construct the B.M. and S.F. diagrams for the beam.*

In order to illustrate the method of addition of component B.M. diagrams, the beam load will be divided up into two systems: (i) the loads on 'AC' and 'BD,' (ii) the load on 'AB.'

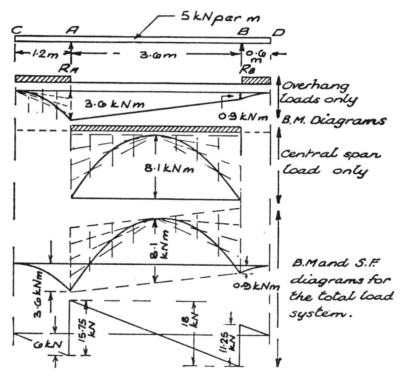

FIG. 285.—OVERHANGING BEAM WITH UNIFORM LOAD.

Taking component system (i) *by itself*:

$$\text{B.M.}_A = [(5 \times 1\cdot2) \times 0\cdot6] \text{ kN m} = 3\cdot6 \text{ kN m (negative).}$$
$$\text{B.M.}_B = [(5 \times 0\cdot6) \times 0\cdot3] \text{ kN m} = 0\cdot9 \text{ kN m (negative).}$$

The B.M. diagram for this component system is shown in Fig. 285 (upper diagram).

Considering component system (ii), i.e. the load on 'AB,'

$$\text{B.M.}_{max} = \frac{Wl}{8} = \frac{5 \times 3\cdot6 \times 3\cdot6}{8} = 8\cdot1 \text{ kN m } (positive).$$

We now have to combine these two B.M. diagrams together. The net B.M. diagram for the complete load system is shown in Fig. 285, which also shows the practical method of obtaining the final B.M. diagram in a more direct manner. In using this construction, care should be taken to plot the value '8·1 kN m' (i.e. 'Wl/8' for the central span) *vertically* from the mid-point of the temporary sloping base line shown in the diagram.

Shear Force Diagram.—We require, first of all, to calculate the support reactions.

$$(R_A \times 3\cdot6) + (3 \times 0\cdot3) = 5 \times 4\cdot8 \times \frac{4\cdot8}{2}$$
$$3\cdot6R_A + 0\cdot9 = 57\cdot6$$
$$3\cdot6R_A = 57\cdot6 - 0\cdot9 = 56\cdot7$$
$$R_A = 15\cdot75 \text{ kN.}$$
$$(R_B \times 3\cdot6) + (6 \times 0\cdot6) = 5 \times 4\cdot2 \times \frac{4\cdot2}{2}$$
$$3\cdot6R_B + 3\cdot6 = 44\cdot1$$
$$3\cdot6R_B = 44\cdot1 - 3\cdot6 = 40\cdot5$$
$$R_B = 11\cdot25 \text{ kN.}$$

From 'C' to 'A' the load on the beam = 6 kN, hence the S.F. diagram must drop by this amount over the length 'CA.' At 'A' there is a vertical jump upwards of '15·75 kN' (i.e. ('R$_A$'). Between 'A' and 'B' the diagram must drop 5 × 3·6 = 18 kN. At 'B' there is a vertical jump upwards of 11·25 kN then a final drop of '3 kN' over the length 'BD.'

Note.—If various portions of a beam, as in the last example, carry uniform loads of similar intensity, the *slope* of the S.F. diagram will be the same in each of these different portions.

Suggested scales:

25 mm = 1 m, 10 mm = 1 kN m (B.M.), 5 mm = 1 kN (S.F.).

Beams with Uniformly Distributed Load Partially Covering the Span

EXAMPLE (i).—'*AB*' (*Fig.* 286) *is a simply supported beam carrying a uniformly distributed load of 6 kN per metre, which partially covers the span. Construct the B.M. and S.F. diagrams for the beam.*

S.M.(M)—8*

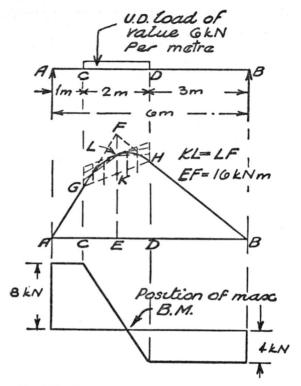

FIG. 286.—UNIFORM LOAD PARTIALLY COVERING SPAN.

The principle of the construction of the B.M. diagram in this case is to regard, *temporarily*, the U.D. load 'CD' as being a concentrated point load of value '12 kN,' acting at 'E,' the mid-point of 'CD.' This is for construction purposes only, and allowance is subsequently made for the fact that the load is actually distributed.

$$\text{B.M.}_\text{E} = \frac{Wab}{6} = \frac{12 \times 2 \times 4}{6} = 16 \text{ kN m.}$$

'EF' is drawn to represent '16 kN m' and the triangular B.M. diagram 'AFB' is completed.

The ordinates 'CG' and 'DH' are drawn and 'GH' is drawn in. 'GH' intersects 'EF' in point 'K.' 'KF' is bisected in point 'L' and the parabolic diagram is constructed as indicated in Fig. 286. The final B.M. diagram is 'AGLHB.'

The bisecting of 'KF' makes the necessary allowance referred to above. The S.F. diagram presents no special difficulty.

$$R_A \times 6 = (12 \times 4) = 48$$
$$R_A = 8 \text{ kN.}$$
$$R_B \times 6 = (12 \times 2) = 24$$
$$R_B = 4 \text{ kN.}$$

Suggested scales:

25 mm = 1 m, 5 mm = 1 kN m (B.M.), 5 mm = 1 kN (S.F.).

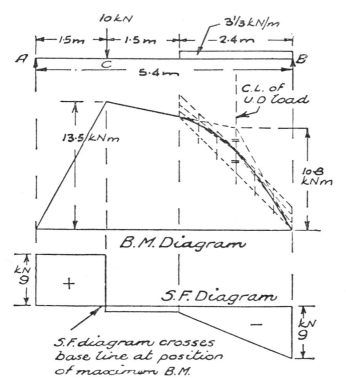

FIG. 287.—COMBINED LOADS.

EXAMPLE (ii).—*The given simply supported beam (Fig. 287) supports a partial uniformly distributed load together with a concentrated load. Construct the B.M. and S.F. diagrams for the beam.*

As in the last example, the U.D. load is replaced temporarily, *for construction purposes,* by a point load of equal value at its centre. The necessary final adjustment is then made as indicated in Fig. 287.

$$R_A \times 5 \cdot 4 = (10 \times 3 \cdot 9) + (8 \times 1 \cdot 2) = 48 \cdot 6$$
$$R_A = 9 \text{ kN.}$$
$$R_B \times 5 \cdot 4 = (10 \times 1 \cdot 5) + (8 \times 4 \cdot 2) = 48 \cdot 6$$
$$R_B = 9 \text{ kN.}$$
$$\text{B.M.}_C = R_A \times 1 \cdot 5 = 9 \times 1 \cdot 5 = 13 \cdot 5 \text{ kN m.}$$

B.M. at centre of U.D. load, *regarding load as concentrated,* $= R_B \times 1 \cdot 2 = 9 \times 1 \cdot 2 = 10 \cdot 8$ kN m. [Care must be exercised not to compute the *correct* B.M. at the centre of the uniform load in this method.]

The 'straight-line' B.M. diagram is completed and the portion corresponding to the uniform load modified as indicated. The S.F. diagram is easily constructed by the rules previously given.

EXAMPLE (iii).—*The cantilever shown in Fig. 288 carries a partial uniformly distributed load and, in addition, one concentrated load. Draw the B.M. and S.F. diagrams for the cantilever.*

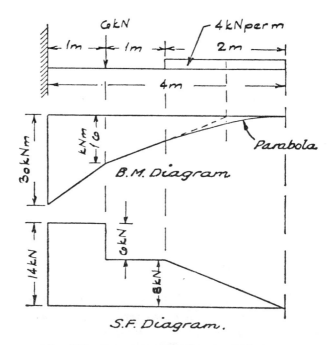

FIG. 288.—CANTILEVER WITH PARTIAL U.D. LOAD.

The treatment is as for the previous examples and the detailed calculations are left as an exercise.

Beams Carrying Complicated Load Systems

Bending Moment Diagrams

If the complete load system may be divided up into two simple systems, i.e. two systems which can be readily dealt with, the method 1 shown in Fig. 289 is sometimes adopted. The diagram for one system is drawn above the base line and for the other below (using the same B.M. scale). The total vertical depth of the diagram will then give the necessary bending moment at any particular beam section.

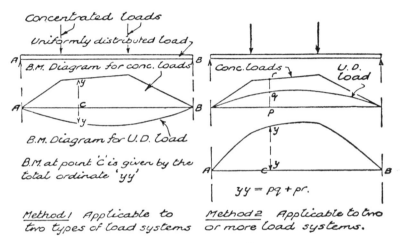

FIG. 289.—SEVERAL LOAD SYSTEMS.

In cases in which there are several systems to be dealt with, the B.M. diagram for each system may be constructed on a common base (or if preferred on separate base lines) and the diagrams added together by the geometrical method of *adding corresponding ordinates* (see Fig. 289). It is essential, of course, that the B.M. scale used for each of the component diagrams shall be the same.

A further practical method for dealing with complicated load systems is to calculate the bending moment values at a suitable number of points in the beam and to draw ordinates to represent these values to a convenient B.M. scale. The reactions will have to be found and the expression '*reaction moment—load moments*' may be used to compute the B.M.

values for the various beam sections. It may be necessary to distinguish carefully between positive and negative bending moments, so that the ordinates may be drawn above or below the base line. The correct *sign* for the bending moment will be obtained if the expression given at the bottom of page 227 be employed.

Shear Force Diagrams

S.F. diagrams do not usually present much difficulty, and the rules already given cover the types of loading considered in this book.

Position of Maximum Bending Moment

If the point in the beam at which the maximum bending moment occurs, in any of the previously worked examples, be examined in conjunction with the corresponding shear force diagram, it will be found that it coincides with the point in which the shear force diagram crosses its base line. As the latter point may be readily obtained *without actually drawing the S.F. diagram*, it is clear that this relationship will give a simple method for determining the position in the beam at which B.M.$_{max}$ occurs. The complete theory underlying this important connection between bending moment and shear force will be found in more advanced books on the theory of structures.

Rule: *The position of maximum bending moment in a beam may be found by determining the point at which the shear force is zero.*

(The case of two, or more, such points is considered later.)

The following steps illustrate the method of calculating a maximum bending moment:

(i) Evaluate the left-end support reaction.

(ii) Proceed across the span from the left end until the load taken up on the beam equals, *in total value*, the left-end reaction. This is clearly the point of zero shear and hence the position of maximum bending moment.

(iii) Calculate the B.M. at this point, by usual methods, to obtain the value of B.M.$_{max}$.

EXAMPLE (i).—*Calculate the maximum bending moment for the given simply supported beam (Fig. 290).*

Step (i).

$$R_A \times 10 = [(4 \times 1\tfrac{1}{4}) \times 8] + [(2 \times 10) \times 5]$$
$$= 40 + 100 = 140$$
$$R_A = 14 \text{ kN.}$$

[R_B should be calculated as a check.]

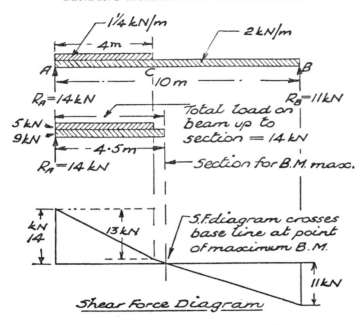

FIG. 290.—POSITION OF MAXIMUM BENDING MOMENT.

Step (ii).—Up to point 'C,' from left end of beam, the load on beam

$$= [(1\tfrac{1}{4} \times 4) + (2 \times 4)] \text{ kN}$$
$$= (5 + 8) \text{ kN} = 13 \text{ kN}.$$

We require another 1 kN to make up 14 kN, which means we have to proceed 0·5 m farther along the beam (as the load is 2 *kN per metre*).

S.F. = 0 and B.M. = maximum at 4·5 m from left-end support.
B.M.$_{4\cdot5m}$ = B.M.$_{max}$ = Reaction moment − Load moments

$$= (14 \times 4\cdot5) - (5 \times 2\cdot5) - \left(9 \times \frac{4\cdot5}{2}\right) \text{ kN m}$$

$$= (63 - 12\cdot5 - 20\cdot25) \text{ kN m}$$
$$= 30\cdot25 \text{ kN m}.$$

EXAMPLE (ii).—*Draw the shear force diagram for the given beam (Fig. 291). Find the position of zero shear* (i) *by means of the S.F. diagram,*

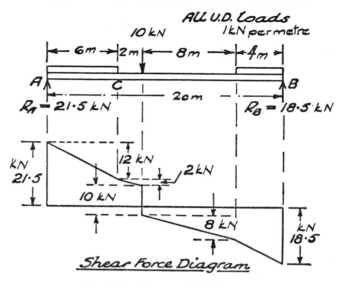

FIG. 291.—MAXIMUM B.M. EXAMPLE.

(ii) *by direct calculation. Calculate the value of the maximum bending moment.*

$$R_A \times 20 = [(1 \times 6) \times 17] + (10 \times 12) + [(1 \times 4) \times 2]$$
$$+ [(1 \times 20) \times 10]$$
$$= 102 + 120 + 8 + 200$$
$$= 430$$
$$R_A = 21\cdot5 \text{ kN.}$$
$$R_B \times 20 = [(1 \times 6) \times 3] + (10 \times 8) + [(1 \times 4) \times 18]$$
$$+ [(1 \times 20) \times 10]$$
$$= 18 + 80 + 72 + 200$$
$$= 370$$
$$R_B = 18\cdot5 \text{ kN.}$$

From left end up to point 'C' the load on beam = 12 kN. Up to the '10 kN' load (but not including it) the load = 12 kN + 2 kN = 14 kN. This is not sufficient. Including the 10 kN point load the total load = 24 kN. This is too great. When the inclusion of a concentrated point load makes the necessary total load exceed the required amount, *the position of B.M.*$_{max}$ *is at the load point.*

B.M.$_{max}$ is therefore at the '10 kN' load point.

B.M.$_{max}$ = Reaction moment — Load moments
$$= (21\cdot5 \times 8) - (6 \times 5) - (8 \times 4)\,\text{kN m}$$
$$= (172 - 30 - 32)\,\text{kN m}$$
$$= 110\,\text{kN m}.$$

The complete S.F. diagram is given in Fig. 291.
Suggested scales: 10 mm = 1 m, 5 mm = 1 kN.

EXAMPLE (iii).—*Calculate the maximum bending moment which the given overhanging beam (Fig. 292) will have to resist. Construct the shear force diagram for the beam.*

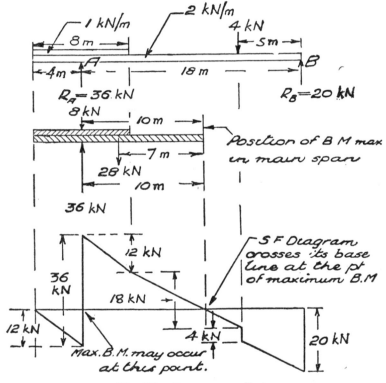

FIG. 292.—OVERHANGING BEAM.

$$R_A \times 18 = [(2 \times 22) \times 11] + [(1 \times 8) \times 18] + (4 \times 5)$$
$$= 484 + 144 + 20 = 648$$

$$R_A = \frac{648}{18}\,\text{kN} = 36\,\text{kN}.$$

$$R_B \times 18 = [(2 \times 22) \times 7] +](1 \times 8) \times 0] + (4 \times 13)$$
$$= 308 + 0 + 52 = 360$$

$$R_B = \frac{360}{18} \, kN = 20 \, kN.$$

In the case of overhanging beams, the maximum bending moment may occur at either of the supports or in the main span of the beam. Wherever the S.F. diagram crosses the base line a *'local'* B.M.$_{max}$ occurs.

$$B.M._A = \text{Moment of load to left of section}$$
$$= (3 \, kN \text{ per metre} \times 4 \, m) \times 2 \, m$$
$$= 24 \, kN \, m \, (negative).$$

Position of B.M.$_{max}$ in span 'AB':
The load on the beam from the *extreme left end* up to the required point must equal 36 kN.

Up to the end of the partial U.D. load, the load on beam $= (8 \times 3)$ kN $= 24$ kN. To obtain the additional $(36 - 24)$ kN $= 12$ kN, we have to proceed 6 m farther. B.M.$_{max}$ therefore occurs at 10 m from the left support.

$$B.M._{max} = \text{Reaction moment} - \text{Load moments}$$
$$= (36 \times 10) - (8 \times 10) - [(2 \times 14) \times 7]$$
$$= 360 - 80 - 196$$
$$= 84 \, kN \, m.$$

This is the absolute max. bending moment in the beam.

Fig. 292 explains the various load jumps by means of which the shear force diagram is constructed.

Suggested scales: 5 mm = 1 m and 25 mm = 10 kN.

Bending Moment Diagram by Link Polygon Method

If the link polygon be drawn for a beam with a concentrated load system and the closer line be inserted, as explained on page 123, the diagram so formed is, to some definite scale, the bending-moment diagram for the beam. This method of construction of B.M. diagrams is not so quick or direct as the methods previously described. It has, however, important applications in more advanced theory of structures, as, for example, in *'moving-load'* problems and in the graphical treatment of *'deflection of beams.'* Fig. 293 shows how to set out the work when both B.M. and S.F. diagrams are required.

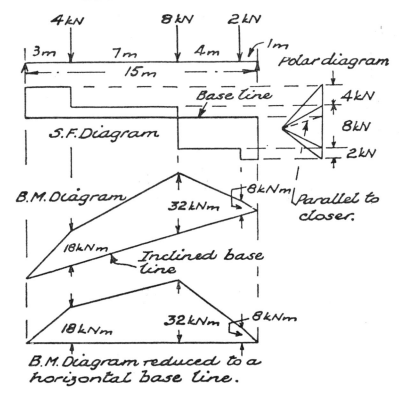

FIG. 293.—B.M. AND S.F. DIAGRAMS BY LINK POLYGON.

The pole is positioned, in the given example, to the left of the load line in order to maintain the convention of sign for bending moment which has been adopted. The position of the pole theoretically is immaterial, but it will be found that it will be advantageous to fix it at some definite *horizontal* distance from the 'load line,' say 100 mm, in order to obtain a suitable final bending moment scale.

Having completed the polar diagram, the '*links*' and the '*closer line*' are drawn in. If it is desired to obtain a horizontal base line for the B.M. diagram, it will be necessary simply to plot the *vertical* ordinates at the load points from a horizontal base as indicated in Fig. 293.

Bending Moment Diagram Scale

Usually, of course, we choose an appropriate B.M. scale according to the magnitude of the calculated B.M. values and the size of diagram that is required. In the present case we have to *calculate* the B.M. scale from the other scales used in the derivation of the diagram.

If 1 mm = 'x' m for '*span*,'
    1 mm = 'y' kN for '*load line*,' and
    'p' mm = actual polar distance, i.e. horizontal distance of pole from load line,

then 1 mm will represent ($x \times y \times p$) kN m in B.M. diagram.

Thus, if 1 mm = 0·1 m for span,
    1 mm = 0·1 kN for load,
and polar distance = 100 mm, the B.M. scale would be
    1 mm = (0·1 × 0·1 × 100) kN m = 1 kN m.

Note that the units for B.M. do not contain 'millimetres' in this case, but are compounded of the '*span*' and '*load*' units.

Shear Force Diagram

It will be clear that the vertical steps, as they represent the corresponding loads, will be given to scale in the polar diagram load line and may be projected across. It only remains to fix the position of the base line. The line drawn in the polar diagram parallel to the 'closer' divides the load line into two parts which respectively represent the support reactions, the upper part representing the left-end reaction. As the first vertical jump upwards of the S.F. diagram represents the left-end reaction, it is obvious that the base line will be obtained by projecting horizontally across in the manner indicated in Fig. 223. The shear force scale will be the same as that used for the *load line* in the polar diagram.

This example was previously solved by the direct calculation method (see Fig. 273).

EXERCISES 11

[Self-weight of beams may be neglected.]

(1) Fig. 294 shows a beam which overhangs its supports at both ends. Calculate, for the loads given, the values of the B.M. and S.F. respectively at a section 1·8 m to the right of the left support. Draw the B.M. and S.F. diagrams for the beam.

(2) Draw the B.M. and S.F. diagrams for the overhanging beam given in Fig. 295. Obtain the value of the maximum bending moment by scaling the B.M. diagram.

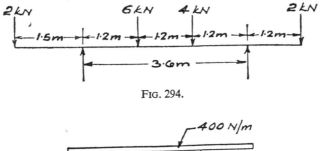

FIG. 294.

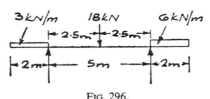

FIG. 295.

(3) Construct the B.M. and S.F. diagrams for the example given in Fig. 296.

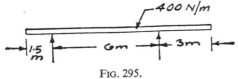

FIG. 296.

(4) A simply supported beam of 6 m effective span carries a uniform load which partially covers the span. The load is 2 m long, commences at 1 m from the left support and has the value of 6 kN per metre run. Construct the B.M. and S.F. diagrams for the beam due to this load.

(5) Find the position and magnitude of the maximum bending moment for the simply supported beam given in Fig. 297. Draw the S.F. diagram for the beam.

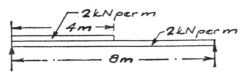

FIG. 297.

(6) Without calculating B.M. values or constructing the B.M. or S.F. diagrams for the beam given in Fig. 298, show that the maximum bending moment occurs at the position of the 8 kN load.

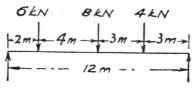

FIG. 298.

(7) Find the position of maximum bending moment for the simply supported beam shown in Fig. 299 (i) by beam load method, (ii) by constructing the S.F. diagram for the beam. Obtain the magnitude of the maximum bending moment.

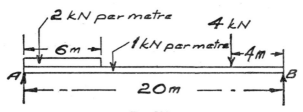

FIG. 299.

(8) Draw the S.F. diagram for the overhanging beam given in Fig. 300. Hence determine the position in the central bay at which a maximum bending moment occurs.

Find the value of the absolute maximum bending moment.

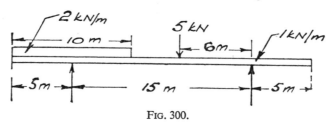

FIG. 300.

(9) Find the position and magnitude of the maximum bending moment in the case of the overhanging beam given in Fig. 301. Construct the S.F. diagram for the beam.

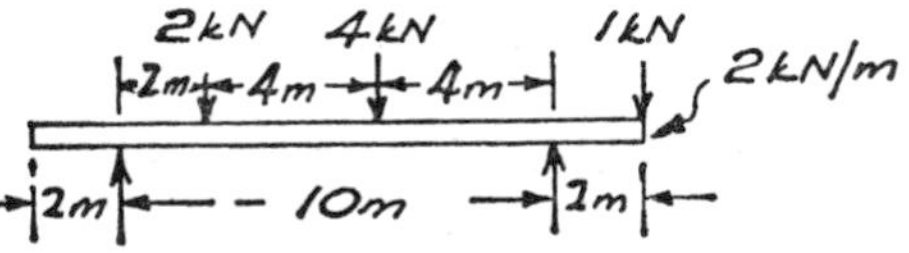

FIG. 301.

(10) Construct the bending moment and shear force diagrams for the overhanging beam shown in Fig. 302.

(11) A simply supported beam of 10 m effective span carries a uniform load of 1 kN per metre which covers the span. In addition there is a point load of 5 kN at 2 m from the left end. Construct the B.M. and S.F. diagrams for the beam.

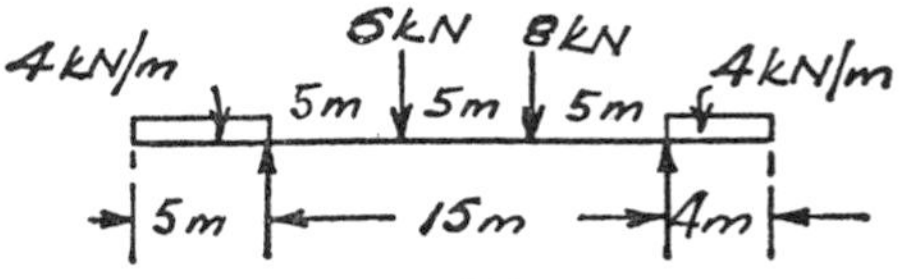

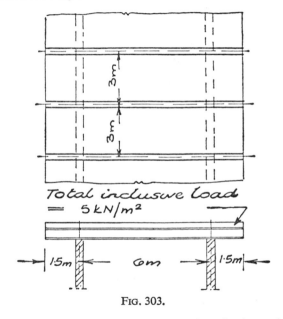

FIG. 302.

(12) Calculate the maximum bending moment which must be resisted by one of the beams shown in the floor diagram (Fig. 303). Draw the B.M. and S.F. diagrams for a beam.

FIG. 303.

(13) A wall 3 m high has to be able to resist a horizontal wind pressure of 500 N/m² for the upper two-thirds of its height. What would be the corresponding bending moment about the base of the wall per metre run of wall, i.e. for every metre length of wall in plan? Construct the B.M. and S.F. diagrams for this metre strip of wall.

(14) Calculate the position and value of the maximum bending moment for the simply supported beam given in Fig. 304.

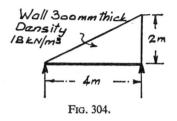

FIG. 304.

[Assume B.M.$_{max}$ to be 'x' m from the left end. The wall height here will be $x/2$ m, by proportion. Equate the weight of the wall, to the left of the section, to the left-end reaction and solve the quadratic for 'x.' Find the B.M. at the calculated section by the usual formula 'Reaction moment − load moment,' treating the load to the left of the section as acting at the C.G. of the triangle.]

(15) Draw the bending moment and shear force diagrams for the mast shown in Fig. 305. Write down the maximum bending moment the mast must be capable of resisting.

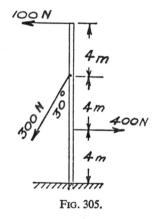

FIG. 305.

[Resolve the '300 N' force horizontally and ignore the vertical component. Treat as a vertical cantilever. Find B.M. at each load point, being careful with the sign of each moment. S.F. diagram is constructed by usual methods, the 'jumps' being horizontal in this case.]

CHAPTER XII

MOMENT OF RESISTANCE. DESIGN OF SIMPLE BEAMS

In the model apparatus shown on page 195, the bending moment on the detached portion of the cantilever was balanced by the couple created by the two equal horizontal forces—the *pull* in the chains and the *thrust* in the bars. The moment of this balancing couple is termed the '*moment of resistance.*' At every beam section subjected to bending moment, the moment of resistance will equal the bending moment, when the beam has deflected to its position of equilibrium. The general problem of beam design is to determine a suitable size and shape for the beam section so that the beam fibres are not excessively stressed and that the material of the beam is used economically.

Definition: *The moment of resistance (M.R.) of a beam section is the moment of the couple which is set up at the section by the longitudinal forces created in the beam by its deflection.*

General Principles of Simple Bending

The upper diagram in Fig. 306 shows the elevation of a simple beam. The lower diagram shows the beam bent under the action of bending

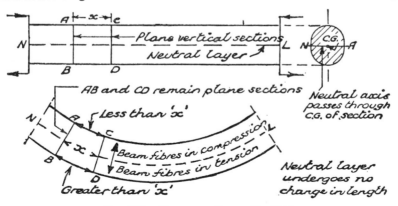

FIG. 306.—THEORY OF SIMPLE BENDING.

moment. In every bent beam there is a layer of material (NL in Fig. 306) which gets neither longer nor shorter when the beam deflects. It is termed the *'neutral layer'* of the beam.

Neutral Axis.—The *'neutral axis'* of a beam section is the straight line in which the neutral layer of the beam cuts that particular section. It represents the level in the beam section at which there is neither stress nor strain.

Variation of Strain and Stress in a Beam Section

An inspection of Fig. 306 indicates that beam layers above the neutral layer 'NL' will be shortened and those below 'NL' will be lengthened. In this case, therefore, all fibres in a beam cross section above the neutral axis will be in compressive strain and hence compressive stress, and those below will be in tensile strain and tensile stress. It will be clear that in negative bending, as in a cantilever, the upper fibres will be in tension and the lower in compression.

The full discussion of the assumptions made in the theory of bending is rather beyond the scope of this book, but one vital assumption must be appreciated. It is assumed that a vertical plane cross section of the beam before bending remains plane (i.e. flat) after bending.

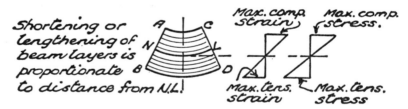

FIG. 307.—STRAIN AND STRESS DISTRIBUTION.

In Figs. 306 and 307, 'AB' and 'CD' represent two cross sections of the beam. They are assumed to be very close together so that the beam layers between them may be assumed to deflect to the arcs of concentric circles.

In Fig. 307 'NA' and 'LC' are straight lines. The beam layers are subjected to a shortening in length from length 'NL' to length 'AC' *in a proportionate manner* as we proceed upwards from 'NL.' The amount of shortening is *directly proportional* to the distance of the particular layer from the neutral layer. Similarly, the amount of lengthening in layers below 'NL' will be directly proportional to their distance from 'NL.'

As all beam layers between sections 'AB' and 'CD' were the same length in the unbent beam, it will be clear that '*strain*' will be proportional to distance from 'NL.' The '*strain variation diagram*' will therefore be linear in character, as shown in Fig. 307. If we assume further that Hooke's law applies to beam fibres, i.e. that the stress in a beam fibre is proportional to its strain, we see that the '*stress variation diagram*' for a beam section will also be linear. These two diagrams are extremely important, and should be thoroughly understood before proceeding further with the theory.

The stress in any fibre in a beam cross section is proportional to its distance from the neutral axis of the section.

Moment of Resistance of a Rectangular Beam Section

The rectangular beam section (Fig. 308) has a breadth b mm and a depth d mm. The applied bending moment at section 'AB' of the beam induces a maximum stress—in the extreme upper and lower fibres of the section—of 'f' N/mm².

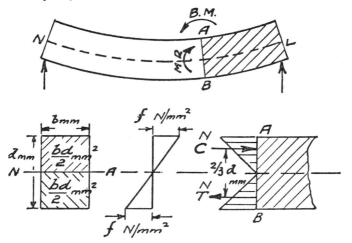

FIG. 308.—RECTANGULAR BEAM SECTION.

We may consider the beam section to be composed of a very large number of thin horizontal strips of equal width and depth. The stress acting upon any given strip will depend upon the position of the strip in the cross section, with reference to the neutral axis of the section. The load carried by a strip will be 'stress × area'. The '*load variation diagram*' will therefore be similar in character to the 'stress variation diagram' as the elemental strips of cross section are equal in area. The system of loads acting on the top half-section of the beam will have a resultant 'C' N, and a resultant pull 'T' N will act upon the bottom half-section.

'C' and 'T' are equal forces and they form a couple of moment 'C' (or 'T') × 'arm of couple.' The moment of this couple is the moment of resistance of the beam section.

Value of 'C' (or 'T')

The stress on the top half-section of the beam section varies uniformly from 'f' N/mm² to 'zero.' The total load 'C' will equal $f/2$ N/mm² × area of half-section.

$$C = \frac{f}{2} \text{ N/mm}^2 \times \frac{bd}{2} \text{ mm}^2 = \frac{fbd}{4} \text{ N.}$$

$$C = T = \frac{fbd}{4} \text{ N.}$$

Value of 'arm of couple'

'C' will act through the centre of gravity of the compression load-variation triangular diagram and 'T' will act through the C.G. of the tensile diagram.

Hence the distance between 'C' and 'T,' i.e. the *'arm of the couple,'* = $[d - (\frac{1}{3} \times d/2) - (\frac{1}{3} \times d/2)]$ mm = $\frac{2}{3}d$ mm.

Moment of couple

$$\text{Moment of couple} = \text{Force} \times \text{arm}$$
$$= C \text{ (or T) N} \times \tfrac{2}{3}d \text{ mm}$$

$$= \frac{fbd}{4} \text{ N} \times \tfrac{2}{3}d \text{ mm}$$

$$= \frac{fbd^2}{6} \text{ N mm.}$$

As explained previously the moment of this couple balances the bending moment at every beam section. We may write 'M' to stand for 'B.M.' or 'M.R.' as required. The proof of the formula has been carried through with the units 'N' and 'millimetres' in order to bring out the nature of the intermediate results. It is obvious that the formula is applicable to any system of units. Hence we may write in general terms:

$$M = \frac{fbd^2}{6}.$$

Allowance for Self-weight of Beams

The self-weight of a beam should always be considered. It is usually a uniform load producing a bending moment of '$Wl/8$' at the centre of span (in the case of simply supported beams). In *problems*, the self-weight of a beam is frequently omitted.

In the case of single timber beams or U.B.s, the beam's own weight is usually small compared with the load carried. Moreover, the choice of a practical beam section will normally leave a margin over the theoretical requirements and this should cover the effect of the weight.

In preliminary calculations the weight of a beam can only be estimated. The estimate should be checked when the beam has been designed and if sufficient allowance has not been made the calculations must be worked through again. Unless stated otherwise, the self-weight of a beam is neglected in the examples in this chapter.

EXAMPLES:

(i) *A timber beam, 50 mm wide × 150 mm deep, carries a uniformly distributed load of total value 4·2 kN. The effective span of the beam is 2·5 m. Calculate the maximum stress induced in the timber.*

$$M = \frac{fbd^2}{6}$$

$$M = \frac{Wl}{8} = \frac{4\cdot2 \times 2\cdot5 \times 1\ 000}{8}\ \text{kN mm} = 1\ 312\cdot5\ \text{kN mm}.$$

[The span must always be in '*millimetres*' in beam stress problems, as the stress contains '*millimetre*' units.]

$$\therefore\quad 1\ 312\cdot5 = \frac{f \times 50 \times 150 \times 150}{6}$$

$$f = \frac{1\ 312\cdot5 \times 6}{50 \times 150 \times 150} \times 1\ 000 = 7\ \text{N/mm}^2.$$

(ii) *Calculate the safe central-point load for a timber beam 50 mm wide × 175 mm deep and of 2·4 m effective span, if the maximum per-missible stress in the timber is 5·5 N/mm².*

Let 'W' N = central load.

$$\text{Max. B.M.} = \frac{Wl}{4}$$

$$M = \frac{fbd^2}{6}$$

$$\therefore \quad \frac{Wl}{4} = \frac{fbd^2}{6}$$

$$\therefore \quad \frac{W \times 2\cdot4 \times 1\,000}{4} = \frac{5\cdot5 \times 50 \times 175 \times 175}{6}$$

$$\therefore \quad W = \frac{4 \times 5\cdot5 \times 50 \times 175 \times 175}{2\cdot4 \times 1\,000 \times 6} = 2\,340 \text{ N.}$$

(iii) *A concrete beam, 150 mm wide × 300 mm deep, has an effective span of 2 m. Assuming the density of the concrete to be 24 kN/m³, calculate the central-load point the beam can carry in addition to its own weight. The maximum tensile stress in the concrete is to be limited to 0·4 N/mm².*

$$\text{Self-weight of beam} = \left[\left(\frac{150}{1\,000} \times \frac{300}{1\,000} \times 2\right) \times 24\right] = 2\cdot16 \text{ kN.}$$

$$\text{B.M.}_{\text{max}} \text{ due to weight of beam} = \frac{Wl}{8}$$

$$= \frac{2\cdot16 \times 2 \times 1\,000}{8} \text{ kN mm} = 540 \text{ kN mm.}$$

$$\text{Moment of resistance of beam section} = \frac{fbd^2}{6}$$

$$= \frac{0\cdot4 \times 150 \times 300 \times 300}{6 \times 1\,000} \text{ kN mm} = 900 \text{ kN mm.}$$

∴   (900 − 540) kN mm = 360 kN mm are available for the B.M. caused by the central load.

Let W kN = value of the central load.

$$\therefore \quad \frac{W \times l}{4} = 360$$

$$W = \frac{4 \times 360}{2 \times 1\,000} \text{ kN} = 0\cdot72 \text{ kN} = 720 \text{ N}.$$

The working stress in tension for concrete is very much smaller than that in compression. To take advantage of the higher compressive strength, the concrete is often relieved of all responsibility for tension and steel bars are embedded in suitable amount and position in order to provide the tensile force in the couple which resists bending.

Such a combination of steel and concrete is known as '*reinforced concrete.*'

(iv) *Design a suitable timber beam section given the following data:*

Effective span = 3 m.
Loads carried: 4·6 *kN, uniformly distributed.*
              1·7 *kN, central-point load.*
Working stress = 6 *N/mm².*

The maximum bending moment will occur at the centre of the span for both load systems.

$$\text{Hence} \quad \text{B.M.}_{\text{max.}} = \left[\frac{4\cdot6 \times 3 \times 1\,000}{8} + \frac{1\cdot7 \times 3 \times 1\,000}{4} \right] \text{kN mm}$$

$$= (1\,725 + 1\,275) \text{ kN mm} = 3\,000 \text{ kN mm}.$$

$$M = \frac{fbd^2}{6}.$$

$$\therefore \quad 3\,000 \times 1\,000 = \frac{6 \times bd^2}{6}. \qquad \therefore \quad bd^2 = \frac{6 \times 3\,000\,000}{6}$$

$$= 3\,000\,000.$$

Theoretically, any beam section which makes 'bd^2' = 3 000 000 will be suitable. However, there is the necessity, in all beams, to limit the maximum deflection. Beams must be '*stiff*' as well as '*strong.*' The depth of a timber floor beam should be from '2' to '3' times its breadth.
Try 'b' = 75 mm. $\therefore$ $75d^2$ = 3 000 000 and d^2 = 40 000.
$$d = \sqrt{40\,000} = 200 \text{ mm}.$$
A beam 75 mm wide × 200 mm deep would be suitable.

(v) *Timber beams, 50 mm wide × 175 mm deep, are to be used in a floor, the effective span of the beams being 2·8 m. The total inclusive floor load is 5·2 kN/m². Assuming the maximum permissible bending stress in the beams to be 6 N/mm², calculate the maximum allowable spacing for the beams, centre to centre.*

Let W N = safe U.D. load for one beam.

$$\frac{W \times 2\cdot8 \times 1\,000}{8} = \frac{fbd^2}{6} = \frac{6 \times 50 \times 175 \times 175}{6}$$

$$W = \frac{8 \times 6 \times 50 \times 175 \times 175}{2\cdot8 \times 1\,000 \times 6}\ \text{N}$$

$$= 4\,375\ \text{N}.$$

If 'x' m = spacing of beams centre to centre, $2\cdot8x$ m² = area of floor supported by one beam.

$$\therefore\quad (2\cdot8x \times 5\,200)\ \text{N} = \text{U.D. load carried by one beam.}$$
$$\therefore\quad 2\cdot8x \times 5\,200 = 4\,375$$
$$\therefore\quad x = 0\cdot301\ \text{m} = \text{say, 300 mm.}$$

Maximum spacing for beams = 300 mm.

(vi) *Fig. 309 shows a portion of the footing of a retaining wall. Assuming the maximum bending moment to occur at the toe of the upper wall, calculate the bending stresses induced in the footing, given a net upward pressure of 225 kN/m².*

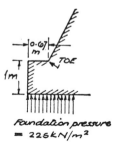

FIG. 309.—FOOTING STRESSES.

Consider 1 m run of footing in a direction perpendicular to the plane of the diagram. We may regard this portion of the footing as a cantilever projecting 0·67 m from its support, of section 1 m wide × 1 m deep and carrying a uniform load of total value = 1 × 0·67 × 225 kN (i.e. 1 m × 0·67 m × 225 kN/m²) = 150 kN.

Max.   B.M. $= 150$ kN $\times$ 0·33 m $= 50$ kN m.

$$M = \frac{fbd^2}{6}$$

$$50 = \frac{f \times 1 \times 1 \times 1}{6} \qquad \therefore \ f = \frac{50 \times 6}{1 \times 1 \times 1} \text{ kN/m}^2.$$

$$\therefore \ f = 300 \text{ kN/m}^2.$$

$$= \frac{300 \times 1\,000}{1\,000 \times 1\,000} = 0·3 \text{ N/mm}^2.$$

Beams with Non-rectangular Sections

It is clear that the formula $M = \dfrac{fbd^2}{6}$ cannot be used for sections which are not rectangular, because the properties of a rectangle were used in its derivation. We cannot employ this formula, for example, to determine the moment of resistance of a standard steel beam section.

The part of the expression $\dfrac{fbd^2}{6}$ which particularly depends upon the assumption of a rectangular section is $\dfrac{bd^2}{6}$. For beams with other than rectangular sections we have to substitute another expression for $\dfrac{bd^2}{6}$.

General Theory of Bending

Fig. 310 shows a portion of a beam in the unbent and in the bent conditions. AB and CD are two vertical cross sections, assumed so close together that the portion of beam between them may be regarded as bending to the arc of a circle.

EF is the part of the neutral layer intercepted between the sections. GH represents a typical layer of material at a distance y from the neutral axis. R is the radius of curvature of the portion of the neutral layer, in the bent beam. The following are the steps in the development of the theory:

(1) Determination of the strain in layer $G'H'$ by principles of geometry.

(2) Evaluation of the stress in this layer by means of Young's modulus.

(3) Determination of the load carried by the little strip of cross section at distance y from the N.A.

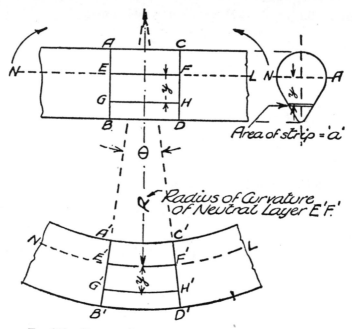

FIG. 310.—DIAGRAM ILLUSTRATING THE THEORY OF BENDING.

(4) Computation of the moment this load has about the N.A. and, by summation, the total moment of all such strip loads.

Step 1.   Extension in layer G'H' = G'H' − GH.

$$\text{Strain in layer G'H'} = \frac{\text{Extension}}{\text{Original length}}$$

$$= \frac{\text{G'H'} - \text{GH}}{\text{GH}}.$$

But GH = EF and EF = E'F' (being on the unstrained layer).

$$\therefore \quad \text{Strain in layer G'H'} = \frac{\text{G'H'} - \text{E'F'}}{\text{E'F'}}.$$

Expressing these distances in terms of R and θ (the angle in radians contained by B'A' and D'C') we have:

$$\text{Strain in layer G'H'} = \frac{(R + y)\theta - R\theta}{R\theta} = \frac{y}{R}.$$

Step 2. $\dfrac{\text{Stress in G'H'}}{\text{Strain in G'H'}} = \text{E}.$

$\therefore$ Stress in layer $= \text{E} \times \text{strain} = \dfrac{\text{E}y}{\text{R}}.$

If $f =$ the stress, $f = \dfrac{\text{E}y}{\text{R}}.$

Step 3. If $a =$ the area of the cross-sectional strip, the load carried $= \text{stress} \times \text{area}$

$$= \dfrac{\text{E}y}{\text{R}} \times a = \dfrac{\text{E}}{\text{R}} \times ay.$$

Step 4. Moment of the load on this strip about NA

$$= \text{Load} \times \text{distance}$$

$$= \left(\dfrac{\text{E}}{\text{R}} \times ay\right) \times y = \dfrac{\text{E}}{\text{R}} \times ay^2.$$

The total 'moment of resistance' of the beam section is made up of all such moments as this.

Total Moment of Resistance $= \sum \dfrac{\text{E}}{\text{R}} \times ay^2$

$$= \dfrac{\text{E}}{\text{R}} \times \sum ay^2.$$

Σay^2 (*sigma ay²*) is a geometrical property of the beam section, with reference to the axis NA. It is termed the *moment of inertia* of the beam section, and is denoted by the letter I.

Writing M for 'moment of resistance,'

$$\text{M} = \dfrac{\text{E}}{\text{R}} \times \text{I}.$$

But $= \dfrac{\text{E}y}{\text{R}}$ or $\dfrac{\text{E}}{\text{R}} = \dfrac{f}{y}$ (step 2).

$$\therefore \text{M} = \dfrac{f\text{I}}{y}.$$

This is the important formula for finding the moment of resistance of a beam section. Writing the results found as a continued ratio we get the complete expression for the theory.

$$\frac{E}{R} = \frac{M}{I} = \frac{f}{y}.$$

In using this expression f will normally represent a maximum stress, so that y, in that case, will be the distance from the neutral axis to an extreme fibre, top or bottom of the section, as the case may be.

M will usually be a bending moment, such as $\dfrac{Wl}{4}$, $\dfrac{Wl}{8}$, etc.

Position of the Neutral Axis.—In the theory we found that the strip load was given by the expression $\dfrac{E}{R} \times ay$.

As long as y is measured downwards all these strip loads will represent tension. If we put y negative, i.e. measured the distance upwards from NA, the load would be compression.

$\sum\dfrac{E}{R} \times ay$ or $\dfrac{E}{R} \sum ay$ (since E and R are constants) will therefore represent a summation of a large number of positive and negative quantities. But, as the total compressive force = the total tensile force (being forces in a couple), $\dfrac{E}{R} \sum ay$ must $= 0$.

This, i.e. $\Sigma ay = 0$, means that the axis, from which y is measured, passes through the centre of gravity of the section.
The neutral axis of a beam section therefore passes through its centre of gravity.

Fig. 311 shows the position of the neutral axis for certain structural sections.

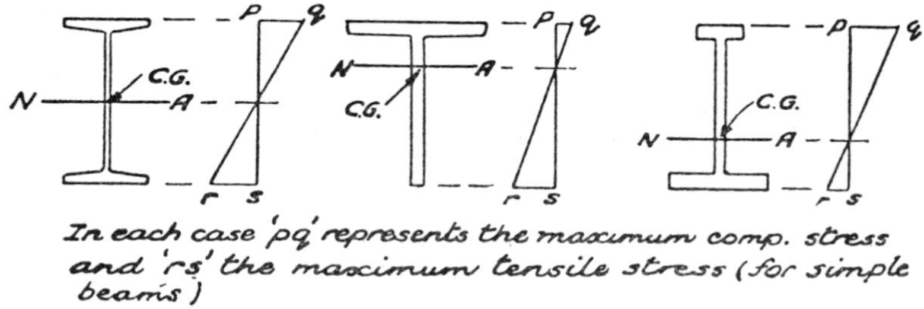

In each case 'pq' represents the maximum comp. stress and 'rs' the maximum tensile stress (for simple beams)

FIG. 311.—POSITION OF NEUTRAL AXIS.

Moment of Inertia.—Fig. 312 shows beam sections divided up into a very large number of thin strips parallel to the neutral axis. If we multiply each strip area by the *square* of its distance from NA and sum up all the quantities obtained, we will obtain the value of 'I_{NA}' for the given beam section.

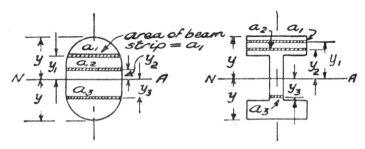

FIG. 312.

$$I_{NA} = (a_1 \times y_1^2) + (a_2 \times y_2^2) + (a_3 \times y_3^2) + \text{etc.}$$

The values for certain structural sections will be found in Fig. 313.

The proofs of 'I' values for geometrical sections such as rectangles and circles involve the use of the calculus.

It will be observed that the value of the moment of inertia of a beam section has nothing to do with the material of the beam. It is a pure geometrical property and is a 'property of section' of the given beam.

Units for 'I'

$$I = \Sigma ay^2.$$

Each term in this summation represents an 'area × distance squared,' e.g. $mm^2 \times mm^2$. However, if the mm^4 unit is used, it will be clear that the arithmetical values of 'moments of inertia' will be very large and unmanageable. Reference to the tables of properties of U.B.s. (see pages 258–61) will show that the B.C.S.A. have appreciated this problem and have given the values for area, moment of inertia and modulus of section in *centimetre* units. This is a departure from the strict SI metric system, but, from a practical point of view, it would appear eminently reasonable.

EXAMPLES:

(i) *A steel beam section has the following dimensions: flange width =*
150 mm, flange thickness = 25 mm, overall depth = 300 mm, web

Section	Moment of Inertia	Section Modulus
Rectangular 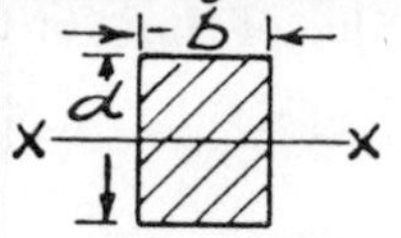	$I_{xx} = \dfrac{bd^3}{12}$	$Z_{xx} = \dfrac{bd^2}{6}$
Solid circular 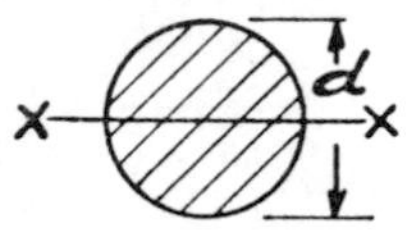	$I_{xx} = \dfrac{\pi d^4}{64}$	$Z_{xx} = \dfrac{\pi d^3}{32}$
Hollow circular 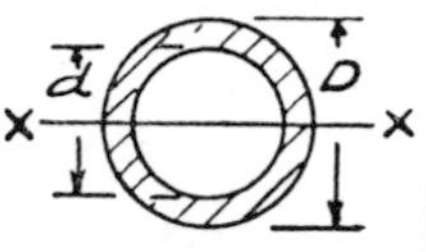	$I_{xx} = \dfrac{\pi}{64}\left[D^4 - d^4\right]$	$Z_{xx} = \dfrac{\frac{\pi}{64}\left[D^4 - d^4\right]}{D/2}$
Steel beam type 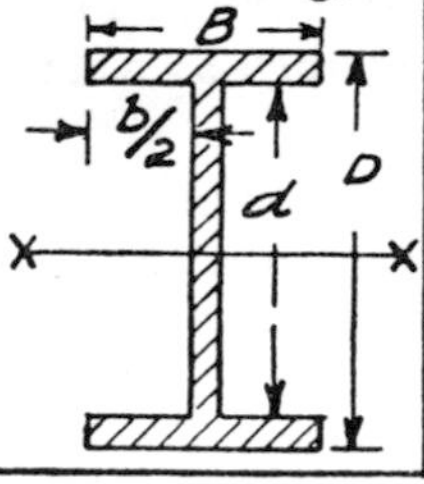	$I_{xx} = \dfrac{BD^3}{12} - \dfrac{bd^3}{12}$ $b = B - \text{web thickness}$	$Z_{xx} = \dfrac{\dfrac{BD^3}{12} - \dfrac{bd^3}{12}}{D/2}$

FIG. 313.—PROPERTIES OF SECTION.

thickness = 12·5 *mm. Calculate the safe uniformly distributed load for this beam if the effective span* = 5 *m, and the maximum permissible stress* = 165 *N/mm²*.

The expression for 'I_{NA}' for the beam section given is

$\dfrac{BD^3}{12} - \dfrac{bd^3}{12}$ (shown as I_{xx} in Fig. 313).

$$\therefore \quad I_{NA} = \left(\frac{15 \times 30^3}{12} - \frac{13 \cdot 75 \times 25^3}{12}\right) \text{cm}^4.$$

$$= (33\,750 - 17\,904)\,\text{cm}^4 = 15\,846\,\text{cm}^4.$$

Let 'W' kN be the safe U.D. load, inclusive of self-weight of beam.

$$\frac{W \times 5 \times 1\,000}{8} = f\frac{I}{y} = \frac{165}{1\,000} \times \frac{15\,846 \times 10^4}{150}$$

$$\left(y = \text{distance of extreme fibre from NA} = \frac{300}{2}\,\text{mm} = 150\,\text{mm}.\right)$$

$$W = \frac{8 \times 165 \times 15\,846 \times 10^4}{5 \times 1\,000 \times 1\,000 \times 150}\,\text{kN} = 279\,\text{kN}$$

(ii) *A solid round steel bar, 50 mm diameter, rests freely on supports 300 mm apart. Calculate the safe central-point load the bar could support if the stress in the steel were limited to 70 N/mm². [Neglect self-weight of bar.]*

'I' for section of bar $= \dfrac{\pi d^4}{64} = \dfrac{\pi \times 50^4}{64}\,\text{mm}^4 = 306\,800\,\text{mm}^4.$

Moment of resistance $= f\dfrac{I}{y} = 70 \times \dfrac{306\,800}{25}$

$$\left[y = \frac{d}{2} = \frac{50}{2} = 25\right].$$

$\therefore$ M.R. $= 859\,000$ N mm.

$$\frac{Wl}{4} = 859\,000$$

$$\frac{W \times 300}{4} = 859\,000$$

$$W = 11\,450\,\text{N}.$$

(iii) *Fig.* 314 *shows a steel chimney subjected to wind pressure. Estimate the stress which will be produced in the steel due to the wind pressure.*

'I' of chimney section $= \dfrac{\pi}{64}(D^4 - d^4)$

$$= \frac{\pi}{64}(750^4 - 700^4)\,\text{mm}^4 = 375 \times 10^7\,\text{mm}^4.$$

The area of chimney elevation at right angles to direction of wind pressure $= (15 \times 0.75)\,\text{m}^2$. The wind is not, however, acting on a flat, but on a curved, surface. It is usual to employ a reduction coefficient to obtain the *effective* area for wind pressure. A coefficient commonly employed for circular chimneys is 0·6.

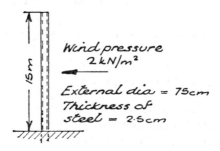

FIG. 314.—STEEL CHIMNEY.

The resultant wind thrust = 0·6 × 15 × 0·75 × 2 kN
= 13·5 kN.

Max. bending moment on chimney (at base)

$$= \left(13\cdot5 \times \frac{15}{2}\right) \text{kN m} = 101\cdot25 \text{ kN m}$$

$$= 101\cdot25 \times 10^6 \text{ N mm}.$$

$$M = f\frac{I}{y}. \quad \left[y = \frac{750}{2} = 375\right]$$

$$101\cdot25 \times 10^6 = f\frac{I}{y}$$

$$= \frac{f \times 375 \times 10^7}{375}$$

$$\therefore \ f = \frac{101\cdot25}{10} \text{ N/mm}^2$$

$$= 10\cdot125 \text{ N/mm}^2.$$

British Standard Sections (Universal Beams).

British Standard (B.S.4) gives a list of beam sections which are known as '*Universal beams*' (U.B.). These sections are also listed in the '*section books*' which are published by steel firms. The various beam properties are evaluated and tabulated for reference purposes. Specimen pages from a section book are given on pages 258–61.

Moments of Inertia.—I_{XX} *or* I_{max}. The 'XX' axis of a beam is always at right angles to the web. It is for this axis that the U.B. section has the highest I-value, hence the term 'I_{max}.' The 'XX' axis is the neutral axis when the beam is used in the normal way, so that the column headed 'I_{max}' is the important one in beam problems.

I_{YY} *or* I_{min}. The 'YY' axis is always parallel to the web of a steel beam. It is for this axis that the section has the minimum moment of inertia. This property is important in the case of a U.B. being used as a column or in the unusual case of a beam being used with the 'YY' axis horizontal.

Description of a 'U.B.'—A U.B. is described as follows: '*overall depth*' × '*width of flange*' × '*weight of section in kg per metre run*.' Thus a 305 mm × 127 mm × 48 kg U.B. has a nominal depth of 305 mm, a nominal flange width of 127 mm, and weighs 48 kg for every metre of length.*

EXAMPLE.—*A 305 mm × 127 mm × 48 kg U.B. is used to carry a total inclusive load of 161·5 kN, uniformly spread over an effective span of 5 m. Calculate the maximum stress in the steel. Look up the necessary properties on page 261.*

'I_{XX}' is given in the tables as 9 485 cm^4.

$$M = f\frac{I}{y}$$

$$M = \frac{Wl}{8} = \frac{161·5 \times 5}{8} \text{ kN m} = 101 \text{ kN m}.$$

y = distance of extreme fibre from neutral axis
$= \frac{1}{2}$ overall depth = 310 mm/2 = 155 mm.

$$\therefore \quad 101 \times 10^6 = \frac{f \times 9\ 485 \times 10^4}{155}$$

$$\therefore \quad f = 165 \text{ N/mm}^2.$$

Section Modulus.—Both 'I' and 'y' in the expression 'I/y' are properties of a given beam section. It is convenient to amalgamate them

* It should be noted that the actual depths and widths of U.B. sections are often slightly different from the nominal depths.

S.M.(M)—9*

into a single property. To this new property the term *'section modulus'* is applied. The usual symbol for section modulus is 'Z.'

$$Z = \frac{I_{NA}}{y} = \frac{\text{Moment of inertia of beam section about NA}}{\text{Distance of extreme fibre from NA}}$$

In most cases (i.e. in all cases in which the N.A. of the beam section is an axis of symmetry) $y = $ half the overall depth of the beam section. In such cases we may write:

$$Z = \frac{I_{NA}}{\frac{1}{2} \text{ overall depth}}.$$

The formula for the moment of resistance of a beam section may now be expressed in the form most convenient for practical design.

$$\mathbf{M} = f\mathbf{Z}.$$

The symbols 'M' and '*f*' may have a variety of special meanings according to the type of beam problem in which they are involved. Usually 'M' will stand for the 'maximum bending moment' in the beam (e.g. 'Wl/8' or 'Wl/4,' etc.) and f will be the permissible working stress.

$$\mathbf{Z} = \mathbf{M}/f.$$

$$\textbf{Necessary section modulus} = \frac{\textbf{Maximum bending moment}}{\textbf{Working stress in bending}}.$$

Values of Section Modulus.—The values of the 'section modulus' for certain geometrical sections are tabulated in Fig. 313. They have been derived from the basic 'I' value by the formula, $Z = I/y$. Thus, for a rectangular section, the section modulus is given by

$$Z = \frac{bd^3}{12} \bigg/ \frac{d}{2} = \frac{bd^2}{6}.$$

Section modulus values for U.B. sections are given on pages 259 and 261. The important column is that headed '*Axis x − x*.'
Units for section modulus:

$$Z = \frac{I}{y} = \frac{cm^4}{cm} = cm^3.$$

Thus a '203 mm × 133 mm × 25 kg' U.B. has a section modulus of 231·1 cm³.

Extract from Steel Section Safe Load Tables.—The Safe Load Tables on pages 262 to 265 inclusive are reproduced by permission of the British Constructional Steelwork Association and the Constructional Steel Research and Development Organisation.

EXAMPLES:

(i) *Given 'I_{XX}' = 20 619 cm^4 and 'I_{YY}' = 768 cm^4 for a 406 mm × 152 mm × 60 kg U.B., calculate 'Z_{XX}' and 'Z_{YY}'.*

$$Z_{XX} = \frac{I_{XX}}{y} = \frac{20\ 619}{20 \cdot 4} \text{ cm}^3 = 1\ 011 \text{ cm}^3.$$

$$Z_{YY} = \frac{I_{XX}}{y} = \frac{768}{7 \cdot 6} \text{ cm}^3 = 101 \text{ cm}^3.$$

These are the values given in the section table (page 261).

(ii) *If 'Z_{XX}' for a 203 mm × 133 mm × 25 kg U.B. = 231 cm^3, calculate 'I_{XX}'.*

$$Z_{XX} = \frac{I_{XX}}{y}.$$

$$y \text{ in this case} = \frac{203 \cdot 2 \text{ mm}}{2} = 101 \cdot 6 \text{ mm} = 10 \cdot 16 \text{ cm}.$$

$$\therefore \quad 231 \cdot 1 = \frac{I_{XX}}{10 \cdot 16}.$$

$$\therefore \quad I_{XX} = 2\ 348 \text{ cm}^4.$$

(iii) *A freely supported steel beam of 5 m effective span carries a uniform load of total value = 168 kN. Calculate the minimum permissible section modulus for the beam, assuming a working stress of 165 N/mm². Select a suitable U.B. section from the tables.*

$$\text{B.M.}_{\cdot \text{max}} = \frac{Wl}{8} = \frac{168 \times 5}{8} = 105 \text{ kN m.}$$

$$M = fZ$$

$$105 \times 10^3 = 165 \times Z(\text{in cm}^3)$$

$$Z = \frac{105 \times 10^3}{165} = 636 \text{ cm}^3$$

UNIVERSAL BEAMS

DIMENSIONS AND PROPERTIES

Serial Size	Mass per metre	Depth of Section D	Width of Section B	Thickness Web t	Thickness Flange T	Root Radius r	Depth between Fillets d	Area of Section
mm	kg	mm	mm	mm	mm	mm	mm	cm²
914 × 419	388	920.5	420.5	21.5	36.6	24.1	791.5	493.9
	343	911.4	418.5	19.4	32.0	24.1	791.5	436.9
914 × 305	289	926.6	307.8	19.6	32.0	19.1	819.2	368.5
	253	918.5	305.5	17.3	27.9	19.1	819.2	322.5
	224	910.3	304.1	15.9	23.9	19.1	819.2	284.9
	201	903.0	303.4	15.2	20.2	19.1	819.2	256.1
838 × 292	226	850.9	293.8	16.1	26.8	17.8	756.4	288.4
	194	840.7	292.4	14.7	21.7	17.8	756.4	246.9
	176	834.9	291.6	14.0	18.8	17.8	756.4	223.8
762 × 267	197	769.6	268.0	15.6	25.4	16.5	681.2	250.5
	173	762.0	266.7	14.3	21.6	16.5	681.2	220.2
	147	753.9	265.3	12.9	17.5	16.5	681.2	187.8
686 × 254	170	692.9	255.8	14.5	23.7	15.2	610.6	216.3
	152	687.6	254.5	13.2	21.0	15.2	610.6	193.6
	140	683.5	253.7	12.4	19.0	15.2	610.6	178.4
	125	677.9	253.0	11.7	16.2	15.2	610.6	159.4
610 × 305	238	633.0	311.5	18.6	31.4	16.5	531.6	303.5
	179	617.5	307.0	14.1	23.6	16.5	531.6	227.7
	149	609.6	304.8	11.9	19.7	16.5	531.6	189.9
610 × 229	140	617.0	230.1	13.1	22.1	12.7	543.1	178.2
	125	611.9	229.0	11.9	19.6	12.7	543.1	159.4
	113	607.3	228.2	11.2	17.3	12.7	543.1	144.3
	101	602.2	227.6	10.6	14.8	12.7	543.1	129.0
610 × 178	91	602.5	178.4	10.6	15.0	12.7	547.1	115.9
	82	598.2	177.8	10.1	12.8	12.7	547.1	104.4
533 × 330	212	545.1	333.6	16.7	27.8	16.5	450.1	269.6
	189	539.5	331.7	14.9	25.0	16.5	450.1	241.2
	167	533.4	330.2	13.4	22.0	16.5	450.1	212.7
533 × 210	122	544.6	211.9	12.8	21.3	12.7	472.7	155.6
	109	539.5	210.7	11.6	18.8	12.7	472.7	138.4
	101	536.7	210.1	10.9	17.4	12.7	472.7	129.1
	92	533.1	209.3	10.2	15.6	12.7	472.7	117.6
	82	528.3	208.7	9.6	13.2	12.7	472.7	104.3
533 × 165	73	528.8	165.6	9.3	13.5	12.7	476.5	93.0
	66	524.8	165.1	8.8	11.5	12.7	476.5	83.6
457 × 191	98	467.4	192.8	11.4	19.6	10.2	404.4	125.2
	89	463.6	192.0	10.6	17.7	10.2	404.4	113.8
	82	460.2	191.3	9.9	16.0	10.2	404.4	104.4
	74	457.2	190.5	9.1	14.5	10.2	404.4	94.9
	67	453.6	189.9	8.5	12.7	10.2	404.4	85.4

Note: These tables are based on Universal Beams with the flange taper shown on page 82.
Universal Beams with parallel flanges have properties at least equal to the values given.
Both Taper and Parallel Flange Beams comply with the requirements of the British Standard
4: Part 1 : 1971 and are interchangable.

UNIVERSAL BEAMS

DIMENSIONS AND PROPERTIES

Serial Size	Moment of Inertia			Radius of Gyration		Elastic Modulus		Ratio $\dfrac{D}{T}$
	Axis x–x		Axis	Axis	Axis	Axis	Axis	
	Gross	Net	y–y	x–x	y–y	x–x	y–y	
mm	cm⁴	cm⁴	cm⁴	cm	cm	cm³	cm³	T
914 × 419	717325	639177	42481	38.1	9.27	15586	2021	25.2
	623866	555835	36251	37.8	9.11	13691	1733	28.5
914 × 305	503781	469903	14793	37.0	6.34	10874	961.3	29.0
	435796	406504	12512	36.8	6.23	9490	819.2	32.9
	375111	350209	10425	36.3	6.05	8241	685.6	38.1
	324715	303783	8632	35.6	5.81	7192	569.1	44.7
838 × 292	339130	315153	10661	34.3	6.08	7971	725.9	31.8
	278833	259625	8384	33.6	5.83	6633	573.6	38.7
	245412	228867	7111	33.1	5.64	5879	487.6	44.4
762 × 267	239464	221138	7699	30.9	5.54	6223	574.6	30.3
	204747	189341	6376	30.5	5.38	5374	478.1	35.3
	168535	156213	5002	30.0	5.16	4471	377.1	43.1
686 × 254	169843	156106	6225	28.0	5.36	4902	486.8	29.2
	150015	137965	5391	27.8	5.28	4364	423.7	32.7
	135972	125156	4789	27.6	5.18	3979	377.5	36.0
	117700	108580	3992	27.2	5.00	3472	315.5	41.8
610 × 305	207252	192203	14973	26.1	7.02	6549	961.3	20.2
	151312	140269	10571	25.8	6.81	4901	688.6	26.2
	124341	115233	8471	25.6	6.68	4079	555.9	30.9
610 × 229	111673	101699	4253	25.0	4.88	3620	369.6	27.9
	98408	89675	3676	24.8	4.80	3217	321.1	31.2
	87260	79645	3184	24.6	4.70	2874	279.1	35.1
	75549	69132	2658	24.2	4.54	2509	233.6	40.7
610 × 178	63970	57238	1427	23.5	3.51	2124	160.0	40.2
	55779	50076	1203	23.1	3.39	1865	135.3	46.7
533 × 330	141682	121777	16064	22.9	7.72	5199	963.2	19.6
	125618	107882	14093	22.8	7.64	4657	849.6	21.6
	109109	93647	12057	22.6	7.53	4091	730.3	24.2
533 × 210	76078	68719	3208	22.1	4.54	2794	302.8	25.6
	66610	60218	2755	21.9	4.46	2469	261.5	28.7
	61530	55671	2512	21.8	4.41	2293	239.2	30.8
	55225	50040	2212	21.7	4.34	2072	211.3	34.2
	47363	43062	1826	21.3	4.18	1793	175.0	40.0
533 × 165	40414	35752	1027	20.8	3.32	1528	124.1	39.2
	35083	31144	863	20.5	3.21	1337	104.5	45.6
457 × 191	45653	40469	2216	19.1	4.21	1954	229.9	23.8
	40956	36313	1960	19.0	4.15	1767	204.2	26.2
	37039	32869	1746	18.8	4.09	1610	182.6	28.8
	33324	29570	1547	18.7	4.04	1458	162.4	31.5
	29337	26072	1328	18.5	3.95	1293	139.9	35.7

Note: One hole is deducted from each flange under 300mm wide (serial size) and two holes from each flange 300mm and over (serial size), in calculating the Net Moment of Inertia about x–x.

UNIVERSAL BEAMS

DIMENSIONS AND PROPERTIES

Serial Size	Mass per metre	Depth of Section D	Width of Section B	Thickness Web t	Thickness Flange T	Root Radius r	Depth between Fillets d	Area of Section
mm	kg	mm	mm	mm	mm	mm	mm	cm²
457 × 152	82	465.1	153.5	10.7	18.9	10.2	404.4	104.4
	74	461.3	152.7	9.9	17.0	10.2	404.4	94.9
	67	457.2	151.9	9.1	15.0	10.2	404.4	85.3
	60	454.7	152.9	8.0	13.3	10.2	407.7	75.9
	52	449.8	152.4	7.6	10.9	10.2	407.7	66.5
406 × 178	74	412.8	179.7	9.7	16.0	10.2	357.4	94.9
	67	409.4	178.8	8.8	14.3	10.2	357.4	85.4
	60	406.4	177.8	7.8	12.8	10.2	357.4	76.1
	54	402.6	177.6	7.6	10.9	10.2	357.4	68.3
406 × 152	74	416.3	153.7	10.1	18.1	10.2	357.4	94.8
	67	412.2	152.9	9.3	16.0	10.2	357.4	85.3
	60	407.9	152.2	8.6	13.9	10.2	357.4	75.8
406 × 140	46	402.3	142.4	6.9	11.2	10.2	357.4	58.9
	39	397.3	141.8	6.3	8.6	10.2	357.4	49.3
381 × 152	67	388.6	154.3	9.7	16.3	10.2	333.2	85.4
	60	384.8	153.4	8.7	14.4	10.2	333.2	75.9
	52	381.0	152.4	7.8	12.4	10.2	333.2	66.4
356 × 171	67	364.0	173.2	9.1	15.7	10.2	309.1	85.3
	57	358.6	172.1	8.0	13.0	10.2	309.1	72.1
	51	355.6	171.5	7.3	11.5	10.2	309.1	64.5
	45	352.0	171.0	6.9	9.7	10.2	309.1	56.9
356 × 127	39	352.8	126.0	6.5	10.7	10.2	309.1	49.3
	33	348.5	125.4	5.9	8.5	10.2	309.1	41.7
305 × 165	54	310.9	166.8	7.7	13.7	8.9	262.6	68.3
	46	307.1	165.7	6.7	11.8	8.9	262.6	58.8
	40	303.8	165.1	6.1	10.2	8.9	262.6	51.4
305 × 127	48	310.4	125.2	8.9	14.0	8.9	262.6	60.8
	42	306.6	124.3	8.0	12.1	8.9	262.6	53.1
	37	303.8	123.5	7.2	10.7	8.9	262.6	47.4
305 × 102	33	312.7	102.4	6.6	10.8	7.6	275.3	41.8
	28	308.9	101.9	6.1	8.9	7.6	275.3	36.3
	25	304.8	101.6	5.8	6.8	7.6	275.3	31.4
254 × 146	43	259.6	147.3	7.3	12.7	7.6	216.2	55.0
	37	256.0	146.4	6.4	10.9	7.6	216.2	47.4
	31	251.5	146.1	6.1	8.6	7.6	216.2	39.9
254 × 102	28	260.4	102.1	6.4	10.0	7.6	224.5	36.2
	25	257.0	101.9	6.1	8.4	7.6	224.5	32.1
	22	254.0	101.6	5.8	6.8	7.6	224.5	28.4
203 × 133	30	206.8	133.8	6.3	9.6	7.6	169.9	38.0
	25	203.2	133.4	5.8	7.8	7.6	169.9	32.3

Note: These tables are based on Universal Beams with the flange taper shown on page 82.
Universal Beams with parallel flanges have properties at least equal to the values given.
Both Taper and Parallel Flange Beams comply with the requirements of the British Standard
4: Part 1 : 1971 and are interchangable.

UNIVERSAL BEAMS

DIMENSIONS AND PROPERTIES

Serial Size	Moment of Inertia			Radius of Gyration		Elastic Modulus		Ratio $\dfrac{D}{T}$
	Axis x–x		Axis	Axis	Axis	Axis	Axis	
	Gross	Net	y–y	x–x	y–y	x–x	y–y	
mm	cm⁴	cm⁴	cm⁴	cm	cm	cm³	cm³	
457 × 152	36160	32058	1093	18.6	3.24	1555	142.5	24.6
	32380	28731	963	18.5	3.18	1404	126.1	27.1
	28522	25342	829	18.3	3.12	1248	109.1	30.5
	25464	22613	794	18.3	3.23	1120	104.0	34.2
	21345	19034	645	17.9	3.11	949.0	84.61	41.3
406 × 178	27279	23981	1448	17.0	3.91	1322	161.2	25.8
	24279	21357	1269	16.9	3.85	1186	141.9	28.6
	21520	18928	1108	16.8	3.82	1059	124.7	31.8
	18576	16389	922	16.5	3.67	922.8	103.8	36.9
406 × 152	26938	23811	1047	16.9	3.32	1294	136.2	23.0
	23798	21069	908	16.7	3.26	1155	118.8	25.8
	20619	18283	768	16.5	3.18	1011	100.9	29.3
406 × 140	15603	13699	500	16.3	2.92	775.6	70.26	35.9
	12408	10963	373	15.9	2.75	624.7	52.61	46.2
381 × 152	21276	18817	947	15.8	3.33	1095	122.7	23.8
	18632	16489	814	15.7	3.27	968.4	106.2	26.7
	16046	14226	685	15.5	3.21	842.3	89.96	30.7
356 × 171	19483	17002	1278	15.1	3.87	1071	147.6	23.2
	16038	14018	1026	14.9	3.77	894.3	119.2	27.6
	14118	12349	885	14.8	3.71	794.0	103.3	30.9
	12052	10578	730	14.6	3.58	684.7	85.39	36.3
356 × 127	10054	8688	333	14.3	2.60	570.0	52.87	33.0
	8167	7099	257	14.0	2.48	468.7	40.99	41.0
305 × 165	11686	10119	988	13.1	3.80	751.8	118.5	22.7
	9924	8596	825	13.0	3.74	646.4	99.54	26.0
	8500	7368	691	12.9	3.67	559.6	83.71	29.8
305 × 127	9485	8137	438	12.5	2.68	611.1	69.94	22.2
	8124	6978	367	12.4	2.63	530.0	58.99	25.3
	7143	6142	316	12.3	2.58	470.3	51.11	28.4
305 × 102	6482	5792	189	12.5	2.13	414.6	37.00	29.0
	5415	4855	153	12.2	2.05	350.7	30.01	34.7
	4381	3959	116	11.8	1.92	287.5	22.85	44.8
254 × 146	6546	5683	633	10.9	3.39	504.3	85.97	20.4
	5544	4814	528	10.8	3.34	433.1	72.11	23.5
	4427	3859	406	10.5	3.19	352.1	55.53	29.2
254 × 102	4004	3565	174	10.5	2.19	307.6	34.13	26.0
	3404	3041	144	10.3	2.11	264.9	28.23	30.6
	2863	2572	116	10.0	2.02	225.4	22.84	37.4
203 × 133	2880	2469	354	8.71	3.05	278.5	52.85	21.5
	2348	2020	280	8.53	2.94	231.1	41.92	26.1

Note: One hole is deducted from each flange under 300mm wide (serial size) and two holes from each flange 300mm and over (serial size), in calculating the Net Moment of Inertia about x–x.

UNIVERSAL BEAMS

BASED ON
BS 449
1969

SAFE LOADS FOR GRADE 43 STEEL

Serial Size mm	Mass per metre in kg	SAFE DISTRIBUTED LOADS IN KILONEWTONS FOR SPANS IN METRES AND DEFLECTION COEFFICIENTS														
		4.00	5.00	6.00	7.00	8.00	9.00	10.00	11.00	12.00	13.00	14.00	15.00	16.00	18.00	20.00
		28.00	17.92	12.44	9.143	7.000	5.531	4.480	3.702	3.111	2.651	2.286	1.991	1.750	1.383	1.120
914 × 419	388		†3958	3429	2939	2572	2286	2057	1870	1714	1583	1470	1372	1286	1143	1029
	343		†3536	3012	2582	2259	2008	1807	1643	1506	1390	1291	1205	1130	1004	904
914 × 305	289	3588	2871	2392	2050	1794	1595	1435	1305	1196	1104	1025	957	897	797	718
	253	3132	2505	2088	1789	1566	1392	1253	1139	1044	964	895	835	783	696	626
	224	2720	2176	1813	1554	1360	1209	1088	989	907	837	777	725	680	604	544
	201	2373	1899	1582	1356	1187	1055	949	863	791	730	678	633	593	527	475
838 × 292	226	2630	2104	1754	1503	1315	1169	1052	957	877	809	752	701	658	585	526
	194	2189	1751	1459	1251	1094	973	876	796	730	674	625	584	547	486	438
	176	1940	1552	1293	1109	970	862	776	705	647	597	554	517	485	431	388
762 × 267	197	2054	1643	1369	1173	1027	913	821	747	685	632	587	548	513	456	411
	173	1773	1419	1182	1013	887	788	709	645	591	546	507	473	443	394	355
	147	1475	1180	984	843	738	656	590	537	492	454	422	393	369	328	295
686 × 254	170	1618	1294	1079	924	809	719	647	588	539	498	462	431	404	360	324
	152	1440	1152	960	823	720	640	576	524	480	443	411	384	360	320	288
	140	1313	1050	875	750	656	584	525	477	438	404	375	350	328	292	263
	125	1146	917	764	655	573	509	458	417	382	353	327	306	286	255	229
610 × 305	238	2161	1729	1441	1235	1081	960	864	786	720	665	617	576	540	480	432
	179	1617	1294	1078	924	809	719	647	588	539	498	462	431	404	359	323
	149	1346	1077	897	769	673	598	538	490	449	414	385	359	337	299	269
610 × 229	140	1195	956	796	683	597	531	478	434	398	368	341	319	299	265	239
	125	1061	849	708	607	531	472	425	386	354	327	303	283	265	236	212
	113	948	759	632	542	474	421	379	345	316	292	271	253	237	211	190
	101	828	662	552	473	414	368	331	301	276	255	237	221	207	184	166
610 × 178	91	701	561	467	400	350	311	280	255	234	216	200	187	175	156	140
	82	615	492	410	352	308	274	246	224	205	189	176	164	154	137	123
533 × 330	212	1716	1372	1144	980	858	762	686	624	572	528	490	457	429	381	
	189	1537	1229	1025	878	768	683	615	559	512	473	439	410	384	342	
	167	1350	1080	900	771	675	600	540	491	450	415	386	360	338	300	
533 × 210	122	922	738	615	527	461	410	369	335	307	284	263	246	231	205	
	109	815	652	543	466	407	362	326	296	272	251	233	217	204	181	
	101	757	605	504	432	378	336	303	275	252	233	216	202	189	168	
	92	684	547	456	391	342	304	273	249	228	210	195	182	171	152	
	82	592	473	394	338	296	263	237	215	197	182	169	158	148		
533 × 165	73	504	404	336	288	252	224	202	183	168	155	144	135	126		
	66	441	353	294	252	221	196	176	160	147	136	126	118	110		
457 × 191	98	645	516	430	368	322	287	258	234	215	198	184	172			
	89	583	467	389	333	292	259	233	212	194	179	167	156			
	82	531	425	354	304	266	236	212	193	177	163	152	142			
	74	481	385	321	275	241	214	192	175	160	148	137	128			
	67	427	341	285	244	213	190	171	155	142	131	122	114			

Generally, tabular loads are based on a flexural stress of 165N/mm², assuming adequate lateral support. Beams without adequate lateral support must not exceed the critical span Lc. unless the allowable compressive stress is reduced in accordance with clause 19.a.(ii) of BS 449 : 1969.
Tabular loads printed in bold face type exceed the buckling capacity of the unstiffened web and the allowance for actual length of bearing; the load bearing capacity should be checked, see page 109.
Tabular loads marked thus † are based on the maximum shear value of the web and are less than the permissible flexural load.

BASED ON

BS 449

1969

UNIVERSAL BEAMS

DIMENSIONS AND PROPERTIES
GRADE 43 STEEL

Actual Size $D \times B$ mm	Critical Span Le metres	Area of section cm²	Moment of Inertia		Radius of Gyration	Elastic Modulus		Ratio $\dfrac{D}{T}$
			Axis x—x	Axis y—y	Axis y—y	Axis x—x	Axis y—y	
			cm⁴	cm⁴	cm	cm³	cm³	
920.5 × 420.5	8.610	493.9	717325	42481	9.27	15586	2021	25.2
911.4 × 418.5	8.456	436.9	623866	36251	9.11	13691	1733	28.5
926.6 × 307.8	5.882	368.5	503781	14793	6.34	10874	961.3	29.0
918.5 × 305.5	5.783	322.5	435796	12512	6.23	9490	819.2	32.9
910.3 × 304.1	5.615	284.9	375111	10425	6.05	8241	685.6	38.1
903.0 × 303.4	5.390	256.1	324715	8632	5.81	7192	569.1	44.7
850.9 × 293.8	5.644	288.4	339130	10661	6.08	7971	725.9	31.8
840.7 × 292.4	5.410	246.9	278833	8384	5.83	6633	573.6	38.7
834.9 × 291.6	5.232	223.8	245412	7111	5.64	5879	487.6	44.4
769.6 × 268.0	5.147	250.5	239464	7699	5.54	6223	574.6	30.3
762.0 × 266.7	4.995	220.2	204747	6376	5.38	5374	478.1	35.3
753.9 × 265.3	4.791	187.8	168535	5002	5.16	4471	377.1	43.1
692.9 × 255.8	4.980	216.3	169843	6225	5.36	4902	486.8	29.2
687.6 × 254.5	4.898	193.6	150015	5391	5.28	4364	423.7	32.7
683.5 × 253.7	4.810	178.4	135972	4789	5.18	3979	377.5	36.0
677.9 × 253.0	4.645	159.4	117700	3992	5.00	3472	315.5	41.8
633.0 × 311.5	6.982	303.5	207252	14973	7.02	6549	961.3	20.2
617.5 × 307.0	6.326	227.7	151312	10571	6.81	4901	688.6	26.2
609.6 × 304.8	6.201	189.9	124341	8471	6.68	4079	555.9	30.9
617.0 × 230.1	4.535	178.2	111673	4253	4.88	3620	369.6	27.9
611.9 × 229.0	4.458	159.4	98408	3676	4.80	3217	321.1	31.2
607.3 × 228.2	4.361	144.3	87260	3184	4.70	2874	279.1	35.1
602.2 × 227.6	4.213	129.0	75549	2658	4.54	2509	233.6	40.7
602.5 × 178.4	3.258	115.9	63970	1427	3.51	2124	160.0	40.2
598.2 × 177.8	3.151	104.4	55779	1203	3.39	1865	135.3	46.7
545.1 × 333.6	7.757	269.6	141682	16064	7.72	5199	963.2	19.6
539.5 × 331.7	7.414	241.2	125618	14093	7.64	4657	849.6	21.6
533.4 × 330.2	7.047	212.7	109109	12057	7.53	4091	730.3	24.2
544.6 × 211.9	4.215	155.6	76078	3208	4.54	2794	302.8	25.6
539.5 × 210.7	4.141	138.4	66610	2755	4.46	2469	261.5	28.7
536.7 × 210.1	4.095	129.1	61530	2512	4.41	2293	239.2	30.8
533.1 × 209.3	4.026	117.6	55225	2212	4.34	2072	211.3	34.2
528.3 × 208.7	3.884	104.3	47363	1826	4.18	1793	175.0	40.0
528.8 × 165.6	3.085	93.0	40414	1027	3.32	1528	124.1	39.2
524.8 × 165.1	2.983	83.6	35083	863	3.21	1337	104.5	45.6
467.4 × 192.8	3.957	125.2	45653	2216	4.21	1954	229.0	23.8
463.6 × 192.0	3.853	113.8	40956	1960	4.15	1767	204.2	26.2
460.2 × 191.3	3.796	104.4	37039	1746	4.09	1610	182.6	28.8
457.2 × 190.5	3.749	94.9	33324	1547	4.04	1458	162.4	31.5
453.6 × 189.9	3.663	85.4	29337	1328	3.95	1293	139.9	35.7

Tabular loads printed in italic type are within the web buckling capacity of the unstiffened web and produce a total deflection not exceeding 1/360th of the span.
Tabular loads printed in ordinary type should be checked for deflection, see page 110.
For explanation of tables, see notes commencing page 102.
1 kilonewton may be taken as 0.102 metric tonne (megagramme) force, but see page 102.

UNIVERSAL BEAMS

SAFE LOADS FOR GRADE 43 STEEL

BASED ON
BS 449
1969

Serial Size mm	Mass per metre in kg	SAFE DISTRIBUTED LOADS IN KILONEWTONS FOR SPANS IN METRES AND DEFLECTION COEFFICIENTS														
		2.00	2.50	3.00	3.50	4.00	4.50	5.00	5.50	6.00	6.50	7.00	7.50	8.00	9.00	10.00
		112.0	71.68	49.78	36.57	28.00	22.12	17.92	14.81	12.44	10.60	9.143	7.964	7.000	5.531	4.480
457 × 152	82	†**995**	**821**	**684**	*586*	*513*	*456*	*411*	*373*	*342*	*316*	*293*	274	257	228	205
	74	†**913**	**741**	**618**	*529*	*463*	*412*	*371*	*337*	*309*	*285*	*265*	247	232	206	185
	67	**823**	**659**	**549**	*471*	*412*	*366*	*329*	*299*	*274*	*253*	*235*	220	206	183	165
	60	†**728**	**591**	**493**	*422*	*370*	*329*	*296*	*269*	*246*	*227*	*211*	197	185	164	148
	52	**626**	**501**	**418**	*358*	*313*	*278*	*251*	*228*	*209*	*193*	*179*	167	157	139	125
406 × 178	74	†**801**	**698**	**582**	*499*	*436*	*388*	*349*	*317*	*291*	*268*	*249*	233	218	194	174
	67	†**721**	**626**	**522**	*447*	*391*	*348*	*313*	*285*	*261*	*241*	*224*	209	196	174	157
	60	†**634**	**559**	**466**	*399*	*349*	*311*	*280*	*254*	*233*	*215*	*200*	186	175	155	140
	54	**609**	**487**	**406**	*348*	*305*	*271*	*244*	*221*	*203*	*187*	*174*	162	152	135	122
406 × 152	74	†**841**	**683**	**569**	*488*	*427*	*380*	*342*	*311*	*285*	*263*	*244*	228	214	190	171
	67	**762**	**610**	**508**	*435*	*381*	*339*	*305*	*277*	*254*	*234*	*218*	203	191	169	152
	60	**667**	**534**	**445**	*381*	*334*	*297*	*267*	*243*	*222*	*205*	*191*	178	167	148	133
406 × 140	46	**512**	**410**	**341**	*293*	*256*	*228*	*205*	*186*	*171*	*158*	*146*	137	128	114	102
	39	**412**	**330**	**275**	*236*	*206*	*183*	*165*	*150*	*137*	*127*	*118*	110	103	92	82
381 × 152	67	**723**	**578**	**482**	*413*	*361*	*321*	*289*	*263*	*241*	*222*	*206*	193	181	161	145
	60	**639**	**511**	**426**	*365*	*320*	*284*	*256*	*232*	*213*	*197*	*183*	170	160	142	128
	52	**556**	**445**	**371**	*318*	*278*	*247*	*222*	*202*	*185*	*171*	*159*	148	139	124	111
356 × 171	67	†**662**	**565**	**471**	*404*	*353*	*314*	*283*	*257*	*236*	*217*	*202*	188	177	157	141
	57	†**574**	**472**	**394**	*337*	*295*	*262*	*236*	*215*	*197*	*182*	*169*	157	148	131	118
	51	†**519**	**419**	**349**	*299*	*262*	*233*	*210*	*191*	*175*	*161*	*150*	140	131	116	105
	45	**452**	**362**	**301**	*258*	*226*	*201*	*181*	*164*	*151*	*139*	*129*	121	113	100	90
356 × 127	39	**376**	**301**	**251**	*215*	*188*	*167*	*150*	*137*	*125*	*116*	*107*	100	94	84	75
	33	**309**	**247**	**206**	*177*	*155*	*137*	*124*	*112*	*103*	*95*	*88*	82	77	69	62
305 × 165	54	†**479**	**397**	**331**	*284*	*248*	*221*	*198*	*180*	*165*	*153*	*142*	132	124	110	99
	46	†**412**	**341**	**284**	*244*	*213*	*190*	*171*	*155*	*142*	*131*	*122*	114	107	95	85
	40	**369**	**295**	**246**	*211*	*185*	*164*	*148*	*134*	*123*	*114*	*106*	98	92	82	74
305 × 127	48	**403**	*323*	*269*	*230*	*202*	*179*	*161*	*147*	*134*	*124*	*115*	108	101	90	81
	42	**350**	*280*	*233*	*200*	*175*	*155*	*140*	*127*	*117*	*108*	*100*	93	87	78	70
	37	**310**	*248*	*207*	*177*	*155*	*138*	*124*	*113*	*103*	*96*	*89*	83	78	69	62
305 × 102	33	**274**	*219*	*182*	*156*	*137*	*122*	*109*	*100*	*91*	*84*	*78*	73	68	61	55
	28	**231**	*185*	*154*	*132*	*116*	*103*	*93*	*84*	*77*	*71*	*66*	62	58	51	46
	25	**190**	*152*	*126*	*108*	*95*	*84*	*76*	*69*	*63*	*58*	*54*	51	47	42	38
254 × 146	43	**333**	*266*	*222*	*190*	*166*	*148*	*133*	*121*	*111*	*102*	*95*	89	83		
	37	**286**	*229*	*191*	*163*	*143*	*127*	*114*	*104*	*95*	*88*	*82*	76	71		
	31	**232**	*186*	*155*	*133*	*116*	*103*	*93*	*84*	*77*	*71*	*66*	62	58		
254 × 102	28	**203**	*162*	*135*	*116*	*102*	*90*	*81*	*74*	*68*	*62*	*58*	54	51		
	25	**175**	*140*	*117*	*100*	*87*	*78*	*70*	*64*	*58*	*54*	*50*	47	44		
	22	**149**	*119*	*99*	*85*	*74*	*66*	*60*	*54*	*50*	*46*	*43*	40	37		
203 × 133	30	**184**	*147*	*123*	*105*	*92*	*82*	*74*	*67*	*61*	*57*	53				
	25	**153**	*122*	*102*	*87*	*76*	*68*	*61*	*55*	*51*	*47*					

Generally, tabular loads are based on a flexural stress of 165N/mm², assuming adequate lateral support. Beams without adequate lateral support must not exceed the critical span Lc, unless the allowable compressive stress is reduced in accordance with clause 19.a.(ii) of BS 449 : 1969.

Tabular loads printed in bold face type exceed the buckling capacity of the unstiffened web without allowance for actual length of bearing; the load bearing capacity should be checked, see page 109.

Tabular loads marked thus † are based on the maximum shear value of the web and are less than the permissible flexural load.

UNIVERSAL BEAMS

BASED ON

BS 449

1969

DIMENSIONS AND PROPERTIES
GRADE 43 STEEL

Actual Size D×B mm	Critical Span Le metres	Area of section cm²	Moment of Inertia		Radius of Gyration	Elastic Modulus		Ratio D — T
			Axis x–x cm⁴	Axis y–y cm⁴	Axis y–y cm	Axis x–x cm³	Axis y–y cm³	
465.1 × 153.5	3.017	104.4	36160	1093	3.24	1555	142.5	24.6
461.3 × 152.7	2.957	94.9	32380	963	3.18	1404	126.1	27.1
457.2 × 151.9	2.894	85.3	28522	829	3.12	1248	109.1	30.5
454.7 × 152.9	3.003	75.9	25464	794	3.23	1120	104.0	34.2
449.8 × 152.4	2.891	66.5	21345	645	3.11	949.0	84.61	41.3
412.8 × 179.7	3.627	94.9	27279	1448	3.91	1322	161.2	25.8
409.4 × 178.8	3.578	85.4	24279	1269	3.85	1186	141.9	28.6
406.4 × 177.8	3.543	76.1	21520	1108	3.82	1059	124.7	31.8
402.6 × 177.6	3.410	68.3	18576	922	3.67	922.8	103.8	36.9
416.3 × 153.7	3.157	94.8	26938	1047	3.32	1294	136.2	23.0
412.2 × 152.9	3.029	85.3	23798	908	3.26	1155	118.8	25.8
407.9 × 152.2	2.954	75.8	20619	768	3.18	1011	100.9	29.3
402.3 × 142.4	2.706	58.9	15603	500	2.92	775.6	70.26	35.9
397.3 × 141.8	2.553	49.3	12408	373	2.75	624.7	52.61	46.2
388.6 × 154.3	3.131	85.4	21276	947	3.33	1095	122.7	23.8
384.8 × 153.4	3.040	75.9	18632	814	3.27	968.4	106.2	26.7
381.0 × 152.4	2.983	66.4	16046	685	3.21	842.3	89.96	30.7
364.0 × 173.2	3.670	85.3	19483	1278	3.87	1071	147.6	23.2
358.6 × 172.1	3.502	72.1	16038	1026	3.77	894.3	119.2	27.6
355.6 × 171.5	3.440	64.5	14118	885	3.71	794.0	103.3	30.9
352.0 × 171.0	3.326	56.9	12052	730	3.58	684.7	85.39	36.3
352.8 × 126.0	2.413	49.3	10054	333	2.60	570.0	52.87	33.0
348.5 × 125.4	2.303	41.7	8167	257	2.48	468.7	40.49	41.0
310.9 × 166.8	3.630	68.3	11686	988	3.80	751.8	118.5	22.7
307.1 × 165.7	3.476	58.8	9924	825	3.74	646.4	99.54	26.0
303.8 × 165.1	3.403	51.4	8500	691	3.67	559.6	83.71	29.8
310.4 × 125.2	2.581	60.8	9485	438	2.68	611.1	69.94	22.2
306.6 × 124.3	2.439	53.1	8124	367	2.63	530.0	58.99	25.3
303.8 × 123.5	2.396	47.4	7143	316	2.58	470.3	51.11	28.4
312.7 × 102.4	1.977	41.8	6482	189	2.13	414.6	37.00	29.0
308.9 × 101.9	1.905	36.3	5415	153	2.05	350.7	30.01	34.7
304.8 × 101.6	1.786	31.4	4381	116	1.92	287.5	22.85	44.8
259.6 × 147.3	3.354	55.0	6546	633	3.39	504.3	85.97	20.4
256.0 × 146.4	3.151	47.4	5544	528	3.34	433.1	72.11	23.5
251.5 × 146.1	2.958	39.9	4427	406	3.19	352.1	55.53	29.2
260.4 × 102.1	2.037	36.2	4004	174	2.19	307.6	34.13	26.0
257.0 × 101.9	1.963	32.1	3404	144	2.11	264.9	28.23	30.6
254.0 × 101.6	1.876	28.4	2863	116	2.02	225.4	22.84	37.4
206.8 × 133.8	2.963	38.0	2880	354	3.05	278.5	52.85	21.5
203.2 × 133.4	2.732	32.3	2348	280	2.94	231.1	41.92	26.1

Tabular loads printed in italic type are within the web buckling capacity of the unstiffened web and produce a total deflection not exceeding 1/360th of the span.
Tabular loads printed in ordinary type should be checked for deflection, see page 110.
For explanation of tables, see notes commencing page 102.
1 kilonewton may be taken as 0.102 metric tonne (megagramme) force, but see page 102.

We must look down the column headed 'Moduli of Section, Axis $x - x$', until we arrive at a value either equal to 636 cm³ or slightly in excess. The nearest value is 646·4 cm³, which is the section modulus of a 305 mm × 165 mm × 46 kg U.B.

(iv) *Select the most economical beam section from the section tables to carry, for an effective span of 4 m, a central point load of 45 kN, together with a uniform load of total value 90 kN. f = 165 N/mm².*
B.M.$_{max}$:

$$\text{Due to central load} = \frac{Wl}{4} = \frac{45 \times 4}{4} = 45 \text{ kN m.}$$

$$\text{Due to uniform load} = \frac{Wl}{8} = \frac{90 \times 4}{8} = 45 \text{ kN m.}$$

$$\text{Total max. B.M.} = \underline{\underline{90 \text{ kN m.}}}$$

$$M = fZ$$

$$Z = M/f = \frac{90 \times 1\ 000}{165} \text{ cm}^3 = 545 \text{ cm}^3.$$

A 356 mm × 127 mm × 39 kg U.B. has a section modulus of 570 cm³. The reader should check, from the section tables, that there is no other section, having a Z-value even higher than 570 cm³, which weighs less than 39 kg per metre.

(v) *Find the necessary section modulus for the simply supported beam shown in Fig. 315. f = 150 N/mm² in this case.*

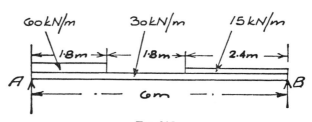

Fig. 315.

$$R_A \times 6 = (108 \times 5·1) + (36 \times 1·2) + (180 \times 3)$$
$$= 550·8 + 43·2 + 540 = 1\ 134$$
$$R_A = 189 \text{ kN.}$$
$$R_B \times 6 = (108 \times 0·9) + (36 \times 4·8) + (180 \times 3)$$
$$= 97·2 + 172·8 + 540 = 810$$
$$R_B = 135 \text{ kN.}$$

Position of max. B.M.

If we proceed along the beam from the left reaction up to a point 2·7 m from the left we take up $[(60 \times 1·8) + (30 \times 2·7)]$ kN = 189 kN of load. This equals the left-end reaction, hence the point of max. bending moment is 2·7 m from the left support.

$$\begin{aligned}
\text{B.M.}_{max} &= \text{Reaction moment} - \text{load moments} \\
&= (189 \times 2·7) - (108 \times 1·8) - (81 \times 1·35)\ \text{kN m} \\
&= (510·3 - 194·4 - 109·4)\ \text{kN m} = 206·5\ \text{kN m} \\
\text{M} &= fZ
\end{aligned}$$

$$\therefore \quad Z = \frac{206·5 \times 1\,000}{150}\ \text{cm}^3 = 1\,377\ \text{cm}^3.$$

A 533 mm × 165 mm × 73 kg U.B. (Z = 1 528 cm³) is the section with the smallest weight which has the necessary modulus.

(vi) *Calculate suitable filler joists for the cantilevered floor example shown in Fig.* 316. *The super. load for the floor* = 7·5 kN/m². *The density of the floor* = 24 kN/m³. f = 180 N/mm².

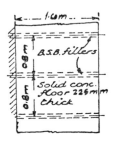

FIG. 316.

For one m² of 225 mm floor the self-weight will be

$$\frac{225\ \text{mm}}{1\,000\ \text{mm}} \times 24 = 5·4\ \text{kN}.$$

$$\begin{aligned}
\text{Dead load} &= 5·4\ \text{kN/m}^2 \\
\text{Super. load} &= 7·5\ \text{kN/m}^2 \\
\text{Total load} &= 12·9\ \text{kN/m}^2.
\end{aligned}$$

Load carried by one filler joist = $(0·8 \times 1·6 \times 12·9)$ kN
$$= 16·5\ \text{kN}.$$

$$\text{B.M.}_{\text{max}} = \frac{Wl}{2} = \frac{16 \cdot 5 \times 1 \cdot 6}{2} \text{ kN m} = 13 \cdot 2 \text{ kN m}.$$

$$M = fZ$$

$$Z = \frac{M}{f} = \frac{13 \cdot 2 \times 1\,000}{180} = 73 \cdot 3 \text{ cm}^3.$$

A 127 mm × 76 mm × 13·36 kg section will be suitable as it provides a section modulus of 74·94 cm³.*

(vii) *A timber beam of 3 m span has a section 75 mm wide × 225 mm deep. It carries two equal concentrated loads, each 3·3 kN. The loads are symmetrically placed 'x' m from either support. Calculate the maximum value of 'x' if the stress in the beam is not to exceed 5·2 N/mm².*

Each reaction = 3·3 kN.

$$\therefore \quad \text{B.M.}_{\text{max}} = (3 \cdot 3 \times x) \text{ kN m}$$
$$= (3 \cdot 3 \times x \times 1\,000) \text{ kN mm} = 3\,300x \text{ kN mm}$$
$$M = fZ$$

$$Z = \frac{bd^2}{6} = \frac{75 \times 225 \times 225}{6} = 632\,800 \text{ mm}^3.$$

$$\therefore \quad 3\,300x \times 1\,000 = 5 \cdot 2 \times 632\,800$$

$$x = \frac{5 \cdot 2 \times 632\,800}{3\,300 \times 1\,000} = 1 \text{ m}.$$

Note. $- M = fZ = \dfrac{fbd^2}{6}$ for rectangular beam sections. This formula was previously obtained by first principles (page 243).

(viii) *A 406 mm × 152 mm × 60 kg U.B., used as a freely supported beam with an effective span of 6 m, carries a uniform load of total value 176·6 kN. Calculate the stress in the beam at a point 102 mm beneath the top of the compression flange, at a beam section 1 m from the left support. Draw the stress variation diagram for the beam at the given section. Z_{XX} for given U.B. = 1 011 cm³.*

$$R_L = R_R = 88 \cdot 3 \text{ kN.}$$

---

* This section is not shown in the tables on pp. 258–65. It is a joist section, the properties of which can be found in any steelmaker's handbook.

Load on beam up to given section = ($\frac{1}{6}$ × 176·6) kN = 29·4 kN.

B.M. at given section = Reaction moment − load moment

$$= [(88·3 \times 1) - (29·4 \times 0·5)] \text{ kN m}$$
$$= (88·3 - 14·7) \text{ kN m} = 73·6 \text{ kN m}.$$

$$M = fZ$$

$$f = \frac{73·6 \times 1\,000}{1\,011} \text{ N/mm}^2 = 72·80 \text{ N/mm}^2.$$

The stress varies from 72·80 N/mm² *at the top of the section* to zero *at the neutral axis*, i.e. 204 mm below (see Fig. 317).

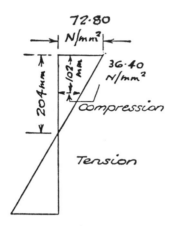

FIG. 317.—STRESS VARIATION.

Therefore at 102 mm below top of beam the stress will be

$$\frac{102}{204} \times 72·80 = 36·40 \text{ N/mm}^2 \text{ (compression).}$$

Example of steel and Timber Floor

A floor, 9 m × 8·25 m, is to be constructed of timber beams supported by steel beams. The beams are to be arranged in the manner indicated in Fig. 318. The total floor load, inclusive of estimated self-weights, is 12 kN/m². A special allowance of 14 kN is to be made for the self-weight of the compound girder. Design suitable timber and steel beams, giving the necessary section modulus in the case of the compound girder.

Working stresses: Timber, 6·5 N/mm². Steel, 165 N/mm².

Timber beams:

All the timber beams have 3 m span and are spaced at 375 mm centres.

Area of floor supported by one beam

$$= (0.375 \times 3) \, m^2 = 1.125 \, m^2.$$

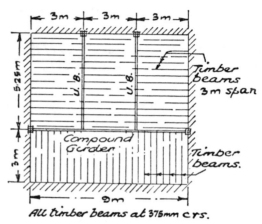

Fig. 318.—Floor Example.

Uniform load carried by one beam

$$= (12 \times 1.125) \, kN = 13.5 \, kN.$$

$$B.M._{max} = \frac{Wl}{8} = \frac{13.5 \times 3}{8} \, kN\,m = 5.0625 \, kN\,m.$$

$$M = \frac{fbd^2}{6}$$

$$5.0625 \times 10^6 = \frac{6.5 \times bd^2}{6}$$

$$bd^2 = \frac{5.0625 \times 10^6 \times 6}{6.5} = 4.67 \times 10^6.$$

If $b = 75$ mm, $d^2 = 62\,320.$ $\therefore$ $d = 250$ mm.

Suitable dimensions for timber beams would be: breadth = 75 mm, depth = 250 mm.

U.B.s:

$$\text{Effective span} = 5.25 \, m.$$

The U.B.s carry the reaction loads of the timber beams. As the distance between the timber beams is small, relative to the effective span of a U.B., we may assume the load carried by a U.B. to be uniformly distributed.

In this case the uniform load to be taken for each U.B. is that corresponding to a floor area $5 \cdot 25 \text{ m} \times \left(\dfrac{3 \text{ m}}{2} + \dfrac{3 \text{ m}}{2} \right)$

$$= (5 \cdot 25 \times 3) \text{ m}^2 = 15 \cdot 75 \text{ m}^2.$$
$$\text{Total uniform load} = (15 \cdot 75 \times 12) \text{ kN} = 189 \text{ kN}.$$
$$M = fZ$$

$$\frac{189 \times 5 \cdot 25 \times 1\,000}{8} = 165 \times Z$$

$$Z = \frac{189 \times 5 \cdot 25 \times 1\,000}{8 \times 165} \text{ cm}^3$$

$$= 752 \text{ cm}^3.$$

A 356 mm × 171 mm × 51 kg U.B. has a section modulus of 794 cm³.

[If no allowance has been made in the floor loading for self-weight of U.B.s in any given case, the excess of supplied section modulus over required section modulus should be checked to ensure the necessary margin.]

Compound girder:

The compound girder carries two reaction loads (from the respective U.B.s) and a uniform load due to the fact that it directly supports the ends of the timber beams in the lower bay. Each U.B. carries a uniform load of 189 kN, and hence transmits a load of 94·5 kN to the compound girder.

The uniform load carried by the compound

$$= [(9 \times 1 \cdot 5) \text{ m}^2 \times (12 \text{ kN/m}^2)] = 162 \text{ kN}.$$

Adding in the self-weight allowance of 14 kN, the total uniform load carried = 176 kN.

B.M.$_{max}$:

The maximum bending moment occurs at the centre of the beam for the uniform load system. For the concentrated load system it remains at the maximum for the middle 3 m of the beam. B.M.$_{max}$ therefore occurs at the centre of the beam.

B.M.$_{max}$ = Reaction moment − load moments.

The total left-end reaction = $\left(94\cdot5 + \dfrac{176}{2}\right)$ kN = 182·5 kN.

B.M.$_{max}$ = (182·5 × 4·5) − (94·5 × 1·5) − $\left(88 \times \dfrac{4\cdot5}{2}\right)$ kN m

$\quad\quad$ = 821·25 − 141·75 − 198 = 481·5 kN m

$$Z = \frac{M}{f} = \frac{481\cdot5 \times 1\,000}{165} = 2\,918 \text{ cm}^3.$$

Necessary section modulus for the compound girder = 2 918 cm³. A list of properties of compounds will suggest a suitable section to employ.

Shear Stress in Beam Webs

The question of shear stress in beam webs becomes important when U.B.s are fully loaded for small spans. As the span decreases the carrying capacity of a beam in kN increases, and hence the support reactions increase. This results in increasing shear stress in the beam web. A point will arise when the shear stress becomes unsafe and the webs will require stiffening. Section books usually indicate on safe-load tables the spans below which, if the beam be fully loaded, the webs will require strengthening.

Limitation of Deflection in Beams

The maximum deflection of a beam has to be limited to an amount which depends upon the particular employment of the beam. For steel beams in steel-framed buildings the limit is usually taken as 1/360th part of the span, when such deflection is caused by loads other than the weight of the beam, its casing and the structural floor.

The theoretical treatment of beam deflection is beyond the scope of this book, but the following may be found useful.

It will be seen from the safe load tables (e.g. on page 262) that the safe loads are printed in three different types. The explanation of the different types is given in the footnotes to the tables, but it should be noted that the loads printed in *ordinary* type must be checked for deflection. This means that the loads as printed in ordinary type, and which have been calculated on a stress criterion, could produce a deflection greater than $\frac{1}{360}$ of the span. If this deflection is not to be exceeded, the tabular load must be correspondingly reduced.

Section and Safe Load Tables.—The 'section and safe load tables' for joists, published in the various handbooks of steel firms, are based on a maximum fibre stress in bending of 165 N/mm². The tables on pages 262–5 are taken from one such handbook. The scope of this book only requires the use of the tables for examples of simple beams—with assumed equal safe bending stresses in compression and tension flanges.

Deflection formulae.—Fig. 319 gives four important deflection formulae. The proofs of these will have to be considered by the reader in his more advanced studies.

Type of beam & nature of load	Maximum B.M.	Maximum deflection
cantilever, point load W at free end, length ℓ	$W\ell$	$\dfrac{1}{3}\dfrac{W\ell^3}{EI}$
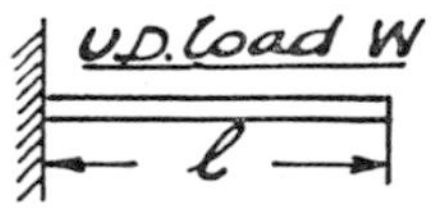 cantilever, U.D. load W, length ℓ	$\dfrac{W\ell}{2}$	$\dfrac{1}{8}\dfrac{W\ell^3}{EI}$
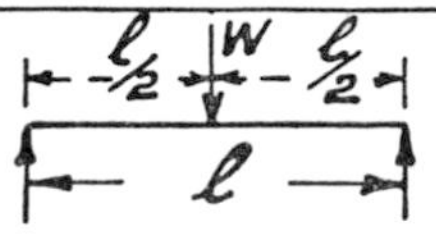 simply supported, point load W at mid-span, $\frac{\ell}{2}+\frac{\ell}{2}$, length ℓ	$\dfrac{W\ell}{4}$	$\dfrac{1}{48}\dfrac{W\ell^3}{EI}$
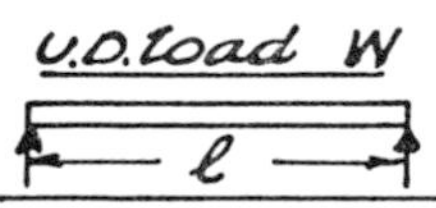 simply supported, U.D. load W, length ℓ	$\dfrac{W\ell}{8}$	$\dfrac{5}{384}\dfrac{W\ell^3}{EI}$

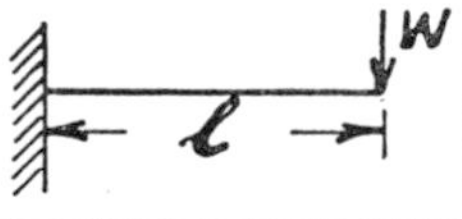

FIG. 319.—DEFLECTION FORMULAE.

EXAMPLE (i).—**Calculate the maximum deflection of a** 406 *mm* × 178 *mm* × 74 *kg U.B. which carries a U.D. load of* 300 *kN. The effective span is* 6 *m.* $E = 210$ *kN/mm²*. I_{XX} *for given beam section* = 27 279 *cm⁴*.

(i) *By formula:*

$$\text{Maximum deflection} = \frac{5}{384} \frac{Wl^3}{EI}$$

$$= \frac{5}{384} \times \frac{300 \times (6 \times 100)^3}{210 \times 100 \times 27\,279} \text{ cm} = 1\cdot473 \text{ cm.}$$

EXAMPLE (ii).—*Calculate the maximum deflection of a timber beam, 75 mm wide × 225 mm deep, which carries a central point load. The effective span is 4 m and the load produces a maximum beam stress of 7 N/mm². E = 8·4 kN/mm².*
[*Neglect self-weight of beam.*]

Let W N. = central load

$$\frac{Wl}{4} = \frac{fbd^2}{6}$$

$$\frac{W \times 4 \times 1\,000}{4} = \frac{7 \times 75 \times 225 \times 225}{6}$$

$$W = 4\,430 \text{ N.}$$

$$\text{Max. deflection} = \frac{1}{48} \frac{Wl^3}{EI}$$

$$I = \frac{bd^3}{12} = \frac{75 \times 225 \times 225 \times 225}{12}$$

$$= 71\cdot19 \times 10^6 \text{ mm}^4.$$

$$\therefore \quad \text{Max. deflection} = \frac{1}{48} \times \frac{4\,430 \times (4 \times 1\,000)^3}{8\,400 \times 71\cdot19 \times 10^6} = 9\cdot88 \text{ mm.}$$

Timber beams carrying plastered ceilings should not deflect more than about 2·75 mm per metre of span.

EXAMPLE (iii).—*If the maximum deflection of a steel beam is limited to 1/325 of the span, check that the maximum span/depth ratio of the beam carrying a U.D. load and stressed to 165 N/mm² is 18·8. E = 210kN/mm².*

$$\text{Max. B.M.} = \frac{Wl}{8}$$

$$\text{Max. deflection} = \frac{5}{384} \times \frac{Wl^3}{EI}$$

$$= \frac{5}{48} \times \frac{Wl}{8} \times \frac{l^2}{EI}$$

$$= \frac{5}{48} \times \frac{M}{I} \times \frac{l^2}{E}$$

But

$$\frac{M}{I} = \frac{f}{y}$$

$$\therefore \quad \text{Max. deflection} = \frac{5}{48} \times \frac{f}{y} \times \frac{l^2}{E}$$

Now $y = \frac{D}{2}$ ($\frac{1}{2}$ depth of beam), $f = 165 \text{ N/mm}^2$ and $E = 210 \text{ kN/mm}^2$.

$$\therefore \quad \text{Max. deflection} = \frac{5}{48} \times \frac{165 \times 2}{D} \times \frac{l^2}{210 \times 1\,000} = \frac{l^2}{6\,109\,D}$$

$$\text{Limiting deflection} = \frac{l}{325}$$

$$\therefore \quad \frac{l}{325} = \frac{l^2}{6\,109\,D}$$

$$\text{or} \quad \frac{l}{D} = \frac{6\,109}{325} = 18 \cdot 8$$

Experiments on Beam Deflection

The experimental laws of deflection may be verified by the apparatus given in Fig. 320. The photograph shows a simply supported beam carrying two concentrated loads. The deflection is being measured in this experiment by means of a recording dial. The dial is mounted on the frame of the testing machine, with a plunger resting directly on the beam.

Modulus of Rupture

Beam specimens of timber are tested to destruction in testing machines in order to determine the relative values of the different varieties of timber for use as beams. One of the standard beam specimens is 20 mm × 20 mm in section. It is supported and loaded in the manner laid down in B.S. 373. If the maximum bending moment the beam can support before fracture be divided by the section modulus of the beam section,

FIG. 320.—BEAM DEFLECTION EXPERIMENT.

a value of 'f' is determined, which is a guide to the value of the particular variety of timber as a beam material. This value of 'f' is often termed the '*modulus of rupture*' of the given timber. The modulus of rupture must be divided by a '*factor of safety*' in order to arrive at a suitable working stress for beam-design calculations. The factor of safety is of the order of '8.' This high 'factor' is a safeguard against excessive beam deflection.

If the modulus of rupture of a given timber be assumed to be 48 N/mm² and a factor of safety of '8' be adopted, the working stress in bending would be $\dfrac{48 \text{ N/mm}^2}{8} = 6 \text{ N/mm}^2$. C.P. 112 gives details of working stresses to be used in timber design (see page 299).

Fig. 321* shows a specimen of timber being tested in a transverse bending machine. The specimen is centrally loaded in this case. An

FIG. 321.—TRANSVERSE BENDING MACHINE.

alternative method is to have two symmetrical and equal point loads, thus subjecting the specimen to a uniform bending moment over the central portion of its length.

* Reproduced by courtesy of the Building Research Station.

EXERCISES 12

[The self-weights of beams are to be neglected unless otherwise stated.]

(1) A timber beam has a section 75 mm wide × 150 mm deep. It is subjected to a bending moment of 2 250 kN mm.

Calculate (i) the maximum fibre stress produced, (ii) the resultant compressive and tensile forces which constitute the couple resisting bending, (iii) the arm of the resistance moment.

(2) Calculate the safe central-point load for a timber beam, 50 mm wide × 150 mm deep × 3 m effective span, if the maximum permissible bending stress is 6·3 N/mm².

(3) What would be the safe point load for the beam of exercise (2), if the load were placed at 1 m from the left end?

(4) Calculate suitable dimensions for a timber beam of 3 m effective span which has to carry a uniform load of 4 280 N. The working stress is 6·3 N/mm².

(5) Calculate the maximum stress induced in the timber beam given in Fig. 322. The beam is 50 mm wide and 150 mm deep.

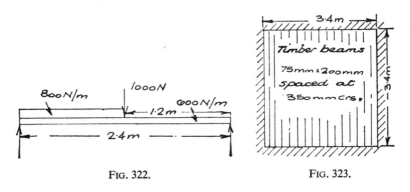

<table>
<tr><td>Fig. 322.</td><td>Fig. 323.</td></tr>
</table>

(6) Obtain the safe load per square metre of floor (inclusive of self-weight of floor) for the timber floor shown in Fig. 323. The maximum stress in the timber is not to exceed 8 N/mm².

(7) The cantilever (Fig. 324) is composed of timber weighing 6·4 kN/m³. It carries a concentrated load of 9 kN at 0·75 m from the support and a uniform load of 400 N per metre run. Assuming a working permissible stress of 5·6 N/mm², find the additional concentrated load the cantilever could safely carry at its free end.

(8) A 254 mm × 146 mm × 31 kg U.B. ($Z_{XX} = 352 \text{ cm}^3$) has an effective span of 2·5 m. Calculate the safe uniform load this beam could carry if the working stress be 165 N/mm². What would be the safe central load in this case?

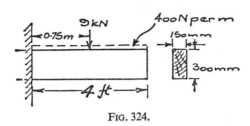

FIG. 324.

(9) (i) I_{XX} for a 203 mm × 133 mm × 25 kg U.B. section is 2 348 cm⁴. Calculate Z_{XX}.

(ii) Z_{XX} for a 457 mm × 191 mm × 74 kg U.B. section is 1 458 cm³. Calculate I_{XX}.

(iii) If I_{XX} for a U.B. is 21 520 cm⁴ and $Z_{XX} = 1 059 \text{ cm}^3$, what is the overall depth of the beam?

(10) Calculate the necessary section modulus for the U.B. shown in Fig. 325. $f = 140 \text{ N/mm}^2$.

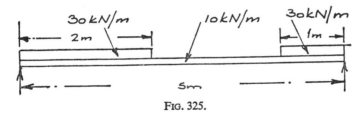

FIG. 325.

(11) Assuming a permissible working stress of 165 N/mm², obtain the necessary section modulus for the steel beam given in Fig. 326.

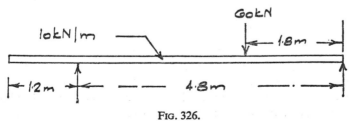

FIG. 326.

(12) Select a suitable U.B. section, from the tables given on pages 258 to 261, given the following data:

Effective span = 6 m.

Loads: 50 kN U.D. together with a point load of 40 kN at mid-span.

Working stress = 165 N/mm².

(13) A balcony floor is composed of concrete with steel filler joists at 0·75 m centres. The total thickness of the floor is 225 mm and its average density is 21 kN/m³. The filler joists are 127 mm × 76 mm × 13·36 kg. Joists [Z_{XX} = 74·94 cm³] and the floor projects 1·2 m from a wall. Assuming a super. load of 10 kN/m² on the floor, calculate the maximum stress in the steel.

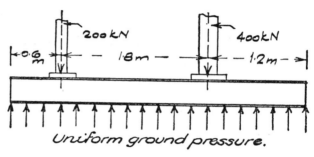

Fig. 327.

(14) Fig. 327 shows a compound grillage which is composed of three U.B.s. Calculate the section modulus required in each U.B., assuming a working stress of 220 N/mm². Neglect self-weight of grillage.

[Invert the diagram and treat as an overhanging beam with a uniformly distributed load of 166⅔ kN per metre run.]

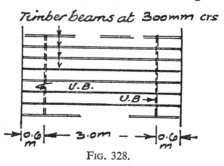

Fig. 328.

(15) Fig. 328 is a plan view of a floor with overhanging timber beams resting on U.B.s. Assuming the floor to weigh 800 N/m² and the live

load to be 7·5 kN/m², find suitable dimensions for the timber beams. 'f' = 8·4 N/mm².

(16) Calculate the safe inclusive load per m² for the floor given in Fig. 329.

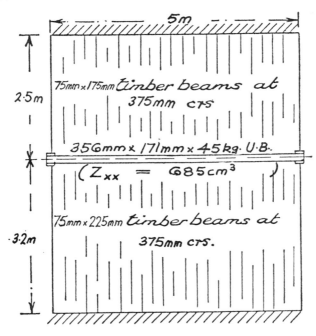

FIG. 329.

Working stresses: Steel   = 165 N/mm².
                  Timber = 8·5 N/mm².

(17) Calculate the stress caused by *bending* in the case of the solid mild steel column given in Fig. 330.

(18) A steel tube, 100 mm external diameter and 3 mm thick, projects 0·375 m horizontally from a wall. It carries a vertical load of 'W' N at the free end. Calculate the maximum permissible value of 'W,' if the bending stress it produces must not exceed 45 N/mm².

(19) A steel beam has the following dimensions:

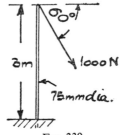

FIG. 330.

Flange width   = 125 mm
Flange thickness = 25 mm
Overall depth   = 250 mm
Web thickness   = 12 mm

Calculate the safe uniform load the beam can carry, in addition to a point load of 100 kN at the centre, if the effective span = 3 m.

The working stress for the steel = 165 N/mm^2.

(20) The effective span of timber floor beams, 75 mm wide × 150 mm deep, is to be 2·8 m. Assuming a total inclusive floor load of 7·65 kN/m^2, calculate the maximum permissible spacing for the beams. The modulus of rupture of the timber is 64 N/mm^2. A factor of safety of 8 must be used.

CHAPTER XIII

INTRODUCTION TO THE PRINCIPLES OF COLUMN CALCULATIONS

IT is intended to outline, in as simple a manner as possible, the fundamental principles underlying the calculation of the strength of a compression member.*

Two factors have to be taken into consideration when dealing with compression members which usually do not enter into the corresponding tension-member problem.

In the case of compression members, (i) the length of the member is very important, (ii) the manner in which the ends are fixed has an important bearing on the safe load the member can carry.

A compression member has a two-fold tendency to failure. Firstly, the applied load tends to crush the fibres of the member. In a *'short'* member this is the important effect produced. Secondly, the load may tend to produce failure by the side-bending or *'buckling'* of the member (Fig. 331). This is an important consideration in a *'long'* compression

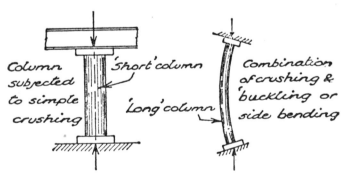

FIG. 331.—SHORT AND LONG COLUMNS.

member. The interpretations of *'short'* and *'long'* will be given a little later.

* For a more advanced treatment of column design, based on B.S. 449, the reader should consult *Structural Steelwork*, by Reynolds and Kent.

If you stand your 300 mm-scale rule on end and subject it to an increasing compressive load, it soon begins to '*buckle*' sideways. A 150 mm-scale rule of the same cross-section would stand up to a greater load, whilst a 600 mm rule would buckle more easily than the 300 mm rule. It is clear, therefore, that, other things being equal, the 'buckling' tendency becomes of greater relative importance as the length of the member increases.

Consider two scale rules, both 300 mm long and of the same sectional area. Assume one of the rules to be of the normal section and the other to be of triangular section, such as is found in the special rules having full-length multiple scales. If the simple compression test previously mentioned were applied to these two rules, the one with triangular section would stand up to a higher load before buckling.

It appears, therefore, that length and 'cross-sectional area' are not the only factors involved in the strength of an axially loaded column of a given material. The *shape* of the section has an important bearing on the value of the safe column load. The precise manner in which the cross-sectional dimensions enter into calculations of column strength introduces another '*property of section*' which involves the '*moment of inertia*' of the column section and its '*sectional area.*' This new property is termed the '*radius of gyration*' of the section. It is denoted by the letter '*r*' in the subsequent calculations.

Sometimes the letters '*k*' and '*g*' are used for denoting 'radius of gyration.'

Radius of Gyration

The radius of gyration of a column section *about* a given axis (i.e. *with reference to* a given axis) is the square root of the result obtained by dividing the moment of inertia of the section, about the axis, by the area of the section.

$$r = \sqrt{\frac{I}{A}}.$$

In Fig. 332, $r_{XX} = \sqrt{\frac{I_{XX}}{A}}$

$$r_{YY} = \sqrt{\frac{I_{YY}}{A}}$$

EXAMPLE.—*Calculate 'r_{XX}' and 'r_{YY}' for a 254 mm × 254 mm × 73 kg*

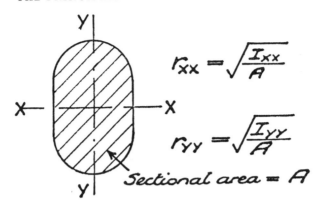

$$r_{XX} = \sqrt{\frac{I_{XX}}{A}}$$

$$r_{YY} = \sqrt{\frac{I_{YY}}{A}}$$

Sectional area = A

FIG. 332.—RADIUS OF GYRATION.

U.C. section given $I_{XX} = 11\ 360\ cm^4$, $I_{YY} = 3\ 873\ cm^3$ *and sectional area* = $92{\cdot}9\ cm^2$.

$$r_{XX} = \sqrt{\frac{I_{XX}}{A}} = \sqrt{\frac{11\ 360}{92{\cdot}9}} = 11{\cdot}1\ cm$$

$$r_{YY} = \sqrt{\frac{I_{YY}}{A}} = \sqrt{\frac{3\ 873}{92{\cdot}9}} = 6{\cdot}46\ cm$$

[Note that the *unit* in which to express radius of gyration is the '*centimetre*.']

See note on page 251. Although the use of '*cm*' is a departure from the strict SI metric system, the steel tables give the values of '*r*' in '*cm*.' These values may be checked by the tables given on page 293.

Least Radius of Gyration.—In the beam calculations of the last chapter, 'I_{max},' i.e. 'I' about the 'XX' axis, was the important value of the moment of inertia. When a U.C. is used as a column, it will clearly tend to side-bend (under concentric loading) so that 'YY' is the neutral axis, because for this axis the resistance to bending is much smaller. The 'YY' axis is therefore the important axis in column calculations, in the majority of cases. As 'I_{YY}' is '*least I*,' the value of '*least r*' will be associated with the 'YY' axis.

$$\text{Least } r = \sqrt{\frac{\text{least I}}{A}} = \sqrt{\frac{I_{YY}}{A}}.$$

The 'YY' axis, for all U.C. sections and for most compound column sections, is the one which is associated with 'least r.' In other cases, if there is any doubt, evaluate both 'I_{XX}' and 'I_{YY}.'

Formulae for Radius of Gyration Values

In Fig. 333 certain geometrical sections are shown, the least 'r' values for which should be memorised.

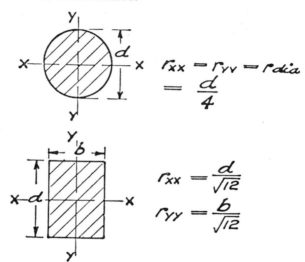

$$r_{XX} = r_{YY} = r_{dia} = \frac{d}{4}$$

$$r_{XX} = \frac{d}{\sqrt{12}}$$

$$r_{YY} = \frac{b}{\sqrt{12}}$$

FIG. 333.—CIRCULAR AND RECTANGULAR SECTIONS.

Solid circular section:

$$r_{XX} = r_{YY} = \sqrt{\frac{I}{A}} = \sqrt{\frac{\pi d^4}{64} \div \frac{\pi d^2}{4}} = \sqrt{\frac{d^2}{16}} = \frac{d}{4}.$$

'Least r' in this case is *one-quarter the diameter.*

Solid rectangular section:

$$r_{XX} = \sqrt{\frac{I_{XX}}{A}} = \sqrt{\frac{bd^3}{12} \div bd} = \sqrt{\frac{d^2}{12}} = \frac{d}{\sqrt{12}}.$$

$$r_{YY} = \sqrt{\frac{I_{YY}}{A}} = \sqrt{\frac{db^3}{12} \div db} = \sqrt{\frac{b^2}{12}} = \frac{b}{\sqrt{12}}.$$

$$\text{Least } r = \frac{\text{Lesser transverse dimension}}{\sqrt{12}}.$$

Steel-beam type section (Fig. 334):

In the example given, 'least *r*' will be associated with the 'YY' axis. Taking the symbols shown in the diagram and treating the section as consisting of three component rectangles,

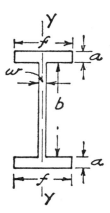

FIG. 334.—JOIST COLUMN SECTION.

$$I_{YY} = \frac{af^3}{12} + \frac{bw^3}{12} + \frac{af^3}{12}.$$

$$A = af + bw + af.$$

$$\text{Least } r = \sqrt{\frac{I_{YY}}{A}}.$$

EXAMPLES:

(i) *Calculate 'least r' for a solid circular column section 150 mm diameter.*

$$\text{Least } r = \frac{d}{4} = \frac{150 \text{ mm}}{4} = 37 \cdot 5 \text{ mm}.$$

(ii) *Find the least radius of gyration for a rectangular section, 100 mm × 150 mm.*

$$\text{Least } r = \frac{100 \text{ mm}}{\sqrt{12}} = 28 \cdot 9 \text{ mm}.$$

(iii) *Obtain the least radius of gyration for the steel column section given in Fig. 335.*

S.M.(M)—10*

Least $r = \sqrt{\dfrac{I_{YY}}{A}}$

$$I_{YY} = 2\left(\dfrac{2\cdot5 \times 20^3}{12}\right) + \dfrac{25 \times 1\cdot25^3}{12}$$

$$= 3\,333\cdot33 + 4\cdot07 = 3\,337\cdot4 \text{ cm}^4.$$
$$A = 2(2\cdot5 \times 20) + (25 \times 1\cdot25) = 131\cdot25 \text{ cm}^2.$$

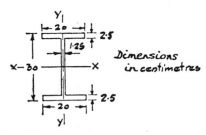

FIG. 335.—30 cm × 20 cm STEEL COLUMN.

Least $r = \sqrt{\dfrac{3\,337\cdot4}{131\cdot25}} = \sqrt{25\cdot43} = 5\cdot04$ cm.

(iv) *A hollow rectangular section is 30 cm × 15 cm, overall dimensions. The material of the section is 2·5 cm thick. Calculate the least radius of gyration of the section.*

Let the YY-axis be parallel to the 30 cm side.

$$I_{\text{least}} = I_{YY} = \dfrac{BD^3}{12} - \dfrac{bd^3}{12} = \left(\dfrac{30 \times 15^3}{12} - \dfrac{25 \times 10^3}{12}\right) \text{cm}^4$$

$$= (8\,437\cdot5 - 2\,083\cdot3) = 6\,354\cdot2 \text{ cm}^4.$$

Sectional area $= 200 \text{ cm}^2$.

$$\therefore \quad \text{Least } r = \sqrt{\dfrac{6\,354\cdot2}{200}} = 5\cdot64 \text{ cm.}$$

End Fixture of Columns

Before explaining the application of the section property 'radius of gyration' in column-strength calculations we must consider the effect on compression members of the manner in which the ends are fixed.

If you place your 300 mm-scale rule upright on a rough table and press vertically with the tip of the finger, the rule will deflect from end to end in one curve. The ends are merely held in position, and there is no restraint at the ends tending to prevent free bending. Columns fixed in this manner are said to be '*position fixed only.*'

Now imagine the rule to be held in two grips as shown in Fig. 336. The bending tendency is not now in one simple curve. The end restraint

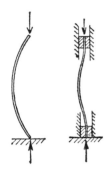

FIG. 336.—EFFECT OF END FIXTURE.

causes a complicated form of bending. A column held in the manner demonstrated is said to be '*direction fixed*' at its ends (see also Fig. 337).

Effective Column Length.—A compression member whose ends are '*position fixed only*' behaves, as has been stated, as if its ends were '*pin-jointed.*' Even when the ends are more rigidly held, a portion of a compression member tends to deflect as if this portion had pin-joints at its

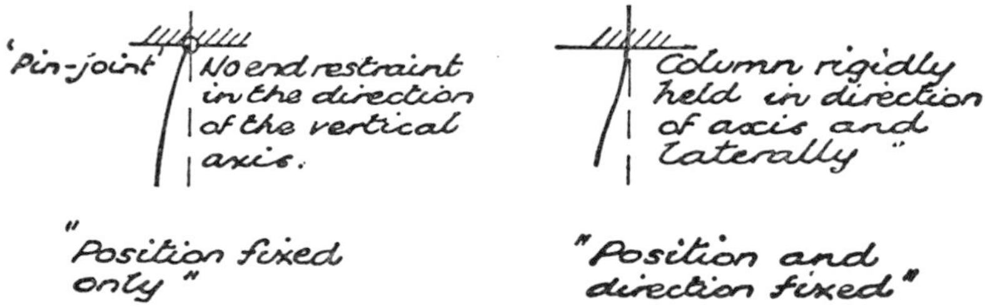

'Pin-joint'   No end restraint in the direction of the vertical axis.

Column rigidly held in direction of axis and laterally "

"Position fixed only "

"Position and direction fixed"

FIG. 337.—END FIXTURE OF COLUMNS.

ends. The length of the member which is thus equivalent to a pin-jointed strut is known as its '*effective length.*' It will be clear that, other things being equal, the less the 'effective length' the stronger the member.

Various building regulations give methods to be adopted in deciding upon the effective length of a column in any given practical case. The diagrams given in Fig. 338 indicate that the 'effective length' may be as low as 0·07 of the actual length and as high as twice that length.

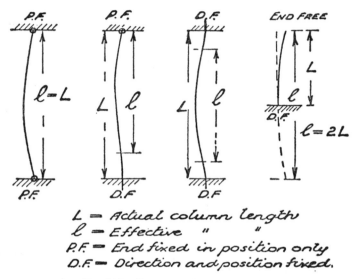

L = Actual column length
l = Effective   "   "
P.F. = End fixed in position only
D.F. = Direction and position fixed.

FIG. 338.—EFFECTIVE COLUMN LENGTH.

Slenderness Ratio.—The actual safe concentric load for a column depends upon a combination of its 'effective length' and the 'least radius of gyration' of its section. The '*slenderness ratio*' of a column is the ratio of its 'effective length' to its 'least radius of gyration.'

$$\text{Slenderness ratio} = \frac{\text{Effective length in centimetres}}{\text{Least } r \text{ in centimetres}} = \frac{l \text{ cm}}{r \text{ cm}} = \frac{l}{r}.$$

EXAMPLES:

(i) *A mild steel column of solid circular section, 125 mm dia., has an effective length of 4 m. Calculate its slenderness ratio.*

$$\text{Least radius of gyration of column section} = \frac{\text{Diameter}}{4} = \frac{125 \text{ mm}}{4}$$

$$= 31·25 \text{ mm.}$$

$$\text{Slenderness ratio} = \frac{l}{r} = \frac{4 \times 1\,000 \text{ mm}}{31·25 \text{ mm}} = 128.$$

(ii) *A steel strut in a truss has a length of 2 m. Its ends are to be regarded as 'position fixed only.' Assuming 'r' (least) from section tables to be given as 17·4 mm, calculate the slenderness ratio of the strut.*

Effective length of strut = actual length = 2 m.

$$\text{Slenderness ratio} = \frac{l}{r} = \frac{2 \times 1\ 000 \text{ mm}}{174 \text{ mm}} = 115$$

A *'short'* compression member is one with a low slenderness ratio value. The higher the value of *'l/r'*, the *'longer'* the member is in this general classification.

The precise importance of the 'slenderness ratio' value in any given case is that it determines the working compressive stress permissible in the member.

Calculation of Column Strength

The value of the estimated safe load for a given column will depend upon the particular *'column formula'* we accept. There are several such formulae in common use. The reader is referred to text-books on the 'theory of structures' for descriptions of formulae known as *'Euler's,'* *'Rankine's,'* etc. It is possible, in the limits set for this chapter, to refer in detail to one method of calculation only. In this method a permissible working stress is given for each $\frac{'l'}{r}$ value. Lists of such stresses are given in B.S. No. 449.

The table shown in Fig. 339 gives the values of permissible axial stresses as laid down for Grade 43 steel.

MILD STEEL COLUMNS

Slenderness Ratio $\frac{\text{Effective Column Length}}{\text{Least Radius of Gyration}} = \frac{l}{r}$	Working Stress. N/mm² of Section for Axial Loading. p_c.	Slenderness Ratio $\frac{\text{Effective Column Length}}{\text{Least Radius of Gyration}} = \frac{l}{r}$	Working Stress. N/mm² of Section for Axial Loading. p_c.
20	147	90	91
30	143	100	79
40	139	110	69
50	133	120	60
60	126	130	52
70	115	140	46
80	104	150	40

FIG. 339.—WORKING STRESSES IN MILD STEEL COLUMNS.

UNIVERSAL COLUMNS
Parallel Flanges

DIMENSIONS AND PROPERTIES

Serial Size	Mass per metre	Depth of Section D	Width of Section B	Thickness		Root Radius r	Depth between Fillets d	Area of Section
				Web t	Flange T			
mm	kg	mm	mm	mm	mm	mm	mm	cm²
356 × 406	634	474.7	424.1	47.6	77.0	15.2	290.1	808.1
	551	455.7	418.5	42.0	67.5	15.2	290.1	701.8
	467	436.6	412.4	35.9	58.0	15.2	290.1	595.5
	393	419.1	407.0	30.6	49.2	15.2	290.1	500.9
	340	406.4	403.0	26.5	42.9	15.2	290.1	432.7
	287	393.7	399.0	22.6	36.5	15.2	290.1	366.0
	235	381.0	395.0	18.5	30.2	15.2	290.1	299.8
Column Core	477	427.0	424.4.	48.0	53.2	15.2	290.1	607.2
356 × 368	202	374.7	374.4	16.8	27.0	15.2	290.1	257.9
	177	368.3	372.1	14.5	23.8	15.2	290.1	225.7
	153	362.0	370.2	12.6	20.7	15.2	290.1	195.2
	129	355.6	368.3	10.7	17.5	15.2	290.1	164.9
305 × 305	283	365.3	321.8	26.9	44.1	15.2	246.6	360.4
	240	352.6	317.9	23.0	37.7	15.2	246.6	305.6
	198	339.9	314.1	19.2	31.4	15.2	246.6	252.3
	158	327.2	310.6	15.7	25.0	15.2	246.6	201.2
	137	320.5	308.7	13.8	21.7	15.2	246.6	174.6
	118	314.5	306.8	11.9	18.7	15.2	246.6	149.8
	97	307.8	304.8	9.9	15.4	15.2	246.6	123.3
254 × 254	167	289.1	264.5	19.2	31.7	12.7	200.2	212.4
	132	276.4	261.0	15.6	25.1	12.7	200.2	167.7
	107	266.7	258.3	13.0	20.5	12.7	200.2	136.6
	89	260.4	255.9	10.5	17.3	12.7	200.2	114.0
	73	254.0	254.0	8.6	14.2	12.7	200.2	92.9
203 × 203	86	222.3	208.8	13.0	20.5	10.2	160.8	110.1
	71	215.9	206.2	10.3	17.3	10.2	160.8	91.1
	60	209.6	205.2	9.3	14.2	10.2	160.8	75.8
	52	206.2	203.9	8.0	12.5	10.2	160.8	66.4
	46	203.2	203.2	7.3	11.0	10.2	160.8	58.8
152 × 152	37	161.8	154.4	8.1	11.5	7.6	123.4	47.4
	30	157.5	152.9	6.6	9.4	7.6	123.4	38.2
	23	152.4	152.4	6.1	6.8	7.6	123.4	29.8

UNIVERSAL COLUMNS

Parallel Flanges

DIMENSIONS AND PROPERTIES

Serial Size	Moment of Inertia			Radius of Gyration		Elastic Modulus		Ratio $\frac{D}{T}$
	Axis x–x		Axis y–y	Axis x–x	Axis y–y	Axis x–x	Axis y–y	
	Gross	Net						
mm	cm⁴	cm⁴	cm⁴	cm	cm	cm³	cm³	
356 × 406	275140	243076	98211	18.5	11.0	11592	4632	6.2
	227023	200312	82665	18.0	10.9	9964	3951	6.8
	183118	161331	67905	17.5	10.7	8388	3293	7.5
	146765	129159	55410	17.1	10.5	7004	2723	8.5
	122474	107667	46816	16.8	10.4	6027	2324	9.5
	99994	87843	38714	16.5	10.3	5080	1940	10.8
	79110	69424	31008	16.2	10.2	4153	1570	12.6
Column Core	172391	152936	68057	16.8	10.6	8075	3207	8.0
356 × 368	66307	57806	23632	16.0	9.57	3540	1262	13.9
	57153	49798	20470	15.9	9.52	3104	1100	15.5
	48525	42250	17470	15.8	9.46	2681	943.8	17.5
	40246	35040	14555	15.6	9.39	2264	790.4	20.3
305 × 305	78777	72827	24545	14.8	8.25	4314	1525	8.3
	64177	59295	20239	14.5	8.14	3641	1273	9.4
	50832	46935	16230	14.2	8.02	2991	1034	10.8
	38740	35766	12524	13.9	7.89	2368	806.3	13.1
	32838	30314	10672	13.7	7.82	2049	691.4	14.8
	27601	25472	9006	13.6	7.75	1755	587.0	16.8
	22202	20488	7268	13.4	7.68	1442	476.9	20.0
254 × 254	29914	27171	9796	11.9	6.79	2070	740.6	9.1
	22416	20350	7444	11.6	6.66	1622	570.4	11.0
	17510	15890	5901	11.3	6.57	1313	456.9	13.0
	14307	12976	4849	11.2	6.52	1099	378.9	15.1
	11360	10297	3873	11.1	6.46	894.5	305.0	17.9
203 × 203	9462	8374	3119	9.27	5.32	851.5	298.7	10.8
	7647	6758	2536	9.16	5.28	708.4	246.0	12.5
	6088	5383	2041	8.96	5.19	581.1	199.0	14.8
	5263	4653	1770	8.90	5.16	510.4	173.6	16.5
	4564	4035	1539	8.81	5.11	449.2	151.5	18.5
152 × 152	2218	1932	709	6.84	3.87	274.2	91.78	14.1
	1742	1515	558	6.75	3.82	221.2	73.06	16.8
	1263	1104	403	6.51	3.68	165.7	52.95	22.4

Note: One hole is deducted from each flange under 300mm wide (serial size) and two holes from each flange 300mm and over (serial size), in calculating the Net Moment of Inertia about x–x.

UNIVERSAL COLUMNS

SAFE LOADS FOR GRADE 43 STEEL

BASED ON
BS 449
1969

Serial Size in mm	Mass per metre in kg	SAFE CONCENTRIC LOADS IN KILONEWTONS FOR EFFECTIVE LENGTHS IN METRES											
		2.0	2.5	3.0	3.5	4.0	5.0	6.0	8.0	10.0	12.0	14.0	16.0
356 × 406	634	**11313**	**11313**	**11313**	**11313**	**11313**	11002	10517	9090	7287	5643	4382	3462
	551	**9826**	**9826**	**9826**	**9826**	**9826**	9527	9087	7801	6206	4782	3705	2923
	467	**8337**	**8337**	**8337**	**8337**	**8337**	8057	7668	6534	5158	3955	3056	2408
	393	**7012**	**7012**	**7012**	**7012**	7001	6755	6414	5427	4253	3247	2503	1970
	340	**6057**	**6057**	**6057**	**6057**	6040	5821	5518	4644	3620	2755	2121	1668
	287	5382	5309	5237	5174	5102	4911	4647	3889	3015	2288	1759	1381
	235	4406	4346	4286	4235	4174	4013	3790	3153	2432	1840	1412	1108
Column Core	477	**8501**	**8501**	**8501**	**8501**	8494	8201	7795	6615	5200	3978	3070	2417
356 × 368	202	3778	3723	3671	3621	3560	3398	3173	2555	1917	1430	1090	852
	177	3305	3257	3211	3168	3114	2970	2770	2224	1665	1241	945	739
	153	2857	2815	2776	2737	2690	2563	2388	1909	1426	1061	807	631
	129	2413	2377	2344	2311	2270	2161	2010	1601	1192	885	673	526
305 × 305	283	**5046**	**5046**	**5046**	4964	4840	4505	4048	2981	2116	1539	1158	
	240	4431	4356	4288	4200	4091	3796	3397	2480	1753	1272	957	
	198	3654	3593	3534	3459	3366	3113	2772	2006	1412	1023	768	
	158	2911	2862	2813	2751	2673	2462	2180	1562	1095	792	594	
	137	2524	2481	2438	2383	2314	2126	1877	1338	935	676	507	
	118	2164	2127	2090	2041	1981	1816	1599	1134	791	571		
	97	1780	1749	1718	1677	1626	1487	1304	920	640	461		

The above safe loads are tabulated for ratios of slenderness up to but not exceeding 180.
Safe loads printed in ordinary type are calculated for the 'effective lengths' of stanchions
in accordance with Table 17 a of BS 449 : 1969. Safe loads printed in bold type are
based on the limiting stress stress of 140 N/mm² for material over 40mm thick.
1 kilonewton may be taken as 0.102 metre tonne (megagramme) force, but see page 102.

UNIVERSAL COLUMNS

BASED ON BS 449 1959

SAFE LOADS FOR GRADE 43 STEEL

Serial Size in mm	Mass per metre in kg	SAFE CONCENTRIC LOADS IN KILONEWTONS FOR EFFECTIVE LENGTHS IN METRES														
		1.0	1.5	2.0	2.5	3.0	3.5	4.0	4.5	5.0	6.0	7.0	8.0	9.0	10.0	11.0
254 × 254	167	3164	3100	3036	2980	2905	2809	2686	2537	2363	1977	1610	1307	1070	886	744
	132	2497	2445	2394	2348	2286	2206	2104	1981	1838	1526	1236	1000	817	676	567
	107	2033	1991	1949	1910	1858	1790	1704	1600	1480	1222	986	796	650	537	450
	89	1695	1659	1624	1592	1547	1490	1417	1328	1227	1010	813	656	535	442	370
	73	1381	1351	1323	1295	1259	1210	1149	1075	991	813	653	525	428	354	296
203 × 203	86	1622	1580	1541	1488	1416	1321	1205	1077	950	728	562	443	356		
	71	1341	1306	1273	1229	1168	1087	990	883	777	594	458	360	290		
	60	1116	1086	1058	1020	966	896	812	721	632	481	370	291	234		
	52	977	951	927	892	845	783	708	628	550	418	321	252	203		
	46	865	842	820	789	745	689	622	550	481	365	280	220	176		
152 × 152	37	685	661	626	574	505	429	357	298	249	180					
	30	552	533	504	460	403	341	283	235	197	142					
	23	429	413	388	351	303	253	208	172	144	103					

The tables on pages 292 to 295 are reproduced by permission of the British Constructional Steelwork Association and the Constructional Steel Research and Development Organisation.

Examples on Steel Columns

(i) *Calculate the safe axial load for a mild steel column of solid circular section, 125 mm diameter, having an effective length of 3·125 m.*

$$\text{Least } r = \frac{d}{4} = \frac{125 \text{ mm}}{4} = 31\cdot25$$

$$\therefore \quad \frac{l}{r} = \frac{3\cdot125 \times 1\,000}{31\cdot25} = 100.$$

The working stress corresponding to this value of the slenderness ratio = 79 N/mm².

$$\text{Sectional area of column} = \frac{\pi d^2}{4} = \frac{\pi \times 125^2}{4} \text{ mm}^2 = 12\,270 \text{ mm}^2.$$

$$\therefore \quad \text{Safe axial load} = (79 \times 12\,270) \text{ N}$$
$$= 969\cdot2 \text{ kN.}$$

(ii) *The safe axial load for a joist column (i.e. a U.C. section used as a column) 305 mm × 305 mm × 97 kg, for an effective height of 4 m, is given on page 294 as 1 626 kN. Taking the necessary properties from the table on page 293, check the safe load given.*

On page 293 we find the least radius of gyration for the column section (axis YY) to be 7·68 cm.

$$\therefore \quad \frac{l}{r} = \frac{4 \times 100 \text{ cm}}{7\cdot68 \text{ cm}} = 52\cdot08.$$

This value lies between the tabular values '50' and '60.' We must therefore '*interpolate*' to find the working stress corresponding to the given $\frac{l}{r}$ value.

$\frac{l}{r}$	*Working Stress*
50	133 N/mm²
60	126 N/mm²
Difference 10	Difference 7

Difference 2·08   Difference $= \frac{7}{10} \times 2\cdot08 = 1\cdot46 \text{ N/mm}^2.$

NOTE.—All calculations are based on the table on page 291.

$\therefore$ The *rise* from 50 to 52·08 in $\dfrac{l}{r}$ value means a *drop* of 1·46 N/mm^2 below the 133 N/mm^2 stress value.

$\therefore$ If $\dfrac{l}{r} = 52\cdot08$, the working stress $= (133 - 1\cdot46)$ N/ mm^2

$$= 131\cdot54 \text{ N/mm}^2.$$

The property tables (page 292) give 123·3 cm^2 as the area of the section.

$\therefore$ Safe axial load $= (131\cdot54 \times 123\cdot3 \times 100)$ N
$$= 1\ 622 \text{ kN.}$$

This answer differs slightly from the tabular value. The reason is that in this answer f_c has been calculated as 131·54 N/mm^2, whereas the value used in the tables has been rounded off to 132 N/mm^2.

(iii) A *U.C. column, 152 mm × 152 mm × 37 kg, has an actual length of 5 m. The end fixture is such that the effective length is 0·85 times the actual length. The least radius of gyration for the column section is 3·87 cm and the area of section = 47·4 cm^2. Calculate the safe axial load for the column.*

Effective column length $= 5 \text{ m} \times 0\cdot85$
$$= 4\cdot25 \text{ m}$$

$$\frac{l}{r} = \frac{4\cdot25 \times 100 \text{ cm}}{3\cdot87 \text{ cm}} = 110.$$

For this value of $\dfrac{l}{r}$, the working stress $= 69$ N/mm^2.

$\therefore$ Safe axial load $= (69 \times 47\cdot4 \times 100)$ N
$$= 327 \text{ kN.}$$

(iv) *A column of effective length 2·4 m has to support an axial load of 1 370 N. Select a suitable U.C.-column section from the list given below.*

Size	Sectional Area	Least Radius of Gyration
(1) 203 mm × 203 mm × 71 kg	91·1 cm^2	5·28 cm
(2) 203 mm × 203 mm × 86 kg	110·1 cm^2	5·32 cm
(3) 254 mm × 254 mm × 73 kg	92·9 cm^2	6·46 cm

Try section No. 1:

$$\frac{l}{r} = \frac{2\cdot4 \times 100 \text{ cm}}{5\cdot28 \text{ cm}} = 45\cdot45.$$

By interpolation:

$$\frac{l}{r} \qquad\qquad p_c \text{ N/mm}^2$$

$$
\begin{array}{cc}
40 & 139 \\
50 & 133 \\
\hline
\text{Difference} = 10 & \text{Difference} = 6
\end{array}
$$

Difference $= 5\cdot45$ Difference $= \dfrac{6}{10} \times 5\cdot45 = 3\cdot27$ N/mm^2.

$\therefore$ for $\dfrac{l}{r} = 45\cdot45$, $p_c = (139 - 3\cdot27)$ N/mm$^2 = 135\cdot73$ N/mm^2.

Safe axial load $= (135\cdot73 \times 91\cdot1 \times 100)$ N $= 1\ 237$ N.
This section is therefore unsuitable.
Try section No. 2:

$$\frac{l}{r} = \frac{2\cdot4 \times 100 \text{ cm}}{5\cdot32 \text{ cm}} = 45\cdot0.$$

$$\therefore \quad p_c = 136 \text{ N/mm}^2.$$

Safe axial load $= (136 \times 110\cdot1 \times 100)$ N $= 1\ 497$ N.
This section is suitable.
It will be found that section No. 3 is not suitable.

Timber Posts and Struts*

The method of calculation of the safe axial load for a timber post or strut is given in the London Building (Constructional) By-laws, 1952. The tables given are published by kind permission of the G.L.C.

In the by-laws the maximum permissible compressive stress for axially loaded posts and struts is dependent upon the lesser of (i) the ratio of the effective length of the member to the least radius of gyration (l/k) or (ii) the ratio of the effective length to the least lateral dimension (l/b). Part (ii) above applies only to posts and struts of solid rectangular cross section. The following limits are laid down:

* Reproduced, by permission and courtesy of G.L.C., from London Building (Constructional) By-Laws, 1952.

The value of 'l/k' must not exceed 200 *and that of 'l/b' must not exceed* 58.

As will be seen in the table given (Table 32 in the G.L.C. By-laws), there are two classes of timber referred to: Class A and Class B.

Structural timber shall be either:

(*a*) Douglas fir (coast) (*Pseudotsuga taxifolia Brit.*), Longleaf pitch pine (*Pinus palustris Mill.*), or Shortleaf pitch pine (*Pinus echinata Mill.*) which for the purpose of these by-laws shall be known as class A timbers; or

(*b*) Canadian spruce (*Picea glauca Voss.*), European larch (*Larix decidua Mill.*), Redwood (*Pinus sylvestris L.*), Western hemlock (*Tsuga heterophylla Sarg.*) or Whitewood (*Picea abies Karst.*), which for the purpose of these by-laws shall be known as Class B timbers.

TABLE 32

Maximum permissible compressive stress in posts and struts in N/mm².

Ratio of Effective Length to		Class of Timber		Ratio of Effective Length to		Class of Timber	
Least radius of gyration. (l/k)	Least lateral dimension. (l/b)	A	B	Least radius of gyration. (l/k)	Least lateral dimension. (l/b)	A	B
0	0	6·895	5·516	80	23	4·826	3·861
10	3	6·757	5·412	90	26	4·206	3·378
20	6	6·619	5·309	100	29	3·654	2·896
30	9	6·481	5·171	120	35	2·758	2·206
40	11	6·274	5·033	140	40	2·137	1·724
50	14	5.998	4·826	160	46	1·655	1·310
60	17	5·723	4·521	180	52	1·379	1·103
70	20	5·309	4·275	200	58	1·103	0·896

The maximum permissible compressive stress for intermediate values of *l/k* or *l/b* shall be obtained by interpolation between the two nearest stresses for the class of timber used.

FIG. 340.

Effective Length of Posts and Struts (Timber)

The importance of end fixture on the 'effective length' of steel columns has been considered earlier in the chapter. Table 33 below is extracted from the London Building (Constructional) By-laws, 1952, by permission and courtesy of the G.L.C., and deals with timber posts and struts.

TABLE 33

Effective length of posts and struts.

Type of post or strut.	Effective length, where L is the length of the post or strut between centres of restraining members.
Properly restrained at both ends in position and direction	0·7L
Properly restrained at both ends in position and at one end in direction	0·85L
Properly restrained at both ends in position but not in direction	L
Properly restrained at one end in position and direction and at the other end partially restrained in direction but not in position .	1·5L
Properly restrained at one end in position and direction but not restrained at the other end .	2·0L

Where a post or strut is of a type not specified in this table, the effective length of that post or strut shall be determined to the satisfaction of the district surveyor.

FIG. 341.

It will be noted that 'L' in Table 33 is, in each case, given as the length of the post or strut, *between centres of restraining members.*

Introduction to the Principles of Eccentric Loading

An eccentric load is one which does not act at the centre of gravity of the section of the member. As the increase in stress due to the fact of eccentricity is usually very marked, eccentric loads are avoided as far as is possible. If unavoidable, the actual amount of eccentricity is kept as low as practical requirements permit.

In foundation work it is desirable to have uniform ground pressure under the base slab. If several columns are supported by one combined grillage, the resultant of the column loads should pass through the centre of gravity of the plan outline of the grillage.

In riveted joints, the resultant load carried does not, sometimes, pass through the centre of gravity of the rivets in the group. In such a case some of the rivets will have to take more than their share of the total load.

In a steel frame, with columns continuing through several floors, a floor beam has to be connected either to the web or to the flange of the column. The beam reaction therefore constitutes an '*eccentric load.*'

Theory of Eccentricity

Fig. 342 shows a short column carrying a load 'W' which acts at a distance 'e' from the centre of gravity of the section. If we introduce, at the C.G. of the section, two vertical forces each equal to 'W' in magnitude but acting in opposite directions to one another, we do not affect the equilibrium conditions, as the two forces introduced merely cancel

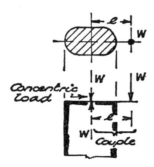

FIG. 342.—ECCENTRIC LOAD.

one another. This procedure, however, enables us to get a clear picture of the real effect of the given eccentric load. The two forces bracketed together (see Fig. 342) form a *couple* which tends to bend the column. The remaining force, acting at the C.G. of the section, tends to crush the column.

We thus have the important result that *an eccentric load is equivalent to a concentric load together with a bending moment.*

The magnitude of the hypothetical concentric load = 'W' and the magnitude of the bending moment = 'W × e,' i.e. '*eccentric load × arm of eccentricity.*'

Stress Produced by an Eccentric Load

The concentric load portion of the equivalent system will produce a direct stress (compressive), uniform in value all over the section.

The bending moment will produce 'bending stresses,' compressive and tensile (Fig. 343). The variation of stress over the column section due to bending will be similar to that produced in a simple beam, i.e. the stress-variation diagram will be a *straight line*.

It should be noted that the neutral axis for *bending* passes through the centre of gravity of the column section and is at *right angles* to the direction of eccentricity. The compressive stress due to the bending moment will occur in the portion of section which lies to the same side of the neutral axis as the eccentric load. The *net* stress distribution over the column section is found by adding the direct and bending stresses in the manner indicated in the following examples.

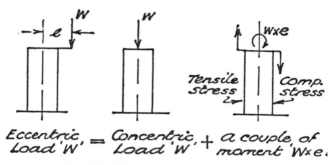

Eccentric = Concentric + a couple of
Load 'W'   Load 'W'   moment 'W × e.

FIG. 343.—EFFECT OF ECCENTRIC LOADING.

EXAMPLES:

(i) *Calculate the safe axial load for a timber post of solid rectangular section,* 100 *mm* × 150 *mm, if the length of the post between the centres of the restraining members* = 2·5 *m. The post is properly restrained at both ends in position and direction and the timber is 'Class A.'*

> Least lateral dimension = 100 mm
> Value of 'L' in mm = 2·5 × 1 000 = 2 500 mm
> Effective length of post = 0·7 × L mm
> = 0·7 × 2 500 mm = 1 750 mm.

Ratio of effective length to least lateral dimension

$$= \frac{1\ 750 \text{ mm}}{100 \text{ mm}} = 17\cdot5$$

Maximum permissible compressive stress.
From Table 32:

> When 'l/b' = 17,   max. stress = 5·72 N/mm²
> When 'l/b' = 20,   max. stress = 5·31 N/mm²
> Increase = 3   · Decrease = 0·41 N/mm²

For increase in 'l/b' of (17·5 − 17·0) i.e. 0·5 the decrease in max.

permissible stress = $\left(\dfrac{0\cdot41}{3} \times 0\cdot5\right)$ N/mm² = 0·068 N/mm².

Hence for $l/b = 17 \cdot 5$, max. stress $= (5 \cdot 72 - 0 \cdot 068)$
$$= 5 \cdot 652 \text{ N/mm}^2.$$

Sectional area of post $= 100 \text{ mm} \times 150 \text{ mm} = 15\,000 \text{ mm}^2$.
$\therefore$ Safe axial load $= (5 \cdot 652 \times 15\,000) \text{ N}$
$$= 84 \cdot 78 \text{ kN}.$$

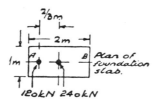

FIG. 344.—FOUNDATION PRESSURE.

(ii) *A concrete foundation slab carries the loads indicated in Fig. 344. Calculate the maximum pressure exerted on the subsoil. Test whether there will be any tendency for uplift of the slab.*

$$\text{Direct stress} = \frac{\text{Total load}}{\text{Area of base}}.$$

If there is an actual *concentric* load in addition to the eccentric load, add the two loads together in computing the direct stress.

$$\text{Direct stress} = \frac{(120 + 240) \text{ kN}}{2\text{m}^2} = \frac{360}{2} \text{ kN/m}^2$$

$$= 180 \text{ kN/m}^2.$$
$$\text{Bending moment} = 120 \text{ kN} \times \tfrac{2}{3} \text{ m} = 80 \text{ kN m}.$$

$$\text{Section modulus of slab} = \frac{bd^2}{6} = \frac{1 \times 2 \times 2}{6} \text{ m}^3$$

$$= \tfrac{2}{3} \text{ m}^3.$$

$$\text{Bending stresses} = \frac{M}{Z} = \frac{80}{2/3} = 120 \text{ kN/m}^2.$$

Max. stress (i.e. pressure) on foundation at 'A' $= (180 + 120)$ kN/m^2 $= 300$ kN/m^2.

Net stress or pressure at 'B' $= (180 - 120) \text{ kN/m}^2 = 60 \text{ kN/m}^2$. The slab is therefore under no tilting tendency as would have occurred if the '*bending tension stress*' had exceeded the '*direct compression stress*.'

(iii) *Figure* 345 *shows a plan of a square masonry pier carrying a single eccentric load W. Determine the maximum value of the eccentricity 'e' if there is to be no tension in the masonry.*

$$\text{Direct stress} = \frac{W}{A} = \frac{W}{B^2}.$$

$$\text{Bending moment} = We.$$

$$\text{Section modulus of pier} = \frac{bd^2}{6} = \frac{B \times B^2}{6} = \frac{B^3}{6}$$

$$\text{Bending Stress} = \frac{M}{Z} = \frac{We}{\dfrac{B^3}{6}} = \frac{6We}{B^3}.$$

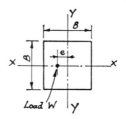

FIG. 345.

For there to be no tension, the maximum value of 'e' will be given when the direct compressive stress equals the bending tensile stress, i.e. when

$$\frac{W}{B^2} = \frac{6We}{B^3}$$

Whence
$$e = \frac{B}{6}.$$

Since this eccentricity could be on either side of the appropriate axis, it is seen that provided that the load falls within the *middle third* there will be no tension in the masonry.

Further Application of the Theory of Eccentricity

The problem of rectangular sections under eccentric loading is very important in masonry work. Formulae may be derived to obtain the maximum stresses without all the detailed working given in the foregoing examples. Such formulae are considered in Chapter XV.

In this introduction to the stresses caused by eccentric loading, it is not possible to consider the method of computation usually adopted in respect of joist sections, since this computation involves something more than the simple addition of direct and bending stresses.

EXERCISES 13

All calculations to be made, where necessary, by use of the section tables on pages 292 to 295.

(1) Calculate the value of the least radius of gyration in each of the following cases: (i) a solid circular section, 100 mm dia., (ii) a rectangular section, 100 mm × 75 mm, (iii) U.C. section, 254 mm × 254 mm × 167 kg (I_{XX} = 29 914 cm⁴, I_{YY} = 9 796 cm⁴, A = 212·4 cm²). (iv) a joist type section, flanges 150 mm wide × 25 mm thick, web thickness 12·5 mm, overall depth = 300 mm.

(2) Calculate the slenderness ratio of a column of 3 m effective length if the section of the column were (i) a circle 150 mm dia., (ii) a U.C. section 305 mm × 305 mm × 97 kg (r_{XX} = 13·4 cm, r_{YY} = 7·68 cm).

(3) Obtain the safe axial load for a U.C. section, 203 mm × 203 mm × 60 kg, if the effective height is 3·75 m. Look up the necessary section properties and 'p_c' value.

(4) Find the maximum permissible effective length for a 152 mm × 152 mm × 30 kg U.C. pillar, which has to carry an axial load of 372 kN. 'Least r' for section = 3·82 cm, sectional area = 38·2 cm². The following table is provided:

$\dfrac{l}{r}$	Permissible Stress
70	115 N/mm²
80	104 N/mm²
90	91 N/mm²

(5) A column has to support an axial load of 2 840 kN. The actual height of the column is 4·5 m. The effective length of the column is to be taken as 0·7 × the actual height. Find which of the following sections would be most suitable: 356 mm × 368 mm × 153 kg, 305 mm × 305 mm × 158 kg, 254 mm × 254 mm × 167 kg. The necessary section properties and the appropriate working stress must be looked up (pages 293 and 291).

(6) Calculate the maximum and minimum compressive stresses in the

eccentrically loaded short masonry pillar shown in Fig. 346 (Neglect self-weight of pillar.)

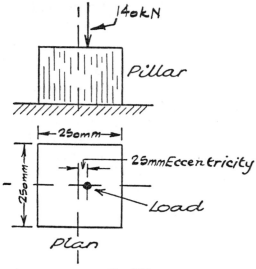

FIG. 346.

(7) In the case of the eccentrically loaded short joist column given in Fig. 347, show that the compressive stress varies from 60 N/mm² to zero.

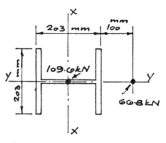

203 mm × 203 mm × 46 kg

FIG. 347.

The column-section properties are:

$$Z_{XX} = 449 \text{ cm}^3, \qquad Z_{YY} = 151 \text{ cm}^3, \qquad A = 58 \cdot 8 \text{ cm}^2.$$

Sketch the stress-variation diagram for the column section.

(8) Fig. 348 shows a short masonry pillar with foundation slab. Taking the conditions given in the figure, calculate the foundation pressures at 'A' and 'B' respectively. The density of the masonry is 21 kN/m³.

(Treat the self-weight of the pillar and base as a concentric load on the foundation.)

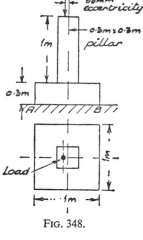

(9) A 203 mm × 203 mm × 71 kg U.C. column has the following properties of section:

$$\text{Sectional area} = 91 \cdot 1 \text{ cm}^2.$$
$$Z_{XX} = 708 \text{ cm}^3.$$

A load is carried on the 'YY' axis of the section at an eccentricity of x mm with respect to the 'XX' axis. Calculate the maximum value of 'x' if no tensile stress is to be developed in the column.

FIG. 348.

(10) Calculate the necessary diameter for a short, circular concrete column which has to support an axial load of 206 kN. The working compressive stress in the concrete is 4·2 N/mm².

(i) Assuming 'E' for the concrete to be 14 000 N/mm², find how much the designed column will shorten on a length of 300 mm.

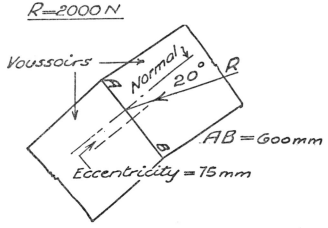

FIG. 349.

(ii) Compute the safe load for the column if the load were placed at a distance of 25 mm from the centre of the column section.

(11) Fig. 349 shows a joint between two adjacent voussoirs in an arch ring. The resultant thrust between the voussoirs is 2 000 N and is positioned as shown in the diagram. The joint 'AB' is 600 mm long and the joint carries the given thrust for 300 mm thickness of arch ring. Calculate the maximum compressive stress at the joint.

[Resolve the thrust normally to the joint 'AB.' Treat as an eccentric load, acting on an area of breadth 300 mm and depth 600 mm.]

Note.

Examples 7 and 9 above have been included purely as examples in the computation of direct and bending stresses. The paragraph on p. 305 in respect of joist sections is relevant.

This remark applies also to revision examples 56 and 57 on p. 375.

CHAPTER XIV

CALCULATION METHODS FOR LOADED FRAMES

IN this chapter an introduction will be made to the calculation of the member forces in a loaded frame. The calculation method is sometimes to be preferred to the construction of a stress diagram.

Method of Sections

Consider the truss shown in Fig. 350 to be cut through at the section plane 'SS,' and the portion to the right of the plane to be removed.

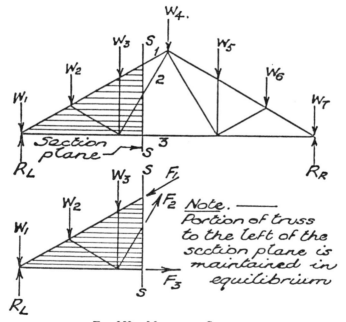

FIG. 350.—METHOD OF SECTIONS.

In order to prevent collapse of the truss certain forces 'F₁,' 'F₂,' and 'F₃' will have to be introduced. These forces are assumed to act in the direction of the lengths of the respective members.

In the uncut state the hatched portion of the truss was in equilibrium under the action of the external forces 'R_L,' 'W_1,' 'W_2,' 'W_3,' and forces exerted on it by bars (1), (2), and (3) at the imaginary section plane 'SS.' The forces 'F_1,' 'F_2,' and 'F_3' which we have to substitute for the respective member forces—to maintain equilibrium when the cut is actually made in the truss—will therefore be equal to these member forces.

We may regard the hatched portion of the truss as being a solid body acted upon by seven forces, viz. 'R_L,' 'W_1,' 'W_2,' 'W_3,' 'F_1,' 'F_2,' and 'F_3' in such a way as to maintain equilibrium. The laws of equilibrium may now be employed to deduce the magnitudes of the unknown forces.

The usual method of solution of an example such as that given in Fig. 350 is to employ the *'principle of moments.'* The method of calculation is then sometimes referred to as the *'method of moments.'*

Method of Moments

The section plane 'SS' is positioned to cut the member whose force is required and not more than two others. The plane need not be vertical— it merely indicates the member forces involved in the calculation. No measurements are actually made to any point in the section plane.

Rule.—*To find the force in any given member of the three cut by the plane, take moments about the intersection point of the other two members.*

By this procedure only one unknown member force will appear in the *'moments equation.'*

Struts and Ties.—The determination of *'struts'* and *'ties'* in the calculation method is effected as follows:

Place arrow heads, indicating the assumed sense of the respective member forces, on the force lines. These arrows, should be placed on the side of the plane away from the portion of truss being considered, i.e. to the right of the plane in the example shown in Fig. 350. If the chosen arrow-head direction is incorrect in the case of any particular member, the numerical value of the force obtained for that member will be prefixed by a negative sign when the 'moments equation' is solved. The arrow head in such a case must be reversed. If (and after the possible reversal just indicated) the arrow head points towards the section plane, a thrust is indicated and the member is a *'strut.'* If, on the other hand, the arrow head points away from the plane, thus indicating a 'pull,' the particular member is a *'tie.'* The numerical value of the force in any member is not affected by an incorrect estimation of the proper arrowhead sense.

Note.—Having drawn-in the section plane, and having decided to consider the portion of the truss, say, to the left of the plane, completely ignore all external loads on the truss to the right of the plane.

EXAMPLE (i).—*Calculate the force in each of the members marked* (1), (2), *and* (3) *respectively in the given roof truss* (*Fig.* 351). *State whether the respective members are struts or ties.*

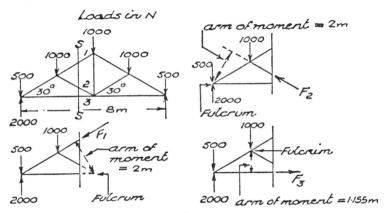

FIG. 351.—CALCULATION OF MEMBER FORCES.

The members given in the question may be solved by introducing a section plane 'SS.'

Member (1).—To find the force in member (1) we have to take moments about the intersection point of members (2) and (3), i.e. the midpoint of the bottom tie (see Fig. 351, bottom left diagram). Only the external loads which act on the portion of truss to the left of the section plane are considered.

$$(F_1 \times 2) + (1\,000 \times 2) + (500 \times 4) = 2\,000 \times 4$$
$$2F_1 + 2\,000 + 2\,000 = 8\,000$$
$$2F_1 = 4\,000$$
$$\therefore \quad F_1 = 2\,000 \text{ N.}$$

The numerical answer is positive. This indicates that the arrow-head direction chosen for member (1) is correct. As the member exerts a thrust *on the section plane*, it is a strut.

Member (2).—To find this member force, moments are taken about the intersection point of members (1) and (3), i.e. the left-end reaction point of the truss (see top right-hand diagram).

$$F_2 \times 2 = 1\,000 \times 2$$
$$\therefore \quad F_2 = 1\,000 \text{ N.}$$

The answer is positive, hence the arrow is correct and the member is a 'strut.'

Member (3).—Moments are taken this time about the mid-point of the left rafter of the truss, this being the point in which member (1) cuts member (2) (see lower right-hand diagram in Fig. 351).

$$(F_3 \times 1{\cdot}155) + (500 \times 2) = 2\,000 \times 2$$
$$1{\cdot}155F_3 = 3\,000$$
$$\therefore \quad F_3 = 2\,600 \text{ N.}$$

Member (3) is a tie.

The '*arms*' of the respective moments, in the foregoing calculations, may be obtained by direct measurement on a scale drawing of the truss or by trigonometrical calculation. The results which would be obtained by drawing a stress diagram for the truss would, of course, agree with those obtained by the calculation method shown.

The '*method of sections*' may be used to find the forces in certain specified members of a frame without the trouble of having to construct a stress diagram. The calculation method may be used to check and confirm a member-force value obtained by a graphical method.

In some cases it is not possible to complete a stress diagram by the usual procedure. Completion may often be effected by calculating one member of the frame, employing a suitable 'section plane' (see page 143).

EXAMPLE (ii).—*Calculate the force in the member marked by a cross in the truss given in Fig. 352.*

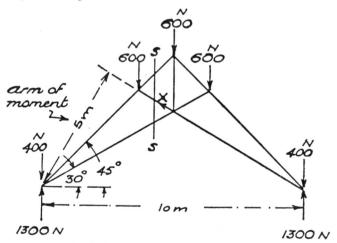

FIG. 352.—CALCULATION OF ONE MEMBER.

'SS' is a section plane cutting the given member and two others. The two latter members intersect at the left-end reaction point of the truss, hence this is the point about which moments must be taken.

Let 'F' N = force in the given member.

The 'arm' of the moment = 5 m

$$\therefore \quad F \times 5 = 600 \times 3.66 \text{ m.}$$

[3·66 m is the horizontal distance from the left-end reaction point to the vertical line of action of the 600 N force.]

$$\therefore \quad F = 439.2 \text{ N.}$$

The arrow head being correct, the member is a strut.

The reader may check the accuracy of the answer by drawing a stress diagram for the truss.

EXAMPLE (iii).—*Obtain the forces in members marked 'C' and 'D' respectively in the cantilever-truss example given in Fig. 353. The 'method of sections' must be used. Find also, by any calculation method, the forces in members 'A' and 'B.'*

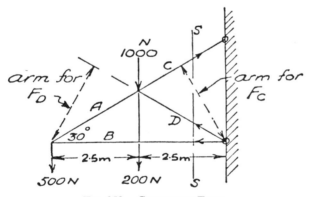

FIG. 353.—CANTILEVER TRUSS.

Member 'C':

Section plane 'SS' cuts member 'C' and two others. The two other members intersect at the lower reaction point.

Taking moments about this point, the force in member 'C' has an 'arm' of (5 sin 30°) m = (5 × ½) m = 2·5 m.

$$\therefore \quad (F_C \times 2.5) = (1\,000 \times 2.5) + (200 \times 2.5) + (500 \times 5)$$
$$2.5F_C = 2\,500 + 500 + 2\,500$$
$$= 5\,500$$
$$F_C = 2\,200 \text{ N.}$$

The arrow head has been correctly placed on member 'C' as the answer is positive, hence member 'C' is a tie.

Member 'D':

In this case the two remaining members intersect at the extreme left point of the truss.

$$\text{The arm for 'F}_D\text{'} = 5 \sin 30° = 2\text{·}5 \text{ m}$$
$$\therefore \quad F_D \times 2\text{·}5 = (1\,000 \times 2\text{·}5) + (200 \times 2\text{·}5)$$
$$\therefore \quad F_D = 1\,200 \text{ N}.$$

Member 'D' is a strut.

Member 'A':

'F_A' is easily found by vertical resolution of the forces acting at the left end of the truss.

$$F_A \cos 60° = 500$$

$$\therefore \quad F_A = \frac{500}{\cos 60°} = \frac{500}{\frac{1}{2}} = 1\,000 \text{ N}.$$

Member 'A' is clearly a 'tie' as it must have an upward vertical component at the bottom end.

Member 'B'.—By horizontal resolution:

$$F_A \cos 30° = F_B$$
$$\therefore \quad F_B = 1\,000 \times 0\text{·}866 = 866 \text{ N}.$$

Member 'B' is a strut, as it clearly must push to the left to balance the effect 'F_A' has towards the right.

Member 'A'.—We could have found 'F_A' by a *'moments'* method. Imagine a vertical section plane cutting members 'A' and 'B.' If a 'fulcrum' be chosen on member 'B,' this 'member force' will vanish from the 'moments equation.' A convenient point is the mid-point of the bottom member of the frame. The 'arm' of the moment for 'F_A' will be 1·25 m ($= 2\text{·}5 \sin 30°$).

$$\therefore \quad F_A \times 1\text{·}25 = 500 \times 2\text{·}5$$
$$\therefore \quad F_A = 1\,000 \text{ N}.$$

Similarly 'F_B' could be found by taking moments about any convenient point on member 'A.'

Braced Girders with Parallel Flanges

EXAMPLE.—*Calculate the force in each member of the braced girder shown in Fig. 354.*

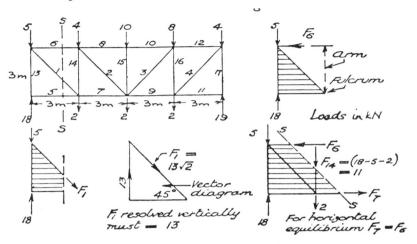

FIG. 354.—SOLUTION OF A BRACED GIRDER.

The first step is to number the members in the order in which they are going to be solved. We will calculate the members in the following order: (i) *diagonal members,* (ii) *horizontal members,* (iii) *vertical members.*

$$R_L \times 12 = (5 \times 12) + (6 \times 9) + (12 \times 6) + (10 \times 3)$$
$$R_L = 18 \text{ kN.}$$
$$R_R \times 12 = (4 \times 12) + (10 \times 9) + (12 \times 6) + (6 \times 3)$$
$$R_R = 19 \text{ kN.}$$

Diagonal Bars

Bar (1).—Imagine a section plane 'SS' cutting bar (1). The portion of the girder to the left of the section plane is in *vertical* equilibrium under the action of (i) the reaction (18 kN) vertically upwards, (ii) the external load (5 kN) vertically downwards, and (iii) the *vertical effect* of the force in bar (1) (see Fig. 354, bottom left diagrams). The total external load which bar (1), *when resolved vertically*, has to balance is $(18 - 5) \text{ kN} = 13 \text{ kN}$, i.e. the *shear force* in the panel in which bar (1) is the diagonal. The force in bar (1) must be more than '13 kN,' so that when resolved vertically it equals the necessary '13 kN.' From the vector

diagram, $\dfrac{13 \text{ kN}}{F_1} = \sin 45°$

$$\therefore \quad F_1 = \frac{13 \text{ kN}}{\sin 45°} = 13\sqrt{2} \text{ kN} \quad \left(\text{Sin } 45° = \frac{1}{\sqrt{2}} \right)$$

We may term '$\sqrt{2}$' in this case the *'reduction coefficient'* for reducing *'shear force values'* to corresponding *'inclined bar forces.'* The coefficient depends simply upon the slope of the inclined bar. If the inclination be 'θ' to the horizontal, the reduction

coefficient $= \dfrac{1}{\sin \theta}$. If the girder dimensions are given, the reduction

coefficient $= \dfrac{\text{Length of diagonal}}{\text{Height of truss}}$.

Rule.—*To find the force in any inclined member, multiply the shear force in the corresponding panel by the appropriate reduction coefficient.*

> *Bar* 1. $F_1 = $ S.F. $\times \sqrt{2} = (18 - 5)\sqrt{2} \text{ kN}$
> $= 13\sqrt{2} \text{ kN (tie)}.$
>
> *Bar* 2. $F_2 = $ S.F. $\times \sqrt{2} = (18 - 5 - 4 - 2)\sqrt{2} \text{ kN}$
> $= 7\sqrt{2} \text{ kN (tie)}.$
>
> *Bar* 3. $F_3 = $ S.F. $\times \sqrt{2} = (19 - 4 - 8 - 2)\sqrt{2} \text{ kN}$
> $= 5\sqrt{2} \text{ kN (tie)}.$
>
> *Bar* 4. $F_4 = $ S.F. $\times \sqrt{2} = (19 - 4)\sqrt{2} \text{ kN}$
> $= 15\sqrt{2} \text{ kN (tie)}.$

Horizontal Bars

Fig. 354 (top right-hand diagram) shows what 'bar 6' does to prevent collapse of the truss. 'Bar 6' must exert sufficient thrust to prevent the girder failing by turning about the opposite joint in the lower flange. The joint concerned is the one in which the diagonal in the panel meets the lower flange of the girder. For equilibrium of the hatched portion of truss:

$$(F_6 \times 3) + (5 \times 3) = 18 \times 3$$
$$F_6 + 5 = 18$$
$$F_6 = 13 \text{ kN}.$$

'Bar 6' is obviously a strut.

To solve '*bar* 7,' moments are taken about the opposite joint in the top flange, i.e. where the '4 kN' load acts. The reader must get accustomed to 'visualising' the portion of truss concerned in the 'moments' equation without having to draw a number of separate diagrams. Until familiarity with the method is acquired, the necessary sketch diagrams had better be constructed.

Bar 5. As there is no other horizontal force at the left-end reaction point, there cannot be any force in bar 5, i.e. $F_5 = 0$.

Bar 6. $F_6 = 13$ kN (as above).   Strut.

Bar 7. $(F_7 \times 3) + (5 \times 3) = 18 \times 3$
$F_7 = 13$ kN.   Tie.   ['Bar 7' must *pull.*]

Bar 8. $(F_8 \times 3) + (4 \times 3) + (2 \times 3) + (5 \times 6) = 18 \times 6$
$F_8 + 4 + 2 + 10 = 36$
$F_8 = 20$ kN.

'Bar 8' is a strut.

Bar 9. $(F_9 \times 3) + (4 \times 3) = 19 \times 3$.
[Note the '8 kN' and '2 kN' pass through the fulcrum.]
$F_9 + 4 = 19$
$\therefore$ $F_9 = 15$ kN.   'Bar 9' is a tie.

Bar 10. $F_{10} = F_8$ by horizontal equilibrium requirement at top of 'bar 15.' Bar 10 is a strut. Force = 20 kN.

Bar 11. $F_{11} = 0$ (as for 'bar 5').

Bar 12. $(F_{12} \times 3) + (4 \times 3) = 19 \times 3$
$F_{12} + 4 = 19$
$F_{12} = 15$ kN.   'Bar 12' is a strut.

Vertical Members

The vertical members are solved by considering vertical equilibrium at one end of the member. Having decid d which end to take, *do not pay any attention to any external loads which may act at the other end of the member.*

Bar 13.   Consider bottom end.
$F_{13} = 18$ kN (strut).

Bar 14.   Bottom end.
$F_1 = 13\sqrt{2}$ kN in the direction of the length of the bar. Its vertical upward component = 13 kN (i.e. the S.F. in the panel).

$\therefore$ Bar 14 must push downwards with a force of '11 kN,' so that with the '2 kN' external load the total downward force = 13 kN.

$\therefore$ $F_{14} = 11$ kN (strut).

Bar 15.   Top end.

$F_{15} = 10$kN (strut).

Bar 16.   Bottom end.

$F_{16} + 2 = 15$ (the upward pull in 'bar 4').

$F_{16} = 13$ kN (strut).

Bar 17.   Bottom end.

$F_{17} = 19$ kN (strut).

Some girder types lend themselves to simple solutions (of certain members) by taking special section planes.

Fig. 354 (bottom right-hand diagram) shows a section plane cutting bars '6,' '14,' and '7.'

Considering the horizontal equilibrium of the hatched portion, the only two horizontal forces are 'F_6' and 'F_7.' 'F_7' must therefore equal 'F_6' in magnitude. Note that if member '14' were not vertical we could not express this equality.

Acting vertically on the portion of truss considered, there are four forces.

Equating the force acting upwards to those acting downwards:

$$F_{14} + 5 + 2 = 18$$
$$\therefore \quad F_{14} = 11 \text{ kN.}$$

Clearly, member '14' is a strut.

It is possible, thus, to write down the forces in several members of the frame by inspection. In examinations the candidate should clearly state the principle upon which the solution of a given member is based.

EXAMPLE (ii).—*Calculate the force in each bar of the Warren girder* (*Fig.* 355).

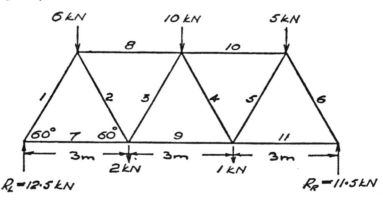

FIG. 355.—WARREN GIRDER.

$$R_L \times 9 = (6 \times 7.5) + (2 \times 6) + (10 \times 4.5) + (1 \times 3) + (5 \times 1.5)$$
$$= 112.5$$
$$R_L = 12.5 \text{ kN.}$$
$$R_R \times 9 = (5 \times 7.5) + (1 \times 6) + (10 \times 4.5) + (2 \times 3) + (6 \times 1.5)$$
$$= 103.5$$
$$R_R = 11.5 \text{ kN.}$$

Inclined Members

$$\text{Reduction coefficient} = \frac{\text{Length of inclined member}}{\text{Height of truss}}$$

$$= \frac{1}{\sin 60°} = 1.155.$$

Bar 1. F_1 = S.F. $\times$ 1.155
$= (12.5 \times 1.155)$ kN = 14.44 kN (strut).

Bar 2. F_2 = S.F. $\times$ 1.155
$= (12.5 - 6) \times 1.155$ kN
$= 7.51$ kN (tie).

Bar 3. F_3 = S.F. $\times$ 1.155
$= (12.5 - 6 - 2) \times 1.155$
$= (4.5 \times 1.155)$ kN
$= 5.20$ kN (strut).

Bar 4.   (Taking forces from right end.)
F_4 = S.F. $\times$ 1.155
$= (11.5 - 5 - 1) \times 1.155$
$= 6.35$ kN (strut).

Bar 5. F_5 = S.F. $\times$ 1.155
$= (11.5 - 5) \times 1.155$
$= (6.5 \times 1.155)$ kN
$= 7.51$ kN (tie).

Bar 6. F_6 = S.F. $\times$ 1.155
$= 11.5 \times 1.155$
$= 13.28$ kN (strut).

Horizontal Bars

Height of truss = $1.5 \tan 60° = 2.598$ m.

Bar 7. $F_7 \times 2.598 = 12.5 \times 1.5$ [Moments about top-flange joint
where the 6 kN load acts.]
$$F_7 = 7.22 \text{ kN (tie).}$$

Bar 8. $(F_8 \times 2{\cdot}598) + (6 \times 1{\cdot}5) = (12{\cdot}5 \times 3)$
$$2{\cdot}598F_8 = 28{\cdot}5$$
$$F_8 = 10{\cdot}97 \text{ kN (strut).}$$

Bar 9. $(F_9 \times 2{\cdot}598) + (2 \times 1{\cdot}5) + (6 \times 3) = (12{\cdot}5 \times 4{\cdot}5)$
$$2{\cdot}598F_9 + 3 + 18 = 56{\cdot}25$$
$$F_9 = \frac{35{\cdot}25}{2{\cdot}598} = 13{\cdot}57 \text{ kN (tie).}$$

Bar 10. $(F_{10} \times 2{\cdot}598) + (5 \times 1{\cdot}5) = 11{\cdot}5 \times 3$
$$2{\cdot}598F_{10} = 27$$
$$F_{10} = 10{\cdot}4 \text{ kN (strut).}$$

Bar 11. $F_{11} \times 2{\cdot}598 = 11{\cdot}5 \times 1{\cdot}5$
$$F_{11} = 6{\cdot}64 \text{ kN (tie).}$$

EXAMPLE (iii).—*Calculate the force in each of the bars marked 'A,'
'B,' 'C,' and 'D' in the loaded frame given in Fig. 356.*

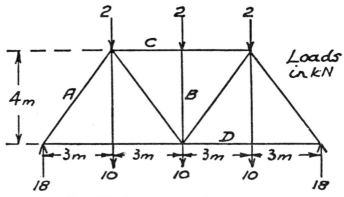

FIG. 356.—SYMMETRICALLY LOADED FRAME.

The inclination of the sloping members is not given in degrees in this example. The length of each diagonal member $= \sqrt{3^2 + 4^2} = 5$ m.

Bar A:

$$\text{Reduction coefficient} = \frac{\text{Length of diagonal}}{\text{Height of girder}}$$

$$= \frac{5 \text{ m}}{4 \text{ m}} = 1{\cdot}25.$$

$$F_A = \text{S.F.} \times 1{\cdot}25 = (18 \times 1{\cdot}25) \text{ kN} = 22{\cdot}5 \text{ kN.}$$

If we imagine a section plane vertically through member 'A' and consider the equilibrium of the girder to left of the plane, it will be clear that member 'A' must be pushing on the section plane, hence it is a strut.

In the usual case of a braced girder, i.e. one in which the loads are not very unsymmetrically disposed, it will be found that diagonal members sloping downwards towards the centre of the girder are 'ties' and those upwards towards the centre are 'struts.' When a girder has to carry a live load which passes on to and off the girder, some diagonal members may be called upon to act as both '*struts*' and '*ties*'—according to the position of the load.

Bar B:

Considering equilibrium at the top end of the bar, $F_B = 2$ kN. The bar is a strut. (Do not worry about the bottom end.)

Bar C:

Taking moments about the mid-point of bottom flange:

$$(F_C \times 4) + (2 \times 3) + (10 \times 3) = 18 \times 6$$
$$4F_C + 6 + 30 = 108$$
$$4F_C = 72$$
$$F_C = 18 \text{ kN.}$$

In the above 'moments equation' bar 'C' was assumed to be exerting a thrust on the left portion of the frame. Hence this bar is a strut. When a truss is supported at its ends, as in this example, all the top flange members are struts.

Bar D:

To solve this bar, moments are taken about the top right-hand joint of the frame. Assuming bar 'D' to be pulling on the portion of the truss to the right:

$$F_D \times 4 = 18 \times 3$$

[The '2 kN' and '10 kN' loads pass through the chosen fulcrum.]

$$\therefore \quad F_D = 54/4 = 13 \cdot 5 \text{ kN.}$$

'Bar D' is a tie.

All bottom flange members will be ties when the frame is supported as in the example.

EXAMPLE (iv).—*Fig. 357 shows a vertical frame carrying wind loads. Determine the nature and the magnitude of the forces in the respective*

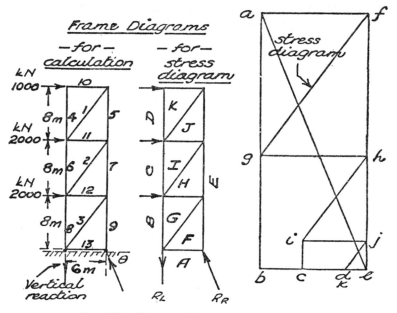

FIG. 357.—VERTICAL FRAME WITH WIND LOADS.

members (i) *by calculation method*, (ii) *by the construction of a stress diagram. Assuming the left-end reaction of the frame to be vertical, determine the frame reactions.*

Only a few members will be solved— the remainder are left as an exercise for the reader. The accuracy of the answers may be checked by means of the stress diagram.

Member 1:

This is a 'diagonal' member and a shear-force 'reduction coefficient' is required.

$$\text{Reduction coefficient} = \frac{\text{Length of diagonal}}{6 \text{ m}}$$

$$= \frac{\sqrt{6^2 + 8^2}}{6} = \frac{10}{6} = 1\tfrac{2}{3}.$$

$$[F_1 = \text{S.F.} \times \text{Reduction coefficient}]$$
$$= (1\,000 \times 1\tfrac{2}{3}) \text{ kN} = 1\,667 \text{ kN}.$$

Imagine a section plane to be taken horizontally through 'member 1.' Considering the portion of truss above the plane to be in equilibrium and acted upon by 'member 1' *from below the plane*, it is clear the member must be *pulling* on the plane so that it may have a component to the left to balance the '1 000 kN' external load. Hence 'F_1' is a tie.

Member 6:

To solve this member moments are taken about the joint in which members '5' and '7' meet. Member '6' must pull on the top panel to prevent collapse about the stated fulcrum, hence it is a 'tie.'

$$F_6 \times 6 = 1\,000 \times 8$$

[The '2 000 kN' load passes through the fulcrum.]

$$\therefore \quad F_6 = \frac{8\,000}{6}\,kN = 1\,333\,kN.$$

Member 12:

This member could be solved by considering horizontal equilibrium at, say, its left end.

It is readily solved by considering a section plane cutting members '8,' '12' and '7' (inclined at about 60° to the horizontal).

Considering horizontal equilibrium of the portion of frame above the section plane:

$$F_{12} = (2\,000 + 2\,000 + 1\,000)\,kN = 5\,000\,kN.$$

'Member 12' is a strut as it has to *push* to the left against the section plane.

Member 3:

$$\begin{aligned} F_3 &= \text{S.F.} \times \text{reduction coefficient} \\ &= (2\,000 + 2\,000 + 1\,000) \times 1\tfrac{2}{3}\,kN \\ &= (5\,000 \times 1\tfrac{2}{3})\,kN = 8\,333\,kN. \end{aligned}$$

'Member 3' is a tie.

Member 13:

Consider horizontal equilibrium at the left end of the bar.

F_{13} = 'F_3' resolved horizontally, as the reaction and member '8' are vertical.

$$F_{13} = F_3 \times 6/10$$

$$= \frac{8\,333 \times 6}{10}\,kN = 5\,000\,kN.$$

'Member 13' is a strut.

Determination of Reactions

To find 'R_L' take moments about the right reaction point.

$$R_L \times 6 = (2\,000 \times 8) + (2\,000 \times 16) + (1\,000 \times 24)$$
$$= 16\,000 + 32\,000 + 24\,000 = 72\,000$$
$$\therefore \quad R_L = 12\,000 \text{ kN.}$$

If 'V' and 'H' be the vertical and horizontal components respectively of the right-end reaction:

$$V = R_L = 12\,000 \text{ kN.}$$
$$H = (2\,000 + 2\,000 + 1\,000) \text{ kN} = 5\,000 \text{ kN.}$$
$$R_R{}^2 = V^2 + H^2 = 12\,000^2 + 5\,000^2 = 169\,000\,000$$
$$\therefore \quad R_R = 13\,000 \text{ kN.}$$

$$\tan \theta = V/H = \frac{12\,000}{5\,000} = 2\cdot4$$

$$\theta = 67° \, 24'.$$

We may now check 'member 13' by considering equilibrium at its right end.

$$F_{13} = R_R \times \cos \theta$$
$$= (13\,000 \times \cos 67° \, 24')$$
$$= 13\,000 \times 0\cdot3843$$
$$= 5\,000 \text{ kN.}$$

Construction of Stress Diagram

Having drawn the load line '$bcde$,' we note that member 'DK' has no force in it, hence points 'd' and 'k' are coincident. The stress diagram Is then completed in the manner indicated on page 142 for the canti-iever-truss example.

Three Concurrent Member Forces in Equilibrium

In the graphical method, the vector lines in a force diagram are scaled in order to obtain the required force values. Thus we scale the '*diagonal*' in a parallelogram of forces example. As explained in page 12, we could express the value of the diagonal in trigonometrical terms. Trigonometry may be usefully employed in problems involving three concurrent forces in equilibrium.

Fig. 358 shows a roof truss carrying a vertical load 'W' at the apex. The corresponding rafter forces are 'P' and 'Q.' 'W,' 'P' and 'Q' form a '*triangle of forces*.' Assuming the angles 'α,' 'β' and 'θ' as indicated,

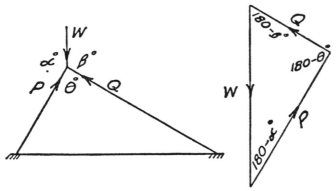

FIG. 358.—LAMI'S THEOREM.

the angles in the triangle of forces will be the supplements of these
angles, at the respective corners shown. Expressing the *sine rule*, used
in the trigonometrical solution of triangles, we have:

$$\frac{W}{\sin(180-\theta)°} = \frac{P}{\sin(180-\beta)°} = \frac{Q}{\sin(180-\alpha)°}.$$

But the sine of an angle = the sine of its supplement

$$\therefore \quad \frac{W}{\sin\theta°} = \frac{P}{\sin\beta°} = \frac{Q}{\sin\alpha°}.$$

Each force is therefore proportional to the sine of the angle between
the other two. This theorem is known as '*Lami's theorem.*'

EXAMPLES:

(i) *A weight of* 100 *N is suspended by two chains in the manner shown
in Fig.* 359. *Calculate the pull in each chain.*

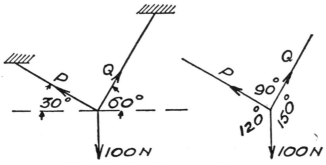

FIG. 359.—THREE CONCURRENT FORCES.

Let 'P' and 'Q' be the pulls in the left and right chains respectively.

$$\frac{100}{\sin 90°} = \frac{P}{\sin 150°} = \frac{Q}{\sin 120°}$$

$$\therefore \quad P = \frac{100 \times \sin 150°}{\sin 90°} = \frac{100 \times \sin 30°}{1} = (100 \times 0·5) \, N = 50 \, N$$

$$Q = \frac{100 \times \sin 120°}{\sin 90°} = \frac{100 \sin 60}{1} = (100 \times 0·866) \, N = 86·6 \, N$$

[Note: sin 150° = sin (180° − 150°) = sin 30°.]

(ii) *Calculate the pull in the tie and the thrust in the rafter in the example given in Fig. 360.*

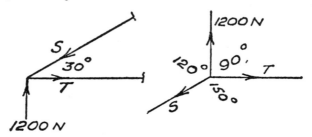

FIG. 360.—FORCES AT END OF TRUSS.

Let 'S' = the thrust in the rafter and 'T' = the pull in the tie.

It will be helpful to visualize (or actually draw out) the three forces as either all acting towards, or all acting from, the common point of concurrence. An error in the assumed arrow-head directions will then be apparent.

$$\frac{1\,200}{\sin 150°} = \frac{T}{\sin 120°} = \frac{S}{\sin 90°}$$

$$\therefore \quad T = \frac{1\,200 \sin 120°}{\sin 150°} = \frac{1\,200 \sin 60°}{\sin 30°} = \frac{1\,200 \times 0·866}{0·5} \, N = 2\,078 \, N$$

$$S = \frac{1\,200 \sin 90°}{\sin 150°} = \frac{1\,200 \times 1}{\sin 30°} = \frac{1200}{0·5} \, N = 2\,400 \, N$$

The two examples given could have been easily solved by resolution method.

Resolving vertically in the latter example:

$$S \cos 60° = 1\ 200$$

$$\therefore \quad S = \frac{1\ 200}{\cos 60°} = \frac{1\ 200}{0\cdot5}\,N = 2\ 400\ N.$$

Resolving horizontally:

$$T = S \cos 30° = (2\ 400 \times 0\cdot866)\,N = 2\ 078\ N.$$

Lami's theorem is useful when the resolution is not so simply effected as in this example.

(iii) *Calculate the pull in the tie and the thrust in the jib in the simple jib-crane example shown in Fig. 361. The resultant pull on the jib head is inclined to the vertical.*

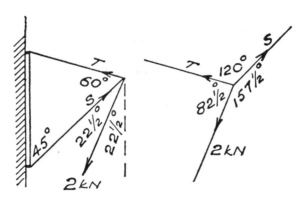

FIG. 361.—SOLUTION BY LAMI'S THEOREM.

Let 'S' = thrust in the jib and 'T' = pull in the tie.

$$\frac{S}{\sin 82\tfrac{1}{2}°} = \frac{T}{\sin 157\tfrac{1}{2}°} = \frac{2}{\sin 120°}$$

$$\therefore \quad S = \frac{2 \sin 82\tfrac{1}{2}°}{\sin 120°} = \frac{2 \sin 82\tfrac{1}{2}°}{\sin 60°} = \frac{2 \times 0\cdot9914}{0\cdot866}\,kN = 2\cdot29\ kN$$

$$T = \frac{2 \sin 157\tfrac{1}{2}°}{\sin 120°} = \frac{2 \sin 22\tfrac{1}{2}°}{\sin 60°} = \frac{2 \times 0\cdot3827}{0\cdot866}\,kN = 0\cdot884\ kN$$

EXERCISES 14

(1) Using the method of sections obtain the force in each of the members marked 'A,' 'B,' and 'C' in the roof-truss example shown in Fig. 362. Distinguish between struts and ties.

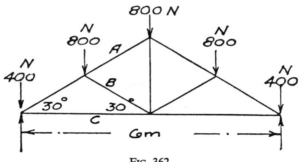

FIG. 362.

(2) Draw a section plane cutting members 'A,' 'B,' and 'C' in the truss given in Fig. 363. Calculate the forces in these members respectively by considering the equilibrium of the portion of the truss: (i) to the left, (ii) to the right, of the section plane.

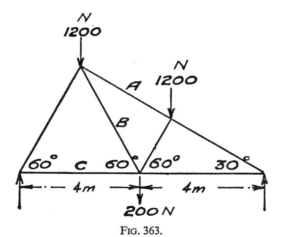

FIG. 363.

(3) Find the force in each member of the loaded frame shown in Fig. 364. State whether the respective members are struts or ties. Check the calculated values by the construction of a stress diagram.

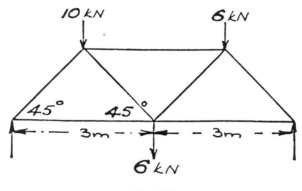

FIG. 364.

(4) Obtain, by the method of sections, the force in member 'A' of the truss given in Fig. 365. State whether the member is in tension or compression.

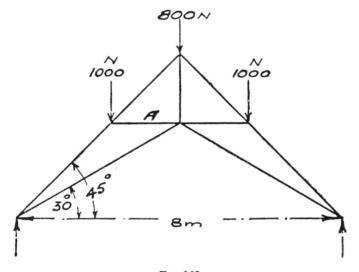

FIG. 365.

(5) Fig. 366 shows a braced girder with symmetrical loading. Calculate the force in each member and state whether 'strut' or 'tie.' Draw up a table of the results.

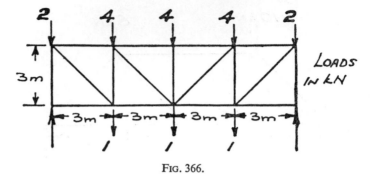

FIG. 366.

(6) The cantilever truss given in Fig. 367 carries wind and dead loads. Using the method of sections, find the forces in members 'A,' 'B,' and 'C' respectively.

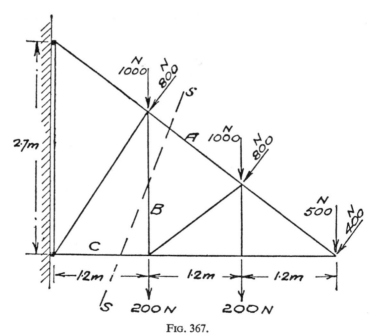

FIG. 367.

['SS' represents a suitable section plane. There will be no necessity to compound the dead and wind loads. The moment of each load (about the appropriate fulcrum) may be taken independently.]

(7) Using Lami's theorem, calculate the force in each of the rafters of the truss given in Fig. 368. Verify the results by drawing a triangle of forces.

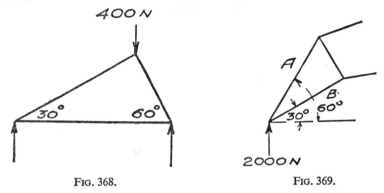

Fig. 368.

Fig. 369.

(8) Calculate the forces in members 'A' and 'B' respectively, in the example given in Fig. 369. Use Lami's theorem. Verify, by resolving vertically and horizontally, that the calculated values are correct.

CHAPTER XV

GRAVITY RETAINING WALLS

THE pressure which a wall has to sustain may be caused by wind, by retained liquid, or by retained earth or other granular material.

In calculations and formulae concerned with the resultant thrust of retained material, it is usual to consider 1 m length (in plan) of the wall. All the forces involved in the stability of a typical retaining wall, i.e. one whose vertical section remains constant throughout its length, are directly proportional to the wall length. It is convenient, therefore, to take 1 m length as a basis for numerical computations.

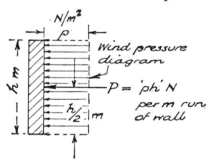

FIG. 370.—WIND PRESSURE.

Wind Pressure

Let   'h' m = Height of wall (Fig. 370).

'p' N/m^2 = Uniform wind pressure.

'P' N = Resultant thrust per metre length of wall.

$\therefore$ P $= p \times (h \times 1) = ph$ N.

The wind-pressure diagram being a rectangle, the resultant thrust will act at $\dfrac{h}{2}$ m from the bottom of the wall.

Overturning moment about base of wall = Force × arm

$$= P \times \frac{h}{2} = \frac{ph^2}{2} \, \text{N m}.$$

EXAMPLE.—*A vertical wall, 2 m high, is subjected to a wind pressure of uniform intensity, 1 500 N/m². Calculate, per metre run of wall,* (i) *the resultant wind thrust,* (ii) *the overturning moment about wall base.*

(i) $P = ph = 1\ 500 \times 2 = 3\ 000\ N = 3\ kN.$

(ii) Overturning moment $= P \times \dfrac{h}{2} = 3\ kN \times \tfrac{2}{2}\ m$

$$= 3\ kN \times 1\ m = 2\ kN\ m.$$

Liquid Pressure

The following is a brief summary of some of the properties of liquid pressure which will be involved in the type of problem considered in this chapter. The liquid in each case is assumed to be at rest.

(i) The intensity of pressure at a given point in a liquid is the same in all directions. For example, at a given depth the pressure is the same horizontally as it is vertically.

(ii) The intensity of pressure increases uniformly with the depth. The pressure-variation diagram will therefore be a straight line (see Fig. 371).

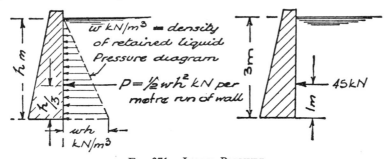

FIG. 371.—LIQUID PRESSURE.

(iii) The value of the pressure intensity at a given point in a liquid will be given by the formula:

Pressure = Density of liquid × Depth of point.

Let   'p' kN/m² = Intensity of pressure.
        'w' kN/m³ = Density of the liquid.
        'h' m = Depth of point considered.
        $\therefore\ \ p = wh$ kN/m².

Thus at a point 4 m deep in water of density 10 kN/m³ the pressure intensity would be $(10 \times 4) = 40$ kN/m².

(iv) The direction of pressure of a retained liquid against the retaining surface is at right angles to that surface. If a retaining wall has a vertical back the resultant liquid thrust will be horizontal.

Magnitude of Resultant Liquid Thrust

Case (i).—**Retaining Wall with Vertical Back.**—In Fig. 371, the area against which the liquid is pressing per metre run of wall $= h$ m $\times$ 1 m $= h$ m². The pressure intensity varies from zero, at the free surface of the liquid, to a value 'wh' kN/m² at the bottom of the wall. The pressure varies uniformly and the average pressure over the total area 'h' m² will be $\dfrac{wh}{2}$ kN/m².

The total thrust per metre length of wall

$$= \text{Average pressure} \times \text{Area under pressure}$$

$$= \left(\frac{wh}{2} \times h \right) \text{kN}.$$

If 'P' represents this resultant thrust.

$$P = \tfrac{1}{2}wh^2 \text{ kN}.$$

Position of the resultant thrust:

The resultant liquid thrust will act horizontally through the centre of gravity of the pressure-variation diagram. It will therefore act at *one-third* the depth of the retained liquid from the bottom of the wall.

EXAMPLE.—*The depth of water behind a retaining wall with a vertical back is 3 m. Calculate the magnitude of the resultant water thrust against the wall per metre run. Obtain the overturning moment this thrust produces about the base of the wall. Density of water* $= 10 \text{ kN/m}^3$.

$$P = \tfrac{1}{2}wh^2$$
$$= \tfrac{1}{2} \times 10 \times 3^2 \text{ kN}$$
$$= 45 \text{ kN per metre run of wall (see Fig. 371).}$$

The resultant water thrust 'P' will act at $\dfrac{h}{3} = \dfrac{3 \text{ m}}{3} = 1$ m above the base of the wall. The overturning moment (sometimes termed '*bending moment*') about any point on the wall base

$$= 45 \text{ kN} \times 1 \text{ m}$$
$$= 45 \text{ kN m per metre run of wall.}$$

Case (ii).—**Retaining Wall with Sloping Back.**

Let 'l' m = slope length of wetted portion of wall in Fig. 372.

Area of wetted surface per metre run of wall = $(l \times 1)$ m^2 = l m^2.

Intensity of pressure at bottom of wall, i.e. at a vertical depth of 'h' m = wh kN/m^2.

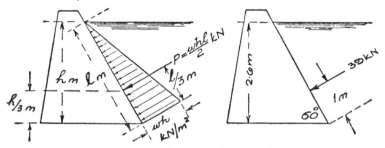

FIG. 372.—LIQUID THRUST ON INCLINED SURFACE.

As before, the average pressure = $\dfrac{wh}{2}$ kN/m^2.

∴ The resultant thrust = $\left(\dfrac{wh}{2} \times l\right)$ kN per metre run of wall.

$$\therefore \ \ P = \frac{whl}{2} \text{ kN}.$$

Position of resultant thrust.—The resultant thrust will act, at right angles to the wall slope, at a point one-third of the length 'l' measured from the wall bottom up the slope.

EXAMPLE.—*The back of a retaining wall is inclined at 60° to the horizontal. The depth of water retained is 2·6 m. Find the resultant thrust on the wall per metre run.*

The slope length of the wetted surface may be obtained by measurement of a scale diagram (see Fig. 372) or by calculation.

By calculation $l = \dfrac{2 \cdot 6 \text{ m}}{\sin 60°} = \dfrac{2 \cdot 6 \text{ m}}{0 \cdot 866} = 3$ m.

Total resultant thrust = $\dfrac{whl}{2}$ kN

$$= \left(\frac{10 \times 2 \cdot 6 \times 3}{2}\right) \text{ kN}$$

$$= 39 \text{ kN per metre run of wall}.$$

The thrust will act at right angles to the wall slope at $\frac{2}{3}$ m = 1 m from the bottom of wall measured up the wall slope.

Earth Pressure

The particles of a granular material like earth or sand, do not move freely amongst themselves when a force, such as the force of gravity, tends to produce relative movement. The surface of a tipped load of earth may be inclined to the horizontal at a considerable angle. This is possible because of the '*friction*' which exists between the particles. In liquids, the particles of which move over one another quite freely without frictional resistance, the free surface is always a horizontal plane.

The existence of friction in granular materials causes differences in properties as between liquid and earth pressures. To appreciate these differences it will be necessary to study some of the '*laws of friction*.'

Friction.—When one body tends to slide over another, the resistance to motion which is experienced at the surface of contact is termed '*frictional resistance*.'

As the force 'F' (Fig. 373) gradually increases in value from zero, the frictional resistance to motion increases just sufficiently to prevent motion. When, however, 'F' has attained a certain '*limiting value*,' the

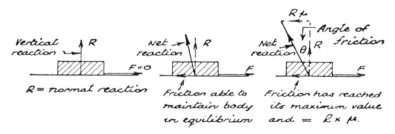

FIG. 373.—FORCE OF FRICTION.

frictional resistance will be unable to prevent the body moving. The value of the frictional resistance when motion is about to take place is termed '*limiting friction*.'

It is found experimentally that two things govern the value of 'limiting friction': (i) the normal reaction between the two surfaces in contact, (ii) the nature of the two surfaces.

Law of limiting friction.—The value of the frictional resistance bears a constant ratio to the normal reaction between the two surfaces. This ratio is a physical constant for any given pair of surfaces in contact.

Let 'F' = the value of the maximum frictional resistance
'R' = the normal reaction between the surfaces
μ = the physical constant referred to above.

The law states that $\dfrac{F}{R} = \mu$, i.e. $F = R\mu$.

'μ' is termed the 'coefficient of friction' for the two surfaces involved. Thus for 'wood on wood' an average value of the coefficient of friction would be 0·35. 'μ' is a Greek letter, and is pronounced 'mu.'

It will be noted that the area of the surfaces in contact does not affect the frictional-resistance value.

EXAMPLE.—*A masonry block weighing* 100 *N rests on a wooden floor. Assuming the coefficient of friction for masonry on timber to be* 0·4, *calculate the least horizontal force which will move the block.*

The normal reaction = 100 N.
$$F = R\mu$$
$$= (100 \times 0\cdot4) = 40 \text{ N.}$$
Necessary horizontal force = 40 N.

Angle of Friction

When there is no force urging the block forward (Fig. 373), i.e. when F = 0, the total reaction between the two surfaces in contact is normal to the surfaces. But when 'F' has a small value, friction is introduced and the total reaction between the surfaces makes a small angle with the normal. When 'F' reaches 'limiting friction' value, the maximum frictional resistance is brought into play and the total reaction makes its maximum angle with the normal. This maximum angle 'θ' is termed the 'angle of friction' for the particular surfaces in contact.

It will be noted that $\tan \theta = \dfrac{R\mu}{R} = \mu$, i.e. the coefficient of friction is equal to the *tangent* of the 'angle of friction.'

Angle of Repose.—Imagine a big open rectangular box full of granular material to be turned upside down suddenly and then the box to be removed vertically. The sides of the granular mass would begin to crumble down but a state of equilibrium would finally be established (Fig. 374). The angle which the surface slope makes with the horizontal in this condition of repose is termed the '*angle of repose*' for the given granular material. This 'angle of repose' may be considered to be the '*angle of friction*' for one portion of the material tending to slide over

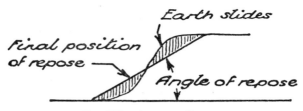

FIG. 374.—ANGLE OF REPOSE.

another (see Fig. 377). In the case of water, in which no friction exists, the angle of repose is zero (see Fig. 375).

CHARACTERISTICS OF BACKING MATERIALS.

Material.	Angle of Repose.	Weight in kN/m³.
Sand (dry)	30°	14–16
(moist)	35°	16–17·5
(wet)	25°	17·5–20
Vegetable earth (dry) . . .	30°	14–16
(moist) . . .	45°	16–17·5
(wet) . . .	15°	17·5–19
Gravel	40°	14·5
Rubble stone	45°	16–17·5
Gravel and sand	25° to 30°	16–17·5
Clay (dry)	30°	19–22
(moist)	45°	19–26
(wet)	15°	
Mud	0°	17–19
Ashes	40°	6·4

FIG. 375.

Earth-pressure Theories

There are various theories of earth pressure resulting in a number of different earth-pressure formulae and graphical constructions. Space permits of reference to two theories only.

Coulomb's Wedge Theory

In this theory a wedge of earth behind the retaining wall (Fig. 376) is assumed to be responsible for the thrust against the wall. This wedge under the action of gravity, tends to slide down its lower boundary plane, which is termed the *'plane of rupture.'* The wall prevents the sliding taking place and is thereby subjected to a thrust.

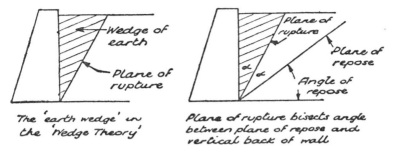

The 'earth wedge' in the 'Wedge Theory'

Plane of rupture bisects angle between plane of repose and vertical back of wall

FIG. 376.—WEDGE OF EARTH.

In the following development of the theory we will assume the back of the wall to be vertical, the surface of the earth to be horizontal and no friction to exist between the earth and the back of the wall. Under these conditions it can be shown that the wall will be subjected to a maximum thrust if the plane of rupture bisect the angle between the back of the wall and the plane of repose.

The wedge of earth (Fig. 377) is in equilibrium under the action of three forces:

(i) Its own weight acting vertically downwards through the C.G. of the wedge.

(ii) The reaction of the wall, which will be horizontal on the no-friction assumption previously laid down.

(iii) The reaction of the earth beneath the plane of rupture. Assuming the angle of friction for earth on earth to be equal to '$\theta°$' (the *'angle of repose'*), this reaction will make an angle $\theta°$ with the normal to the plane of rupture—over which sliding is assumed to be just taking place.

We can therefore draw a *triangle of forces* for these three forces and graphically find force (ii) above. This force, reversed in direction, will be the required earth thrust.

Just as in the case of water, earth pressure increases uniformly with the depth, giving a straight-line pressure-variation diagram. The resultant earth thrust will act at one-third the height of earth retained, from the bottom of the wall.

EXAMPLE.—*Using the graphical method for the wedge theory, find the*

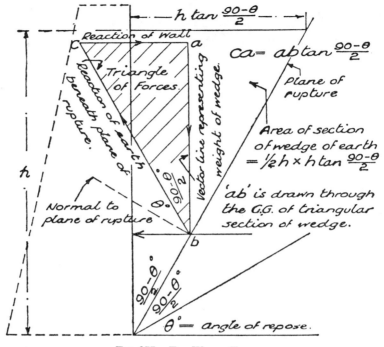

FIG. 377.—THE WEDGE THEORY.

resultant earth thrust against a retaining wall per metre of length, under the following circumstances:

Height of wall (vertical back) = 4 *m.*
Horizontal earth, 0·4 *m from top of wall.*
Density of earth = 14 *kN/m².*
Angle of repose of the earth = 30°.

The graphical solution is given in Fig. 378.

The plane of rupture makes $30° \left(= \dfrac{60°}{2} \right)$ with the back of the wall.

The horizontal dimension of the triangular section of the wedge may be obtained by measurement or calculation. It is (3·6 tan 30°) m = (3·6 × 0·5774) = 2·08 m.

∴   Area of section of wedge = ($\frac{1}{2}$ × 3·6 × 2·08) m² = 3·744 m².

Volume of wedge per metre run = (3·744 × 1) = 3·744 m³.
Weight of wedge per metre run = (3·744 × 14) kN = 52·42 kN.

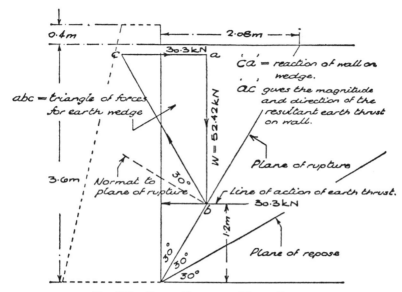

FIG. 378.—EARTH THRUST BY WEDGE THEORY.

Vector line '*ab*' is drawn to represent 52·42 kN. It is drawn through the C.G. of the wedge section.

Vector line '*bc*' is drawn at '30°' to the normal to the plane of rupture (30° = angle of friction), and '*ac*' is drawn horizontally.

The triangle '*abc*' is the triangle of forces for the wedge of earth. Vector line '*ca*' scales 30·3 kN.

The resultant earth thrust *per metre run of wall* is 30·3 kN, acting horizontally at 1·2 m above the bottom of the wall.

Formula for Resultant Thrust

Using the symbols given in Fig. 377, the volume of earth in the wedge $= \frac{1}{2}\left(h \times h \tan \frac{90 - \theta}{2}\right) = \frac{1}{2}h^2 \tan \frac{90 - \theta}{2}$.

∴ As '*w*' = density of the earth, the weight of the wedge per metre run $= \frac{wh^2}{2} \tan \frac{90 - \theta}{2} = W$.

From the triangle of forces for the wedge we find P = earth thrust

$= W \tan \frac{90 - \theta}{2} = \frac{wh^2}{2}\left(\tan \frac{90 - \theta}{2}\right)^2$.

It can be shown by trigonometrical deduction that

$$\left(\tan\frac{90-\theta}{2}\right)^2 = \frac{1-\sin\theta}{1+\sin\theta}$$

$$\therefore \quad P = \tfrac{1}{2}wh^2\frac{1-\sin\theta}{1+\sin\theta}.$$

In this form the formula for the earth thrust is usually known as 'Rankine's formula.' With the particular assumptions made in the present investigation the 'wedge theory' and 'Rankine's theory' yield the same result. Applying Rankine's formula to the previous numerical example:

$$P = \tfrac{1}{2}wh^2\frac{1-\sin\theta}{1+\sin\theta} = \tfrac{1}{2}wh^2\frac{1-\sin 30°}{1+\sin 30°}$$

$$= \tfrac{1}{2} \times 14 \times 3\cdot6 \times 3\cdot6 \times \frac{1-\tfrac{1}{2}}{1+\tfrac{1}{2}}\,\text{kN}$$

$$= (\tfrac{1}{2} \times 14 \times 3\cdot6 \times 3\cdot6 \times \tfrac{1}{3})\,\text{kN}$$

$$= 30\cdot24\,\text{kN (see Fig. 380).}$$

This value agrees with the result obtained by the graphical construction (Fig. 378).

Rankine's Theory

Imagine a very small cube of earth, with horizontal and vertical faces, to be subjected to a normal pressure 'p' on the two horizontal faces. If the cube were composed of liquid instead of earth we would require to exert horizontal pressure $= p$ on the vertical faces to maintain equilibrium (Fig. 379).

Rankine showed that, in the case of an earth cube, a pressure less than 'p' on the vertical faces would be sufficient to prevent collapse of the cube. In the case of a granular material, owing to the frictional forces which the grains can exert on one another, disruption of the cube by slipping along internal planes is considerably hampered.

If 'q' be the necessary pressure applied horizontally to the vertical faces of the cube to prevent 'spreading' when a pressure of 'p' is applied vertically to the top and bottom faces, Rankine proved that 'p' and 'q' are connected by the following relationship:

$$q = p\frac{1-\sin\theta}{1+\sin\theta}, \text{ where } \theta = \text{angle of repose for the earth.}$$

[If $\theta = 0$, as in liquids, $q = p$.]

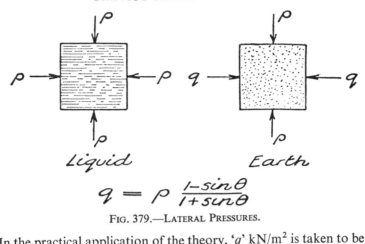

$$q = p \frac{1 - \sin \theta}{1 + \sin \theta}$$

Fig. 379.—Lateral Pressures.

In the practical application of the theory, 'q' kN/m² is taken to be the pressure exerted horizontally by earth which would exert a pressure 'p' kN/m² vertically. The value of 'p' is found as in liquid-pressure calculations, i.e. by multiplying the earth density by the depth of the point considered.

$$\therefore \quad q = wh\frac{1 - \sin \theta}{1 + \sin \theta}.$$

The pressure-variation diagram will be as indicated in Fig. 380. Taking 1 m length of wall, the total earth thrust will be

$$\tfrac{1}{2}wh\frac{1 - \sin \theta}{1 + \sin \theta} \times (h \times 1)\,\text{kN}$$

$$\therefore \quad P = \tfrac{1}{2}wh^2\frac{1 - \sin \theta}{1 + \sin \theta}\ \text{kN per metre run of wall.}$$

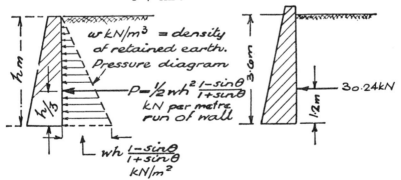

Fig. 380.—Rankine's Formula.

The resultant thrust will act through the C.G. of the triangular pressure diagram, i.e. '$\frac{1}{3}h$' from the bottom of wall.

The overturning moment of the resultant thrust about the base of

$$\text{wall} = \tfrac{1}{2}wh^2 \frac{1 - \sin\theta}{1 + \sin\theta} \times \frac{h}{3} = \frac{wh^3}{6} \frac{1 - \sin\theta}{1 + \sin\theta}.$$

Stability of Gravity Retaining Walls

(i) Graphical Method

The resultant thrust which acts across the base of a retaining wall (and which has to be balanced by the force exerted by the foundation) is compounded of two forces;

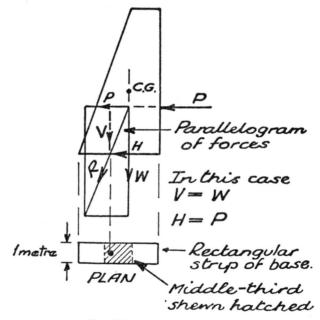

FIG. 381.—RESULTANT THRUST.

(i) The thrust of the retained material behind the wall, and (ii) the self-weight of the wall. The resultant of forces (i) and (ii) is found by the usual vector-diagram method (Fig. 381).

It is upon the character of this resultant thrust—its *position*, *magnitude*, and *direction*—that the suitability of the wall design depends.

Position.—Assuming the wall to merely rest on its base, the resultant must obviously cut the base inside the toe of the wall if overturning is not to take place. If the resultant falls outside the base altogether, equilibrium is not possible. But it is necessary that the resultant thrust shall fall well inside the toe if all tendency to tilt is to be eliminated. It is usually assumed that the line of action of the resultant must cut the base inside the '*middle-third*' portion of its width if all risk of uplift is to be prevented. The law—known as the *middle-third law*—states that when two rectangular surfaces are in contact the resultant thrust must not act outside the middle-third of the width of contact of the surfaces if tension is to be avoided. The base (or any horizontal section) of a typical retaining wall, taking 1 m run as hitherto, is a rectangular area, and hence the middle-third law is applicable. As a tendency for tension in masonry, or at mortar joints, is undesirable, the law is a test of satisfactory design. The law is considered mathematically a little later.

Magnitude.—The resultant is resolved into its vertical and horizontal components (Fig. 381). The problem of the vertical component usually is that of an eccentric load acting on a rectangular section. The stresses produced—sometimes referred to as '*normal stresses*'—are therefore computed by the method given on page 301. A more detailed consideration of these stresses, with the derivation of suitable formulae for their calculation, is given in a subsequent paragraph.

The horizontal component of the resultant tends to cause the wall to slide over its base. Unless other forces are brought into play to assist in the stability from the sliding point of view, the frictional resistance at the base must be able to balance the horizontal component thrust.

The earth in front of the base of the wall assists in lateral stability and the base may be tilted upwards towards the toe to reduce the possibility of horizontal movement.

Direction.—The direction of the resultant thrust provides an alternative method of checking sliding tendency. If 'R' makes a bigger angle with the normal to the base than the appropriate *angle of friction*, the wall will slide, assuming friction to be the only resisting force

Foundation Pressures

The theory given below is applicable to any horizontal section of the wall. We may have to investigate the stresses at the junction of a wall with its base. Similarly, stress computations may have to be effected for a section, say, at mid-depth of the wall to test the suitability of the width provided there. The formulae developed below will be just as applicable to the calculations of such stresses as they will be to the determination of the pressures produced at the wall base.

Let 'V' kN = vertical component of resultant thrust
    'b' m = width of base (or section considered)
    'e' m = eccentricity of resultant, i.e. the distance, from the centre of base, at which 'V' acts (Fig. 382).
Taking 1 m length of wall:

$$\text{Area} = (b \times 1) = b \text{ m}^2.$$

$$\text{Section modulus (Z)} = \frac{1 \times b^2}{6}, \quad \text{i.e.} \quad \left(\frac{breadth \times depth^2}{6} \right) = \frac{b^2}{6}.$$

$$\text{Direct stress} = \frac{\text{Load}}{\text{Area}} = \frac{V}{b} \text{ kN/m}^2.$$

$$\text{Bending stress} = \frac{M}{Z} = \frac{V \times e}{b^2/6} = \frac{6Ve}{b^2} \text{ kN/m}^2.$$

$$\text{Total compressive stress at A} = \frac{V}{b} + \frac{6Ve}{b^2} = \frac{V}{b} \left(1 + \frac{6e}{b} \right)$$

$$\text{Total compressive stress at B} = \frac{V}{b} - \frac{6Ve}{b^2} = \frac{V}{b} \left(1 - \frac{6e}{b} \right).$$

The stress units are 'kN/m^2' in this case.

Middle-third Law.—Considering the stress at 'B', i.e. at the edge of section remote from the eccentric load, if there is to be no tension $\frac{V}{b} \left(1 - \frac{6e}{b} \right)$ must not yield a negative value,

$$\therefore \quad \frac{6e}{b} \text{ must not exceed } 1$$

$$\text{i.e.} \quad e \text{ must not exceed } \frac{b}{6}$$

i.e., 'V' must not deviate to either side of the centre of the section more than $\frac{'b'}{6}$ which confines its position to the '*middle-third*' portion of the width of the section.

The usually accepted forms of stress or pressure-variation diagrams

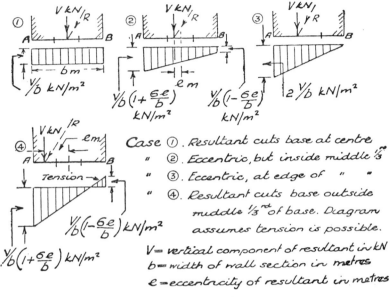

FIG. 382.—NORMAL STRESS DISTRIBUTION.

are given in Fig. 382. These follow directly from the application of the formula:

$$\frac{V}{b}\left(1 \pm \frac{6e}{b}\right).$$

Special case.—At the base of a wall resting freely on its foundation no possibility of tension exists. As previously stated, tension anywhere in a masonry structure is undesirable. If unavoidable it should be limited to about 0·4 kN/m². Where no tension is possible we have to amend the previous theory in order that the laws of equilibrium shall be satisfied.

The resultant up-thrust of the foundation must balance 'V' (see Fig. 383) and therefore must act in the same line. If 'V' act at distance 'a' from the toe of the wall, the form of the pressure triangle must be as indicated in order that its C.G. (through which the up-thrust acts) shall be positioned vertically below the point of application of 'V.'

If 'p' kN/m² be maximum pressure, the average pressure will be 'p/2' kN/m².

Area under pressure per metre run of wall

$$= (3a \times 1) = 3a \ \text{m}^2.$$

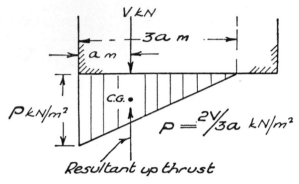

FIG. 383.—No Tension Possible.

$$\therefore \quad \frac{p}{2} \times 3a = V$$

$$\therefore \quad p = \frac{2V}{3a} \text{ kN/m}^2.$$

Safe foundation pressures:

Typical values of safe foundation pressures are given in Fig. 384. These values must be taken only as a general guide to safe bearing capacities.

	Load on ground kN/m²
Alluvial soil, made ground, very wet sand . . .	55
Soft clay, wet or loose sand	110
Ordinary fairly dry clay, fairly dry fine sand, sandy clay .	220
Firm dry clay	330
Compact sand or gravel, London blue or similar hard compact clay	440

FIG. 384.—Safe Bearing Pressures on Subsoils.

Example of Simple Trapezoidal Wall

A retaining wall, 4 m high, has a vertical back. It is 0·6 m wide at the top and 2·4 m wide at the bottom. It retains water which is at a level of 0·4 m from the top of the wall. The average density of the wall material is 17 kN/m³. Find whether there will be any tendency for up-lift to be developed at the base. Calculate the pressures on the foundation subsoil at the toe and heel of the wall respectively.

Water thrust per metre length of wall

$$= \tfrac{1}{2}wh^2 = \tfrac{1}{2} \times 10 \times 3\cdot6^2 \text{kN} = 64\cdot8 \text{ kN}.$$

Weight of wall per metre length

$$= \text{Volume} \times \text{Density} = \frac{0.6 + 2.4}{2} \times 4 \times 17 \, \text{kN} = 102 \, \text{kN}.$$

The centre of gravity of the wall section is conveniently found by the graphical method given on page 106.

The line of action of the resultant water thrust cuts the vertical line, drawn through the C.G. of the wall section, in the point 'O' (see Fig. 385). From 'O' a vector line is drawn horizontally to the left to represent

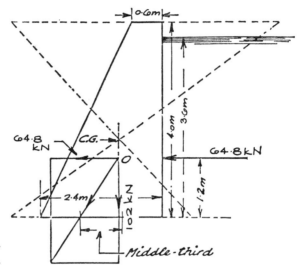

Fig. 385.—Stability of Retaining Wall.

64·8 kN. From 'O' vertically downwards is drawn a vector line to represent 102 kN. The parallelogram of forces is completed and the resultant of the two forces obtained. The resultant will be found to be a force of 121 kN. Its line of action cuts the base at the outer middle-third point. As the resultant does not come *outside the 'middle-third'* there will be no tendency for tension or up-lift at the base.

Foundation pressures.—The vertical component of the resultant is 102 kN, i.e. the weight of the wall.

Eccentricity of loading $(e) = 0.4$ m.

$$\text{Pressure at toe of wall} = \frac{V}{b}\left(1 + \frac{6e}{b}\right) = \frac{102}{2.4}\left(1 + \frac{6 \times 0.4}{2.4}\right) \text{kN/m}^2$$

$$= \frac{102}{2.4} \times 2 = 85 \, \text{kN/m}^2.$$

Pressure at heel of wall $= \dfrac{V}{b}\left(1 - \dfrac{6e}{b}\right) = \dfrac{102}{2\cdot4}\left(1 - \dfrac{6 \times 0\cdot4}{2\cdot4}\right)$

$$= \dfrac{102}{2\cdot4} \times 0 = 0, \text{ as we would expect.}$$

Note.—When the resultant thrust cuts the base at the outer middle-third point, the toe pressure is twice the average pressure, i.e. $2 \times \dfrac{V}{b}$, and the heel pressure is zero.

Example of Wall with Base

Show that the given retaining wall (Fig. 386) is stable from the point of view of (i) *tension at the base,* (ii) *sliding on its foundation. Assume the following data:*

Density of earth $= 14 \text{ kN/m}^3$
Angle of repose $= 30°$
Density of wall $= 21 \text{ kN/m}^3$
Coefficient of friction $= 0\cdot4$ (*between base and foundation*).

The first step in this example is to determine the centre of gravity of the wall section.

Let $\bar{x} =$ distance of C.G. from back of wall. Treating the upper wall as a rectangle plus a right-angled triangle

$$(2\cdot16 + 2\cdot16 + 2\cdot16)\bar{x} = (2\cdot16 \times 0\cdot3) + (2\cdot16 \times 1\cdot0) + (2\cdot16 \times 1\cdot2)$$
$$6\cdot48\bar{x} = 5\cdot4$$
$$\bar{x} = \tfrac{5}{6} \text{ m.}$$

The line of action of 'W' will therefore act at $\tfrac{5}{6}$ m from back of wall. *Earth thrust.*—The earth thrust per metre run of wall

$$= \tfrac{1}{2}wh^2\dfrac{1 - \sin\theta}{1 + \sin\theta}$$

$$\therefore \quad P = \tfrac{1}{2} \times 14 \times 4\cdot5^2 \times \dfrac{1 - \sin 30°}{1 + \sin 30°}$$

$$= (\tfrac{1}{2} \times 14 \times 20\cdot25 \times \tfrac{1}{3}) \text{ kN} = 47\cdot25 \text{ kN.}$$

Weight of wall.—The weight of wall per metre run (W)

$$= 6\cdot48 \text{ m}^3 \times 21 \text{ kN/m}^3 = 136\cdot1 \text{ kN.}$$

The resultant of 'P' and 'W' cuts the base inside the middle-third of the width, therefore there is no tendency for tension. The wall will exert a positive pressure all along the base.

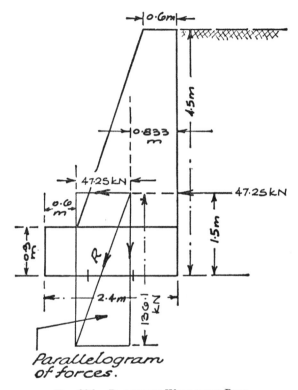

FIG. 386.—RETAINING WALL WITH BASE.

Test for sliding.—In all cases of walls subjected to horizontal thrusts from retained materials, the *vertical* component of the resultant thrust equals the *weight* of the wall.

∴ V = 136·1 kN, i.e. normal reaction = 136·1 kN.

Frictional resistance to sliding = 136·1 kN × 0·4 = 54·44 kN.

Horizontal component of resultant = 47·25 kN. As the frictional resistance exceeds the horizontal component, the wall is theoretically stable from the point of view of sliding.

The resistance to horizontal movement which is provided by the earth in front of the toe of the base (assuming the wall base to be sunken in the ground), and the increased normal pressure which the earth resting on the base projection would provide, are neglected in this problem. When all the resistances to sliding are added together the total resistance should not be much less than twice the horizontal component of the resultant thrust. For a full discussion of all the factors involved in the stability of retaining walls, the reader is referred to a publication issued by the Institution of Structural Engineers (see page 362).

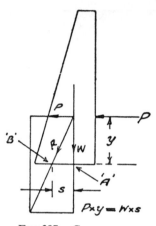

FIG. 387.—CALCULATION METHOD.

Calculation Method

Consider the wall in Fig. 387, which is subjected to a thrust 'P.' If 'P' were not acting, the line of action of 'R' would pass through the point 'A'. The effect of 'P,' therefore, is to displace or 'shift' the resultant a distance 's,' i.e. from point 'A' to point 'B.'

'W' and 'P' are the components of the resultant 'R.' The moment of force 'R' about 'B' is zero, therefore the moment of 'P' about point 'B' must be equal and opposite to that of force 'W.'

$$\therefore \quad P \times y = W \times s$$

$$\therefore \quad s = \frac{P \times y}{W}$$

's' is sometimes referred to as the 'shift' of the resultant force on the wall, due to the force 'P.'

EXAMPLE (i).—*Find the maximum height to which a 225 mm brick wall may be built if it is subjected to a wind pressure of 720 N/m² over its whole height. No tension is permissible in the wall. The density of the brickwork is 20 kN/m³.*

Let 'h' m = maximum height of wall.

Consider 1 m length of wall.

$$\text{Weight of wall} = \left(h \times 1 \times \frac{225}{1\,000} \right) \text{m}^3 \times 20 \text{ kN/m}^3 = 4{\cdot}5h \text{ kN.}$$

Total wind pressure on wall = $(720 \times h \times 1)$ N = $720h$ N.

The resultant wind force acts at $\dfrac{h}{2}$ m from the bottom.

$\therefore$ Overturning moment $= \left(720h \times \dfrac{h}{2}\right) = \dfrac{720}{2}h^2 = 360h^2$ N m.

In order that there shall be no tension, the eccentricity must not exceed $\dfrac{b}{6} = \dfrac{225 \text{ mm}}{6} = 37 \cdot 5$ mm.

Shift of 'W' $= \dfrac{P \times y}{W} = \dfrac{\text{Overturning moment}}{\text{Weight of wall}} = \dfrac{360h^2}{4\ 500h}$

$\dfrac{37 \cdot 5}{1\ 000}$ m $=$ maximum permissible shift

$\therefore \quad \dfrac{37 \cdot 5}{1\ 000} = \dfrac{360h^2}{4\ 500h}$

$$h = \dfrac{4\ 500 \times 37 \cdot 5}{360 \times 1\ 000} = 0 \cdot 47 \text{ m.}$$

EXAMPLE (ii).—*A retaining wall is to be built with a vertical back 6 m high. The widths at the top and bottom are to be 1 m and 2 m respectively. The earth to be retained weighs 18 kN/m³ and has an angle of repose of 30°. The earth is to be level with the top of the wall, Assuming the wall material to weigh 19 kN/m³, show that the proposed wall is liable to overturn.*

The following method may be found convenient for determining centre of gravity positions.

Let $\bar{x} =$ Distance of centre of gravity from back of wall.

Area		*Lever arm*	*Moment*
$6 \times 1 = 6$	$\times$	$0 \cdot 5$	$= 3$
$6 \times \frac{1}{2} = 3$	$\times$	$(1 + \frac{1}{3})$	$= 4$
9	$\times$	$\bar{x}$	$= 7$
		$\bar{x}$	$= \frac{7}{9} = 0 \cdot 78$ m.

Total earth thrust (P) $= \frac{1}{2}wh^2\dfrac{1 - \sin \theta}{1 + \sin \theta} = \frac{1}{2} \times 18 \times 6^2 \times \frac{1}{3}$

$= 108$ kN per metre of wall.

Overturning moment $=$ P $\times \dfrac{h}{3} = 108 \times \dfrac{6}{3}$ kN m $= 216$ kN m.

Area of wall section per metre = 9 m².

∴ Weight of wall per metre run = (9 × 19) kN = 171 kN.

$$\text{Shift of 'W'} = \frac{\text{Overturning moment}}{\text{Weight of wall}} = \frac{216}{171} \text{ m} = 1 \cdot 264 \text{ m.}$$

∴ The point at which the resultant cuts the base = (0·78 + 1·264) m = 2·044 m from back of wall.

This is outside the base, therefore the proposed wall is not stable against overturning.

EXAMPLE (iii).—*A retaining wall 4·5 m high, 1·2 m wide at the top, and 2·7 m wide at the base, retains water which is level with the top of the wall. If the wall material weighs 24 kN/m³, investigate the stability of the wall. Draw a diagram of soil pressure distribution at the base of the wall.*

Position of C.G. of wall ($\bar{x}$ = distance from back of wall):

Area		Lever arm		Moment
4·5 × 1·2 = 5·4	×	0·6	=	3·24
$4 \cdot 5 \times \dfrac{1 \cdot 5}{2} = 3 \cdot 375$	×	$\left(1 \cdot 2 + \dfrac{1 \cdot 5}{3}\right)$	=	5·74
$\overline{8 \cdot 775}$	×	$\bar{x}$	=	$\overline{8 \cdot 98}$
		$\bar{x}$	=	1·02 m.

$$\text{Resultant water thrust} = \tfrac{1}{2}wh^2 = \tfrac{1}{2} \times 10 \times 4 \cdot 5^2 \text{ kN}$$
$$= 101 \cdot 25 \text{ kN.}$$

$$\text{Overturning moment} = 101 \cdot 25 \times \frac{4 \cdot 5}{3} = 151 \cdot 87 \text{ kN m.}$$

Area of wall section = 8·775 m².

∴ Weight of wall per metre run = (8·775 × 24) kN = 211 kN.

$$\text{Shift of 'W'} = \frac{151 \cdot 87}{211} \text{ m} = 0 \cdot 72 \text{ m.}$$

∴ Point at which the resultant cuts the base = (1·02 + 0·72) m = 1·74 m from the back of wall, i.e. the resultant lies within the middle-third of the base width. No tendency for tension will occur at the wall base.

$$\text{Maximum soil pressure (at toe of wall)} = \frac{V}{b}\left(1 + \frac{6e}{b}\right).$$

'V' = 'W', the weight of the wall = 211 kN.

'b' = breadth of base = 2·7 m.

'e' = eccentricity of resultant = (1·74 − 1·35) m = 0·39 m.

$$\therefore \quad \text{Maximum soil pressure} = \frac{211}{2\cdot7}\left(1 + \frac{6 \times 0\cdot39}{2\cdot7}\right) \text{kN/m}^2$$

$$= \frac{211}{2\cdot7}\left(1 + \frac{2\cdot34}{2\cdot7}\right) = 146 \text{ kN/m}^2.$$

$$\text{Minimum soil pressure (at heel)} = \frac{V}{b}\left(1 - \frac{6e}{b}\right)$$

$$= \frac{211}{2\cdot7}\left(1 - \frac{2\cdot34}{2\cdot7}\right) \text{kN/m}^2 = 10\cdot4 \text{ kN/m}^2.$$

The pressure-variation diagram is shown in Fig. 388.

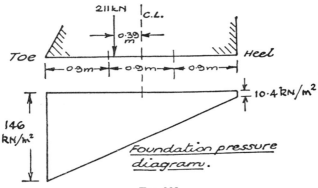

FIG. 388.

EXAMPLE (iv).—*Show by calculation method that the given retaining wall (Fig. 389) is just stable from the point of view of tension at the base. Earth density = 14·4 kN/m³. Angle of repose = 30°. Wall density = 20·8 kN/m³. Check the stability also by the graphical method. The foundation must not be subjected to a pressure exceeding 220 kN/m².*

The reader should verify that the following preliminary results are correct.

C.G. of wall is 0·864 m from the back.

P = 70 kN.   W = 171 kN.

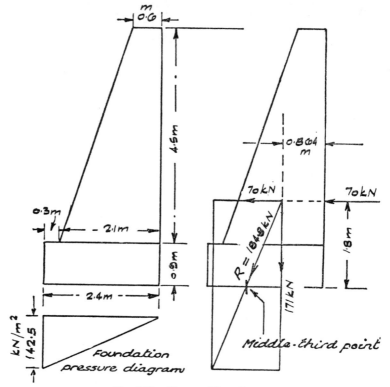

FIG. 389.—TEST OF WALL DESIGN.

Taking these worked values:

Shift of 'W' = $\dfrac{\text{Overturning moment}}{\text{Weight of wall}}$

$$= \frac{70 \times 1\cdot8}{171} \text{ m} = \frac{126}{171} \text{ m} = 0\cdot736 \text{ m}.$$

∴ Resultant thrust cuts base at $(0\cdot736 + 0\cdot864)$ m $= 1\cdot6$ m from heel of wall.

But outer middle-third point is $\frac{2}{3} \times 2\cdot4$ m $= 1\cdot6$ m from the heel.

∴ Resultant cuts base at outer middle-third point, i.e. the wall is just stable from the point of view of tension at base.

The maximum foundation pressure under the given conditions

$$= \frac{2V}{b} \text{ kN/m}^2 = \frac{2 \times 171}{2\cdot4} = 142\cdot5 \text{ kN/m}^2.$$

This is less than the maximum permissible ground pressure.

EXAMPLE (v).—*An 'L-shaped' retaining wall is 2·8 m high (Fig. 390). The base is 1·1 m wide and the wall is 0·5 m thick in each leg. The earth being retained weighs 17·5 kN/m³, and has an angle of repose of 35°. The earth is level with the top of the wall. The wall weighs 24 kN/m³. Test the wall for up-lift at the base. Draw the diagram of soil pressure distribution for the base of the wall. Construct the diagram of earth pressure variation for the back of the wall.*

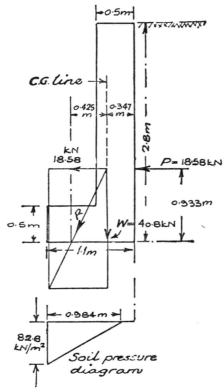

FIG. 390.—L-SHAPED RETAINING WALL.

C.G. of wall section—from vertical back:

Area		Lever arm		Moment
$2·8 \times 0·5 = 1·4$	×	0·25	=	0·35
$0·6 \times 0·5 = 0·3$	×	0·8	=	0·24
$1·7$	×	$\bar{x}$	=	0·59

$$\bar{x} = \frac{0·59}{1·7} = 0·347 \text{ m.}$$

$$\text{Resultant earth thrust} = \tfrac{1}{2}wh^2\frac{1 - \sin\theta}{1 + \sin\theta}$$

$$= \tfrac{1}{2} \times 17\cdot5 \times 2\cdot8^2 \times \frac{1 - \sin 35°}{1 + \sin 35°}$$

$$= \tfrac{1}{2} \times 17\cdot5 \times 7\cdot84 \times \frac{1 - 0\cdot5736}{1 + 0\cdot5736} = 18\cdot58 \text{ kN.}$$

Overturning moment = $(18\cdot58 \times 0\cdot933)$ kN m = $17\cdot34$ kN m.
Area of wall section = $1\cdot7$ m^2.
$\therefore$ Weight of wall per m run = $1\cdot7 \times 24 = 40\cdot8$ kN.

$$\text{Shift of 'W'} = \frac{17\cdot34}{40\cdot8} \text{ m} = 0\cdot425 \text{ m.}$$

$\therefore$ Point at which the resultant cuts base is $(0\cdot347 + 0\cdot425)$ m = $0\cdot772$ m from back of wall, i.e. outside the middle-third.

The appropriate formula (proved on page 348) will be used for the soil pressure.

Distance from point at which the resultant cuts the base to toe of wall = $a = (1\cdot1 - 0\cdot772)$ m = $0\cdot328$ m.

$$\therefore \quad 3 \times a = 3 \times 0\cdot328 = 0\cdot984 \text{ m.}$$

$$\text{Maximum soil pressure} = \frac{2V}{3a}$$

$$= \frac{2 \times 40\cdot8}{0\cdot984} = 82\cdot8 \text{ kN/m}^2 \text{ (see Fig. 390).}$$

$$\text{Maximum earth pressure behind the wall} = wh\frac{1 - \sin\theta}{1 + \sin\theta}$$

$$= \left(17\cdot5 \times 2\cdot8 \times \frac{1 - 0\cdot5736}{1 + 0\cdot5736}\right) \text{ kN/m}^2 = 13\cdot27 \text{ kN/m}^2.$$

The corresponding variation of pressure diagram is linear, of the type illustrated in Fig. 380. It is not shown in Fig. 390.

Retaining Walls Subjected to Inclined Thrusts

In Fig. 391 the thrust behind the wall is not horizontal. The line of action of the thrust is produced to cut the line of action of the weight and the parallelogram of forces is completed in the usual manner. A point to note in this case is that the value of 'V,' the vertical component of the resultant thrust is not equal to 'W,' the weight of the wall. It is greater than 'W' and will be found by resolving 'R' vertically in the usual manner. When friction between the retained earth and the back of the wall is taken into account, we get the type of inclined thrust indicated in Fig. 391.

Buttresses or Piers

The principles underlying the calculations in retaining wall problems may be applied to buttresses with horizontal or inclined thrusts. In such cases 1 m length of wall may not be representative, and it may be necessary to consider a portion of wall plus the buttress as a whole for calculation of weight and applied thrust.

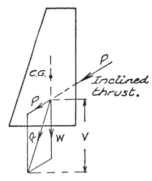

FIG. 391.—WALL WITH INCLINED THRUST.

Retaining Wall with Stepped Back.—In the example given in Fig. 392, the earth resting on the steps at the back of the wall assists in the stability of the wall. We may assume AB to be the *virtual* back of the wall. The value of 'W' in such a case would be the combined weight of the wall material and the supported earth. The line of action of 'W' would pass through the centre of gravity of the combined wall and earth masses.

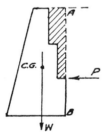

FIG. 392.—WALL WITH STEPPED BACK.

Retaining Wall with Superimposed Loading at the Back

In Fig. 393 the net pressure diagram behind a wall in such a case is illustrated.

AB = the head of earth which would produce the given superimposed pressure of 'p_1' kN/m².

If h_1 m = the head, $p_1 = wh_1$,

$\therefore \quad h_1 = p_1/w$ m.

The pressure diagram ACD indicates that the pressure behind the wall is represented by the trapezoidal diagram BEDC. This may be directly constructed by drawing BE = $p_1 \dfrac{1 - \sin \theta}{1 + \sin \theta}$ and

$$CD = wh\frac{1 - \sin \theta}{1 + \sin \theta} + p_1\frac{1 - \sin \theta}{1 + \sin \theta}.$$

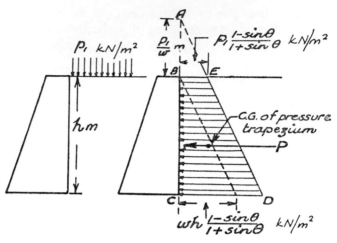

FIG. 393.—SUPERIMPOSED LOAD ON EARTH.

The effect of the super-imposed load is to increase the pressure everywhere on the back of the wall by an amount

$$p_1\frac{1 - \sin \theta}{1 + \sin \theta} \text{ kN/m}^2.$$

EXAMPLE.—*A retaining wall is* $3 \cdot 6$ *m high (Fig. 394). The earth retained has a density of* 15 kN/m^3 *and angle of repose* $30°$, *and is subjected to a superimposed load of* $7 \cdot 5 \text{ kN/m}^2$. *Calculate the overturning moment on the wall about the base per metre run of wall (i) by constructing the pressure diagram for the wall, (ii) by adding the component overturning moments due to retained earth and superimposed load respectively.*

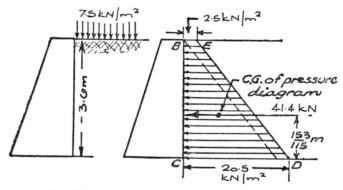

FIG. 394.—EXAMPLE INVOLVING SUPERIMPOSED LOAD.

Method (i):

The net pressure-variation diagram is shown in Fig. 394. The horizontal pressure equivalent to the superimposed load of 7.5 kN/m^2

$$= p_1 \times \frac{1 - \sin \theta}{1 + \sin \theta} = 7.5 \times \frac{1 - \frac{1}{2}}{1 + \frac{1}{2}} = 7.5 \times \frac{1}{3} = 2.5 \text{ kN/m}^2.$$

$$\therefore \quad \text{BE} = 2.5 \text{ kN/m}^2.$$

The pressure at wall base against the wall, due to retained earth,

$$= wh\frac{1 - \sin \theta}{1 + \sin \theta} = 15 \times 3.6 \times \frac{1}{3} = 18 \text{ kN/m}^2.$$

$$\therefore \quad \text{CD} = (18 + 2.5) = 20.5 \text{ kN/m}^2.$$

Employing the formula given on page 107 for the C.G. of a trapezium $\left(\bar{y} = \dfrac{a + 2b}{3(a + b)} \times 3.6 \text{ m} \right)$, the height of C.G. of trapezium above the base of wall

$$= \frac{20.5 + (2 \times 2.5)}{3(20.5 + 2.5)} \times 3.6 \text{ m} = \frac{25.5 \times 3.6}{3 \times 23} = \frac{30.6}{23} \text{ m.}$$

Total thrust against wall = Average pressure $\times$ (3.6 m)
$$= \frac{1}{2} (20.5 + 2.5) \times 3.6 \text{ kN} = 41.4 \text{ kN.}$$

Arm of overturning moment $= \dfrac{30.6}{23}$ m.

$\therefore$ Overturning moment about base

$$= \left(41.4 \times \frac{30.6}{23} \right) \text{ kN m} = 55.08 \text{ kN m.}$$

Method (ii):

Overturning moment due to retained earth alone

$$= \frac{1}{2}wh^2\frac{1 - \sin \theta}{1 + \sin \theta} \times \frac{h}{3} = \left(\frac{1}{2} \times 15 \times 12.96 \times \frac{1}{3} \times \frac{3.6}{3} \right) \text{ kN m}$$

$$= 38.88 \text{ kN m.}$$

The superimposed load causes a uniform pressure of 2·5 kN/m² over the wall back.

∴   Total thrust per metre of wall = (2·5 × 3·6) = 9 kN.

Arm of overturning moment = $\dfrac{3·6}{2}$ = 1·8 m.

∴   Overturning moment = 9 kN × 1·8 m = 16·2 kN m.
Total overturning moment = (38·88 + 16·2) = 55·08 kN m as before.

The second method is the quicker. If the position of the resultant thrust were required by this method, we would have to compound the two parallel thrusts due to retained earth and superimposed loads, acting respectively at $\frac{1}{3}$ and $\frac{1}{2}$ height of wall above the bottom.

Equivalent Liquid Pressure

Readers who take up the more advanced study of retaining walls will come across a method of treatment of earth pressures in which a liquid of calculated density is assumed to replace the given earth. By this means the horizontal pressure at any depth is obtained by the simple formula: pressure = density × depth. Taking the case of earth of density 16 kN/m³ and angle of repose 35° we have:

$$p = wh\frac{1 - \sin\theta}{1 + \sin\theta} = w\frac{1 - \sin\theta}{1 + \sin\theta} \times h$$

$$= 16 \times \frac{1 - \sin 35°}{1 + \sin 35°} \times h = 16 \times \frac{1 - 0·5736}{1 + 0·5736} \times h$$

∴ $p = 4·34h$ kN/m².

Horizontal pressures and thrusts may now be computed on the basis of a liquid of density 4·34 kN/m³.

Further treatment of retaining-wall problems is beyond the scope of this book. *Civil Engineering Code of Practice No. 2. Earth Retaining Structures* may be obtained from the Institution of Structural Engineers, 11, Upper Belgrave Street, London, SW1X 8BH.

The report deals with the question of friction between the retained material and the back of the wall. Walls are sometimes 'surcharged,' i.e. have to support material which is not level with the top of the wall. The surface inclination of the material to the horizontal may be any angle up to the 'angle of repose.' The practical consideration of this case and that of retaining walls with backs inclined to the vertical form part of the report.

EXERCISES 15

(1) A vertical wall, 2 m high, is subjected to a uniform wind pressure of 1 kN/m². Calculate the overturning moment about the wall base per metre run of wall.

(2) A retaining wall with a vertical back is 3 m high. It retains water the level of which is 0·2 m from the top of the wall. Taking the density of water to be 10 kN/m³, calculate the water thrust against the wall per metre run of wall.

(3) Earth, of density 15 kN/m³ and angle of repose 30°, is retained by a wall with a vertical back. The wall is 3·6 m high and the earth is level with the top. Calculate, on the basis of 1 m length of wall, (i) the resultant earth thrust, (ii) the overturning moment about a point in the wall base.

Draw a diagram showing the variation of horizontal pressure on the back of the wall.

(4) A retaining wall has a section in the form of a right-angled triangle. The vertical back is 3 m high and the base is 'x' m. The wall retains water which reaches to the top of the wall. Assuming the densities of the water and the wall to be 10 kN/m³ and 20 kN/m³ respectively, obtain the minimum permissible value of 'x' if there is to be no tendency for tension at the wall base.

(5) A retaining wall with a vertical back has a trapezoidal section. Taking the data given, obtain the resultant thrust at the wall base.

State whether or not the wall presses firmly on its foundation, from the toe to the heel of the base.

Width of wall at the top = 0·6 m.
Width of wall at the bottom = 1·2 m.
Height of wall = 2·8 m.
Density of retained gravel, which is level with top of wall, = 15 kN/m³.
Angle of repose = 40°.
Density of wall material = 22·5 kN/m³.
Note.—Sin 40° = 0·6428.

(6) (i) A wall of rectangular section, 0·45 m thick, is 1·5 m high. The density of the wall material is 16 kN/m³. Calculate the maximum permissible value of uniform horizontal wind pressure to which the wall may be subjected if no tension is to be developed at the wall base.

(ii) Calculate the maximum height to which a wall, of constant thick-

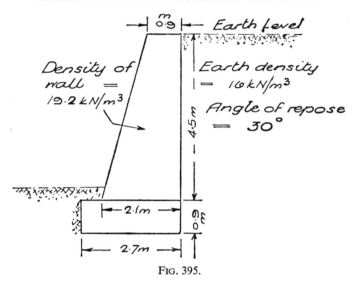

FIG. 395.

ness 0·6 m, may be built, without tension being developed, if the wind
pressure be 1 kN/m² normal to the wall. The wall density is 19·2 kN/m³.

(7) Given that the subsoil under the retaining wall shown in Fig. 395
can safely take a pressure of 165 kN/m² and that a coefficient of friction
of 0·5 may be assumed at the wall base, investigate the stability of the
wall.

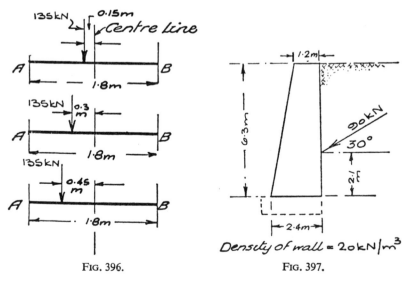

FIG. 396.   FIG. 397.

(8) Fig. 396 shows a rectangular base 'AB' subjected to a resultant thrust which has a vertical component of 135 kN per metre of wall. Obtain the foundation pressure at point 'A' of the base in each of the three cases given. No tension is possible at 'B.'

(9) Friction between the retained earth and the back of the wall is sometimes assumed. In this case the earth thrust is inclined to the horizontal.

Taking the example given in Fig. 397, obtain the maximum compressive stress at the junction of the upper wall and the base. Show that no tensile stress will be developed at this level in the wall.

[The earth thrust shown has been computed for a wall height of 6·3 m, i.e. for the portion of wall *above the horizontal section being considered*. Find 'V' as in Fig. 391 and treat as a vertical force acting at the point in which the resultant 'R' cuts the junction of upper wall and base.]

CHAPTER XVI

REVISION EXAMPLES WITH ABRIDGED SOLUTIONS

In the examples given in this chapter, unless otherwise stated or suitable data be given, the self-weight of members is to be neglected. All beams are assumed to be simply supported.

The solutions are given in abridged form, with sufficient detail to enable the reader to check the intermediate stages in his calculations. Readers unacquainted with trigonometry should use graphical methods where the calculation method involves this branch of mathematics.

The page numbers given in the solutions will enable a reference to be made to examples involving similar principles, previously worked in greater detail in the text.

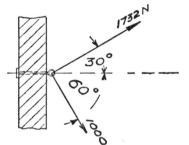

FIG. 398.

(1) A vertical brick wall, 1·5 m high, is 0·45 m thick. The wall is subjected to a horizontal wind pressure of 1·5 kN/m² of wall. Find the magnitude and direction of the resultant thrust at the wall base, taking the density of the brickwork to be 16 kN/m³. Calculations are to be based on 1 m length of wall.

(2) Verify that the withdrawal tendency on the bolt shown in Fig. 398 is wholly horizontal. Find the force of withdrawal.

(3) A pier is subjected to a vertical thrust of 800 N. It also supports a horizontal thrust of 'H' N. Calculate 'H' if the resultant thrust on the pier is 1 000 N.

(4) Assuming a slate weighing 30 N to become loose on a roof of 30° pitch, obtain the gravitational force urging it down the roof slope.

(5) Fig. 399 shows a bracket connection to a steel column. The position of the bracket load is such as to cause rivet 'A' to be subjected to a resultant load of 22·5 kN in the manner indicated. Find (i) the tensile load, (ii) the shear load, which the rivet has to carry.

(6) A simple triangular truss has the following dimensions: Bottom

tie = 15 m, left-hand rafter = 9 m, right-hand rafter = 12 m. It carries a vertical load of 'P' N at the apex. Assuming the left-hand rafter to be exerting a thrust of 1 000 N, obtain (i) the value of 'P', (ii) the thrust in the right-hand rafter.

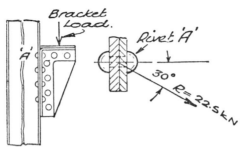

FIG. 399.

(7) The vertical reaction at the left end of a truss is 2 000 N. The main rafter and bottom tie make angles of 30° and 10° respectively with the horizontal. Using Lami's theorem, calculate the given member forces. Check the results by constructing a triangle of forces.

(8) At a certain top boom joint in a braced girder the centre lines of four members meet. The two horizontal boom members exert thrusts, the left-hand member 120 kN, and the right-hand member 150 kN. Find the force in each of the other two members, one of which is vertical and the other inclined downwards towards the right, at 45° to the horizontal. There is no external load at the joint.

(9) A vertical wall, 12 m high, is subjected to horizontal wind pressure as follows: 750 N/m², up to a level 9 m above ground, and 1 000 N/m² above this level. Calculate the overturning moment of this wind pressure about the base of the wall, per metre length of wall.

(10) A bolt, with its axis horizontal, is turned by a straight spanner. Assuming the maximum safe turning moment which may be applied to the bolt to be 20 kN m, calculate the greatest permissible distance, measured along the spanner from the centre of the bolt, at which a vertical force of 100 N may be applied, when the spanner makes an angle of 30° with the horizontal.

(11) In a testing machine a uniform straight bar with the fulcrum at the centre swings as a pendulum in a vertical plane. The bar carries two adjustable weights, 20 N near the bottom end and 10 N near the top. Calculate the moment of the couple which must be imposed on the pendulum to maintain it at an angle of 30° with the vertical, given that

the bottom and top weights have their C.G.s respectively at 200 mm and 100 mm from the fulcrum.

(12) A beam, 4·5 m overall length, is simply supported at two points 'A' and 'B.' 'A' is 0·6 m from the left end and 'B' is 0·9 m from the right end of the beam. The beam carries a 225 mm brick wall, of uniform height 1·2 m, throughout its length. Taking the density of the brickwork to be 18 kN/m³, calculate the reactions at 'A' and 'B' due to the weight of the wall.

(13) A beam (whose self-weight may be neglected) is simply supported at the left end and at a point 1·2 m from the right end. The overall length of the beam = 3·6 m. The beam carries a point load of 20 kN at 0·6 m from the left end. What additional load per metre run can the beam support, if the left-end reaction is not to exceed 45 kN?

(14) A simply supported beam 'AB,' of 6 m effective span, carries a point load of 80 kN at 1·2 m from 'A.' A uniform load of 33⅓ kN per metre extends 3·6 m into the span from the end 'B.'

Determine the support reactions due to these loads:

(i) by calculation;
(ii) by link polygon.

(15) The reactions for a beam, 4·5 m long, are at the left end and 1·5 m from the right end. The beam carries three loads: (i) 1·5 kN per metre over the whole beam length; (ii) 3 kN per metre for a distance of 1·5 m from the left end; and (iii) a point load of 15 kN at the extreme right end of the beam. Calculate the beam reactions.

(16) Fig. 400 shows a wall bracket. Calculate the magnitude of the reaction at 'A' and the magnitude and direction of the reaction at 'B.'

Check the results by a graphical method.

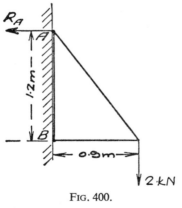

FIG. 400.

(17) A king-post truss of 8 m span is subjected to a uniform positive wind pressure on the left slope of the truss. The length of the principal rafter of the truss is 4·8 m and the trusses are spaced at 3 m centres. The reaction at the right end of the truss is vertical. If this reaction have a value of 3 240 N due to the wind, calculate the intensity of the wind pressure normal to the roof slope.

(18) A trapezoidal retaining-wall section with a vertical back is 1 m wide at the top, 2 m wide at the bottom and 5 m high. Obtain the distance of the centre of gravity of the wall section from the back of the wall

    (i) by calculation;
    (ii) by link polygon construction:
    (iii) by any other graphical construction.

(19) Find the centre of gravity of each of the structural sections given in Fig. 401.

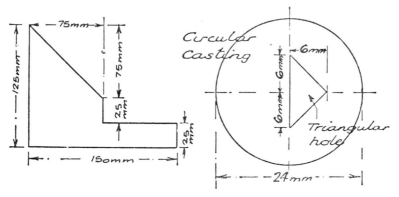

FIG. 401.

(20) A compound girder, 320 mm overall depth, is composed of an I-section with a plate 200 mm × 10 mm on the top flange, and a plate 300 mm × 10 mm on the bottom flange. The dimensions of the I-section are:

Flange width	= 150 mm
Flange thickness	= 20 mm
Depth	= 300 mm
Web thickness	= 10 mm.

Calculate the position of the neutral axis of the section, making no allowance for rivet holes.

(21) Draw the stress diagram for the frame given in Fig. 402. Draw up a table showing the force in each member of the frame, and the type of member.

(22) Forces of 1 kN, 5 kN, and 4 kN act vertically downwards at points 'A,' 'B,' and 'C' respectively on a horizontal beam. 'AB' = 4 m and 'BC' = 6 m.

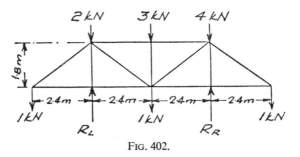

FIG. 402.

Find the magnitude, direction, and position (with respect to point 'A') of the resultant of the three forces:

      (i) by calculation:
      (ii) by means of a link polygon.

(23) In a travelling crane three wheel loads 'A,' 'B,' and 'C' traverse a girder. 'A' = 1 kN, 'B' = 3 kN, and 'C' = 2 kN. AB = 1 m and BC = 2 m. Assuming the fact that the load system will produce its maximum bending moment when the '3 kN' load and the 'C.G. of the three loads' lie equidistant on either side of the centre of the girder, find how far load 'A' is from the near end of the girder when the critical load setting for maximum bending moment is attained. The girder has a span of 10 m.

(24) A truncated right pyramid of square section is 100 mm × 100 mm at one end and 50 mm × 50 mm at the other and is used as a tie member. It is subjected to an axial pull of 1 800 N. Calculate the stress at the mid-section of the pyramid and also determine the maximum stress produced.

(25) (i) How many 20 mm dia. rivets in 22 mm dia. holes in single shear will be required to transmit a shear load of 660 kN ($f = 110$ N/mm²)?

(ii) A tie-bar in a frame is connected to the boom angles by two gusset plates, one on either side of the tie-bar. Four 24 mm dia. turned and fitted bolts are used. Calculate the safe axial load for the tie-bar from the point of view of the shear in the bolts in its end connection ($f_s = 95$ N/mm²).

(26) A timber post, 75 mm × 75 mm in section and 1·2 m in length, is subjected to an axial thrust of 9 kN. Taking Young's modulus for the timber to be 11·2 kN/mm², calculate the contraction in length of the post.

(27) An extensometer, employing a 200 mm gauge length, is attached to a mild steel tie-bar in a bridge truss. As a train passes over the bridge the extensometer registers a maximum extension of 0·05 mm. Assuming the cross-sectional area of the tie-bar to be 3 750 mm², estimate the live load created in it by the passing of the moving load. A suitable value for 'E' must be chosen.

(28) The concrete in a reinforced concrete circular water tank, on drying out after casting, has a shrinkage of 0·01 %, thus causing compressive stress in the steel reinforcement which is in the form of horizontal hoops. Calculate the value of the tensile stress which the filling-in of the water may tend to cause in the steel hoops without any residual tensile stress being created. E for steel = 210 kN/mm². It is assumed that there is no slip between the steel and surrounding concrete.

(29) A simply supported beam, of effective span = 'l,' carries a point load = 'W' at one-quarter of the span from one end. Prove that the maximum bending moment in the beam is given by the expression $\dfrac{3Wl}{16}$.

(30) A simply supported beam 'AB' has an effective span of 8 m. At points 'C,' 'D,' and 'E' on the beam there are concentrated loads of 2 kN, 4 kN, and 1 kN respectively. AC = 1 m, CD = 3 m, DE = 2 m.

Construct the bending moment and shear force diagrams for the beam.

(31) A simply supported beam of 10 m effective span carries a uniform load which develops a bending moment of 16 kN m at a section 2 m from the left end. Calculate the maximum bending moment and maximum shear force for the beam.

(32) A timber beam of 2·4 m effective span carries 0·9 m² of flooring. The inclusive floor load is 5·7 kN/m². Design a suitable section for the beam, making the depth three times the breadth. The maximum fibre stress in the timber is not to exceed 8·4 N/mm².

(33) A timber beam of 3 m effective span has to carry a point load of 2·46 kN at 1·2 m from the left end. Find a suitable depth for the beam assuming a breadth of 50 mm. The safe stress in bending for the timber is 7 N/mm².

(34) A timber beam has a breadth of 75 mm, a depth of 150 mm, and an effective span of 3 m. In addition to a point load of 2·5 kN at 1·2 m from the left end, it carries a uniform load of 1·2 kN. Calculate the maximum fibre stress in the timber.

(35) Calculate the maximum permissible span for a 75 mm × 225 mm

timber beam which carries 1·5 kN per metre run, given that the maximum stress must not exceed 8·4 N/mm² and that the maximum deflection is not to be greater than 1/360th of the span.

'E' for the timber = 11 kN/mm².

(36) A balcony is supported outside a building by timber cantilevers. The width of the balcony is 1·2 m and the total inclusive load carried is 10·5 kN/m² of floor area. If the cantilevers be 75 mm wide × 175 mm deep, calculate their maximum permissible spacing, if the fibre stress in the timber be limited to 7 N/mm².

(37) A concrete footing to a retaining wall projects 0·45 m beyond the toe of the wall. The footing is 0·6 m deep. The net upward pressure at the edge of the footing is 220 kN/m² and vertically beneath the toe of the upper wall the foundation pressure has fallen to 160 kN/m². Calculate the maximum tensile stress in the concrete.

(38) A U.B. has an effective span of 4·2 m. It carries a central point load of 35 kN and a uniform load of total value 70 kN. Taking 'f' = 165 N/mm², find the necessary section modulus for the beam.

(39) A 305 mm × 165 mm × 54 kg U.B. having an effective span of 6 m carries point loads of 'W' kN at 1·5 m from each end, and, in addition, a uniform load of 80 kN. Calculate the value of 'W' if the maximum fibre stress in the steel = 165 N/mm².

$$Z_{XX} \text{ for the U.B} = 752 \text{ cm}^3$$

(40) A steel beam carries three equal point loads of 64 kN each, at the quarter-points of an effective span of 5 m. Assuming a working stress of 165 N/mm², select a suitable section from the following list:

356 mm × 171 mm × 57 kg: Z_{XX} = 894 cm³.
381 mm × 152 mm × 60 kg: Z_{XX} = 968 cm³.
406 mm × 152 mm × 60 kg: Z_{XX} = 1 011 cm³.

(41) Taking the case of a 203 mm × 133 mm × 30 kg U.B., verify the rule that if a U.B. be fully loaded with uniform load so that f = 165 N/mm², the effective span must not exceed 16·97 times the beam depth, if the maximum deflection is to be limited to $\frac{1}{360}$th part of the span. E = 210 kN/mm². I_{XX} for U.B. = 2 880 cm⁴.

(42) A compound girder, weighing 2 kN per metre run, has an effective span of 6 m. The girder carries a superimposed uniform load of total value 28 kN. In addition the girder carries the reactions of secondary beams which are equivalent to two point loads of 240 kN each,

respectively, at 2 m from the end supports. The girder is 500 mm overall depth and $I_{XX} = 83\,385$ cm⁴. Calculate the maximum stress induced in the girder flanges.

(43) The steel corner post in a wire fence is of I-section. The flanges are 40 mm wide and 6 mm thick. The web thickness is 6 mm and the overall depth is 50 mm. The fence wires contain an angle of 120° at the post and the 'YY' axis of the steel section bisects the angle between the wires. Calculate the maximum permissible tension in the wires, assuming there are three horizontal wire circuits at 0·3 m, 0·6 m and 0·9 m respectively above the ground support. The post has no stays and the maximum stress must not exceed 120 N/mm². The tension in each wire is the same.

(44) A beam, simply supported at the ends, carries a uniform load of 1 kN per metre run. The effective span = 3 m.

Calculate the uniform load, in kN per metre run, which the beam can carry, in addition to the given load, assuming this new load to extend from the left support up to the centre of the span. The maximum bending moment in the beam is to be limited to 3·6 kN m.

Draw the B.M. and S.F. diagrams for the beam when carrying the calculated and given loads simultaneously.

(45) A beam of rectangular section, 50 mm wide × 150 mm deep, has an effective span of 2·4 m. The load carried by the beam is uniformly distributed and of 'W' N total value. At 25 mm beneath the top of beam, at a section 0·6 m from the left end, the compressive stress is 4·2 N/mm². Calculate the value of 'W.'

(46) An overhanging beam extends 2 m over the supports at both ends. It carries three equal concentrated loads, one at each end and one in the middle. Calculate the magnitude of the central span if the negative support moment is twice the maximum positive central-span moment.

(47) An overhanging beam of 5 m total length is supported at 1 m from each end. The beam carries a wall 225 mm thick and of average density 16 kN/m³. The elevation of the wall is an equilateral triangle with 5 m base. Calculate (i) the support bending moment, (ii) the maximum central-span bending moment, due to the weight of the wall.

(48) What is the least radius of gyration of a solid circular column section 140 mm diameter? Calculate the safe concentric load for a mild steel column of this section, assuming its effective length to be 4·2 m, given the following 'p_c' values:

l/r	'p_c' (N/mm^2)
100	79
110	69
120	60

(49) A column has to support a concentric load of 450 kN. The effective length of the column is estimated to be 3·75 m.

Select, from the sections given below, the most suitable U.C. column.

152 mm × 152 mm × 37 kg $I_{YY} = 709$ cm^4 Area = 47·4 cm^2.
203 mm × 203 mm × 46 kg $I_{YY} = 1\,539$ cm^4 Area = 58·8 cm^2.

The following 'p_c' values are supplied:

l/r	p_c (N/mm^2)	l/r	p_c (N/mm^2)
70	115	90	91
80	104	100	79

(50) A short masonry pillar of square section, 225 mm side, supports a vertical concentrated load of 'W' kN, having a single eccentricity of 37·5 mm with respect to one of the principal axes of the section. Calculate the value of 'W' if the maximum compressive stress it induces in the masonry is 4·2 N/mm^2.

(51) The resultant thrust at the base of a retaining wall is 120 kN per metre run of wall. It makes an angle of 30° with the vertical, and cuts the base at 1·1 m from the toe. The width at the base = 2·8 m. Draw a diagram showing the distribution of normal pressure on the subsoil under the base.

(52) A retaining wall with a vertical back is 4·5 m high. Calculate the resultant thrust on the back of the wall, per metre run of wall, under each of the given circumstances:

(i) Uniform wind pressure of intensity 1·5 kN/m^2.

(ii) Water, level with the top.

(iii) Earth, level with the top. The earth density is 14·4 kN/m^3 and angle of repose is 30°.

(53) The height of a retaining wall with a vertical back is 3·6 m. It retains earth, level with the top of the wall. The earth density is 16 kN/m^3 and its angle of repose is 30°. There is a superimposed load of 10 kN/m^2 on the earth at the back of the wall. Calculate the resultant overturning moment about the base of the wall.

(54) A boundary wall of rectangular section, 2 m high, is subjected

to a uniform horizontal wind pressure of 750 N/m². The density of the wall material is 20 kN/m³. Calculate the necessary thickness of the wall if no tension is to be developed at the base.

(55) Fig. 403 shows a braced girder carrying two uniform load systems, 8 kN per metre for the length of the girder and 16 kN per metre extending over two bays. The loads are carried by a floor system which transmits the given loads to the panel points of the girder. Calculate the force in each of the members marked 'A' and 'B' respectively.

[*Note.*—The load at a given panel point will be the sum of the loads taken up by proceeding half-way to the adjacent panel points on either side.]

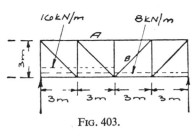

FIG. 403.

(56) A 254 mm × 254 mm × 89 kg U.C. section is used as a short column and is subjected to the following load system:

100 kN on the 'YY' axis, 200 mm eccentric with respect to XX axis.
 80 kN on the 'XX' axis, 75 mm eccentric with respect to YY axis.
480 kN concentric load.

Given $Z_{XX} = 1\,099$ cm³, $Z_{YY} = 379$ cm³, area of column section = 114 cm², calculate the maximum stress induced in the steel.

[For direct stress, add together the three loads and divide by area of section. For maximum bending stress, find the bending stress due to each eccentric load (with reference to the appropriate axis) and add the two stresses. Finally add the direct and total of bending stresses.]

(57) The column section given in Question 56 is used to carry a single load of 200 kN, which has an eccentricity of 150 mm with respect to the 'XX' axis and 50 mm with respect to the 'YY' axis. Calculate the maximum stress in the column.

[Add the direct stress to the combined bending stresses due to the two eccentricities.]

Solutions

[Page references are to examples involving similar principles.]

(1) Wind load = 2·25 kN. Weight of wall = 10·8 kN.
Resultant = $\sqrt{2\cdot25^2 + 10\cdot8^2} = \mathbf{11\cdot03}$ **kN**. If 'θ' = inclination to vertical, $\tan \theta = 2\cdot25/10\cdot8 = 0\cdot208$, $\theta = \mathbf{11° \ 45'}$. [*Page* 13.]

S.M.(M)—13

(2) $1\,732 \times \cos 60° = 1\,000 \cos 30°$.   ∴ No vertical component.
Horizontal force on bolt = $1\,732 \cos 30° + 1\,000 \cos 60° = \mathbf{2\,000\ N}$.
[*Page* 13.]
   (3) $H^2 + 800^2 = 1\,000^2$.   ∴ $H = \mathbf{600\ N}$. [*Page* 13.]
   (4) $30 \cos 60° = 30 \times \frac{1}{2} = \mathbf{15\ N}$. [*Page* 21.]
   (5) (i) Tensile load = $22·5 \cos 30° = \mathbf{19·5\ kN}$.
     (ii) Shear load = $22·5 \cos 60° = \mathbf{11·25\ kN}$. [*Page* 22.]
   (6) Solution by 'triangle of forces' or by 'Lami's theorem.' Lami's
theorem reduces to:

$$\frac{P}{\sin 90°} = \frac{1\,000}{12/15}.$$

$$\therefore\quad P = 1\,250\ N.$$

Also $\dfrac{P}{\sin 90°} = \dfrac{Q}{9/15}$ Which gives Q (force in right-hand rafter) = **750N**.
[*Pages* 34 *and* 325.]
   (7) Reducing to three forces acting away from the point of concurrence, the contained angles are 120°, 80°, and 160°.

$$\frac{2\,000}{\sin 160°} = \frac{\text{Rafter force}}{\sin 80°} = \frac{\text{Tie force}}{\sin 120°}.$$

From this, we get

Rafter force = 5 759 N.   Tie force = 5 064 N.   [*Page* 326.]
   (8) Inclined member is a tie. $T \cos 45° = 3$. $T = \mathbf{42·42\ kN}$.
Vertical member is a strut. $S = T \cos 45° = \mathbf{30\ kN}$ [*Pages* 22 *and* 41.]
   (9) Wind load on lower portion = **6·75 kN**.
     Wind load on upper portion = **3·0 kN**.
Overturning moment = $(6·75 \times 4·5) + (3 \times 10·5) = \mathbf{61·88\ kN\ m}$.
[*Page* 103.]
   (10) x mm = distance.   ∴ $100 \times x \times \cos 30° = 20\,000$.
∴ $x = \mathbf{231\ mm}$. [*Page* 47.]
   (11) $(20 \times 200 \sin 30°) - (10 \times 100 \sin 30°) = \mathbf{1\,500\ N\ mm}$. [*Page*
48.]
   (12) Total weight of brickwork = 21·87 kN, acting at C.G., which is
1·65 m from 'A.'

$R_A \times 3 = 21·87 \times 1·35.$ $R_A = \mathbf{9·84\ kN}.$
$R_B \times 3 = 21·87 \times 1·65.$ $R_B = \mathbf{12·03\ kN}.$ [*Page* 107.]

(13) If 'w kN' per metre be load
$45 \times 2 \cdot 4 = (20 \times 1 \cdot 8) + (3 \cdot 6w \times 0 \cdot 6).$ $\therefore$ $w = 33\frac{1}{3}$ kN **per metre**
[*Page* 59.]
(14) $R_A \times 6 = (80 \times 4 \cdot 8) + (120 \times 1 \cdot 8).$ $\therefore$ $R_A = $ **100 kN.**
   $R_B \times 6 = (80 \times 1 \cdot 2) + (120 \times 4 \cdot 2).$ $\therefore$ $R_B = $ **100 kN.**

[Treat uniform load as concentrated at the C.G. for link polygon construction.] [*Page* 124.]
(15) $(R_L \cdot 3) + (15 \times 1 \cdot 5) = (6 \cdot 75 \times 0 \cdot 75) + (4 \cdot 5 \times 2 \cdot 25).$
   $R_L = - $ **2·44 kN**, i.e. a downward force.
   $(R_R \times 3) = (6 \cdot 75 \times 2 \cdot 25) + (4 \cdot 5 \times 0 \cdot 75) + (15 \times 4 \cdot 5)$
   $R_R = $ **28·69** kN. [*Page* 108.]
(16) $R_A \times 1 \cdot 2 = 2 \times 0 \cdot 9.$ $\therefore$ $R_A = $ **1·5 kN.**
   $H_B = R_A = 1 \cdot 5$ kN. $V_B = 2$ kN.
   $R_B = \sqrt{1 \cdot 5^2 + 2^2} = $ **2·5 kN.**
   $\theta = $ inclination of 'R_B' to horizontal.
   $\tan \theta = V_B / H_B = 2/1 \cdot 5 = 1 \cdot 33$
      $\theta = $ **53°** **(nearly).** [*Page* 83.]
(17) $P = $ resultant wind load.
$P \times 2 \cdot 4 = 3\,240 \times 8.$ $\therefore$ $P = 10\,800$ N.

Wind pressure $= \dfrac{P}{4 \cdot 8 \times 3} = \dfrac{10\,800}{4 \cdot 8 \times 3} = 750$ N/m^2.

[*Page* 84.]
(18) $\bar{x}$ from back of wall.
$7 \cdot 5\bar{x} = (5 \times 0 \cdot 5) + (2 \cdot 5 \times 1\frac{1}{3}).$ $\therefore$ $\bar{x} = $ **0·78 m.** [*Page* 100.]
(19) (i) Dividing into a triangle and two rectangles of respective areas 2 812·5, 3 750, and 1 875 mm^2 and assuming C.G. at $\bar{x}$ from left edge and $\bar{y}$ from bottom:

$8\,437 \cdot 5 = (2\,812 \cdot 5 \times 25) + (3\,750 \times 37 \cdot 5) + (1\,875 \times 112 \cdot 5).$
   $\therefore$ $\bar{x} = $ **50 mm.**

$8\,437 \cdot 5\bar{y} = (2\,812 \cdot 5 \times 75) + (3\,750 \times 25) + 1\,875 \times 12 \cdot 5.$
   $\therefore$ $\bar{y} = $ **38·9 mm.** [*Page* 100.]

(ii) $\bar{x} = $ distance of C.G. from left edge:

$416 \cdot 4\bar{x} = (452 \cdot 4 \times 12) - (36 \times 14)$
$\bar{x} = $ **11·82 mm.** [*Page* 101.]

C.G. is 12 mm from bottom of section.

(20) $\bar{y}$ = height of C.G. above bottom, i.e. height of NA.
Area of I-section = 8 600 mm². Total area = 13 600 mm².
13 600$\bar{y}$ = (8 600 × 160) + (2 000 × 315) + (3 000 × 5)
    $\bar{y}$ = **148·6 mm**. [*Page* 102.]

(21)

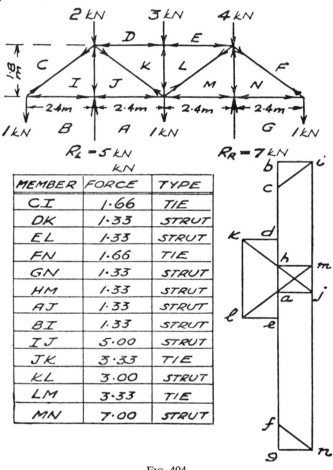

FIG. 404.

MEMBER	FORCE kN	TYPE
CI	1·66	TIE
DK	1·33	STRUT
EL	1·33	STRUT
FN	1·66	TIE
GN	1·33	STRUT
HM	1·33	STRUT
AJ	1·33	STRUT
BI	1·33	STRUT
IJ	5·00	STRUT
JK	3·33	TIE
KL	3·00	STRUT
LM	3·33	TIE
MN	7·00	STRUT

(22) Magnitude = (1 + 5 + 4) = **10 kN**.
Direction   = **vertically downwards.**
Position   = x m from 'A'
        $10x$ = (5 × 4) + (4 × 10)
        x = **6 m**. [*Pages* 71 *and* 118.]

(23) C.G. of load system is 1·5 m from 'A,' i.e. 0·5 m from 'B.' Therefore centre of span is 0·25 m from 'B' and 0·25 m from C.G. of system, measured in opposite directions. This setting will fix load 'A' as being **3·75 m** from the end of the girder nearer to it. [*Page 92.*]

(24) Area at mid-section   = 5 625 mm^2.

$$\text{Stress at mid-section} = \frac{1\,800}{5\,625} = 0\cdot32 \text{ N/mm}^2.$$

Smallest sectional area = 2 500 mm^2.

$$\therefore \quad \text{Maximum stress} = \frac{1\,800}{2\,500} = \textbf{0·72 N/mm}^2. \text{ [}\textit{Page 161.}\text{]}$$

(25) (i) Value of one rivet   = 41·8 kN.

$$\text{Number of rivets required} = \frac{660}{41\cdot8} = 16.$$

(ii) Value of one bolt   = 86 kN.
∴   Safe load for 4 bolts = 344 kN. [*Page 185.*]

(26) Stress $= \dfrac{9\,000}{5\,625} \simeq \textbf{1·6 N/mm}^2.$

$$\text{Strain} = \frac{x \text{ mm}}{1\,200 \text{ mm}}.$$

$$\text{E} = \frac{\text{Stress}}{\text{Strain}} = \frac{1\cdot6}{x/1\,200} = 11\,200 \text{ N/mm}^2.$$

x = contraction in length = **0·171 mm**. [*Page 176.*]

(27) $\text{Strain} = \dfrac{0\cdot05 \text{ mm}}{200 \text{ mm}} = 0\cdot00025.$

Stress = E × strain = 210 × 0·00025
          = 0·0525 kN/mm^2.
Load created = (0·0525 × 3 750) kN = **197 kN**. [*Page 176.*]

(28) Strain = 0·01/100 = 0·0001.
Stress in   steel = E × strain = (210 × 0·0001) = **0·021 kN/mm**^{2}. This compressive stress must be neutralised before any tension is developed. It is assumed there is no relative slip between steel reinforcing bars and the concrete. [*Page 176.*]

(29) $R_L = \dfrac{W \times \frac{3}{4}l}{l} = \frac{3}{4}W.$ B.M.$_{max} = \frac{3}{4}W \times \dfrac{l}{4} = \dfrac{3Wl}{16}.$

[*Page* 207.]

(30) $R_A = 4 \text{ kN.}$ $R_B = 3 \text{ kN.}$
B.M.$_C = (4 \times 1) = \textbf{4 kN m.}$
B.M.$_D = (4 \times 4) - (2 \times 3) = \textbf{10 kN m.}$
B.M.$_E = (3 \times 2) = \textbf{6 kN m.}$ [*Page* 210.]

(31) '*w*' **kN** per metre = uniform load.

$$R_A = \dfrac{10w}{2} = 5w \text{ kN.}$$

$\therefore$ $(5w \times 2) - (2w \times 2/2) = 16.$ $\therefore$ **w = 2 kN per metre.**
$W = 2 \times 10 = 20 \text{ kN.}$

$$\text{B.M.}_{max} = \dfrac{Wl}{8} = \dfrac{20 \times 10}{8} = \textbf{25 kN m.}$$

$$\text{S.F.}_{max} = \pm \dfrac{W}{2} = \pm \textbf{10 kN.}   [\textit{Page } 213.]$$

(32)   B.M.$_{max} = \dfrac{5 \cdot 13 \times 2 \cdot 4 \times 1\,000}{8} = 1\,539 \text{ kN m.}$

$8 \cdot 4 \times \dfrac{bd^2}{6} = 1\,539 \times 1\,000$

$b = \textbf{50 mm}$ and $d = \textbf{150 mm.}$ [*Page* 245.]

(33) $R_A = 1 \cdot 48 \text{ kN.}$ B.M.$_{max} = (1 \cdot 48 \times 1 \cdot 2) = 1 \cdot 78 \text{ kN m.}$

$1 \cdot 78 \times 10^6 = \dfrac{7 \times 50 \times d^2}{6}.$

$b = 50 \text{ mm.}$ $d = 175 \text{ mm.}$
Suitable depth = **175 mm.** [*Page* 245.]

(34) $R_A = 2 \cdot 1 \text{ kN.}$
B.M.$_{max}$ occurs at the point-load position.
B.M.$_{max} = (2 \cdot 1 \times 1 \cdot 2) - (0 \cdot 48 \times 0 \cdot 6) = 2 \cdot 232 \text{ kN m.}$

$2 \cdot 232 \times 10^6 = \dfrac{f \times 75 \times 150 \times 150}{6}.$ $\therefore$ $= \textbf{7·94 N/mm}^2.$

[*Page* 266.]

(35) '*l*' m = max. permissible span.

$$Strength: \frac{1 \cdot 5l \times l \times 10^6}{8} = \frac{8 \cdot 4 \times 75 \times 225^2}{6}.$$

$$\therefore \quad l = 5 \cdot 385 \text{ m.}$$

$$Deflection: y = \frac{5}{384} \frac{Wl^3}{EI}.$$

$$\frac{l \times 1\,000}{360} = \frac{5}{384} \times \frac{1 \cdot 5l \times l^3 \times 1\,000^3}{11 \times 71 \cdot 2 \times 10^6}. \qquad \therefore \quad l = 4 \cdot 82 \text{ m.}$$

$$\therefore \quad \text{Max. span} = \textbf{4·82 m.} \; [\textit{Page } 274.]$$

(36) Safe total load per cantilever = W N.

$$W \times 0 \cdot 6 \times 1\,000 = \frac{7 \times 75 \times 175 \times 175}{6}.$$

$$\therefore \quad W = 4\,466 \text{ N.} \qquad x \text{ m} = \text{spacing (centres).}$$

$$\therefore \quad \frac{4\,466}{1 \cdot 2x} = 10\,500.$$

$$x = 0 \cdot 354 \text{ m} = \textbf{354 mm} \text{ (centres). } [\textit{Page } 268.]$$

(37) Distance of C.G. of pressure diagram from toe of upper wall, i.e. the arm of bending moment,

$$= \frac{a + 2b}{3(a + b)} \times l = \frac{160 + 440}{3(380)} \times 0 \cdot 45 = 0 \cdot 2368 \text{ m.}$$

$$\text{Upward thrust} = \left[\left(\frac{220 + 160}{2} \right) \times (0 \cdot 45 \times 1) \right] \text{kN} = 85 \cdot 5 \text{ kN.}$$

$$\therefore \quad 85 \cdot 5 \times 0 \cdot 2368 \times 10^6 = f \times \frac{1\,000 \times 600 \times 600}{6}.$$

$$\therefore \quad f = \textbf{0·3375 N/mm}^2. \; [\textit{Pages } 106 \textit{ and } 246.]$$

(38) Max. B.M. $= \left(\dfrac{35 \times 4 \cdot 2}{4} \right) + \left(\dfrac{70 \times 4 \cdot 2}{8} \right) = 73 \cdot 5 \text{ kN m.}$

$$\therefore \quad Z = \frac{73 \cdot 5 \times 1\,000}{165} = 446 \text{ cm}^3. \; [\textit{Page } 266].$$

(39) Available B.M. $= \dfrac{165 \times 752}{1\,000} = 124 \cdot 1$ kN m.

B.M.$_{\text{max}}$ due to uniform load $= \dfrac{80 \times 6}{8} = 60$ kN m.

$\therefore$ The two loads 'W' must create a B.M. $= (124 \cdot 1 - 60) = 64 \cdot 1$ kN m.

$\therefore$ W $\times 1 \cdot 5 = 64 \cdot 1$.   W $= \mathbf{42 \cdot 7}$ **kN.** [*Page* 268.]

(40) R$_A$ = 96 kN.

B.M.$_{\text{max}}$ $= (96 \times 2 \cdot 5) - (64 \times 1 \cdot 25) = 160$ kN m.

$Z = \dfrac{160 \times 1\,000}{165} = 970$ cm^3.

Use 406 mm $\times$ 152 mm $\times$ 60 kg U.B. which is the lightest section that is adequate. [*Page* 266.]

(41) $16 \cdot 97 \times$ beam depth $= 16 \cdot 97 \times 207$ mm $= 3\,513$ mm.

Safe load $=$ W kN. $M = f\dfrac{I}{y}$.

$\dfrac{W \times 3\,513}{8} = \dfrac{165 \times 2\,880 \times 10\,000}{1\,000 \times 103 \cdot 5}$.

$\therefore$ W $= 104 \cdot 6$ kN. [*Page* 255.]

Max. deflection $= \dfrac{5}{384} \times \dfrac{104 \cdot 6 \times 1\,000 \times 3\,513^3}{210 \times 2\,880 \times 10^7} = 9 \cdot 76$ mm.

Max. deflection permissible $= \dfrac{3\,513}{360} = \mathbf{9 \cdot 76}$ **mm.** [*Page* 274.]

(42) Self-weight of girder $= 6 \times 2 = 12$ kN.

Total U.D. load $= (12 + 28) = 40$ kN.

Total B.M.$_{\text{max}}$ $= \left(\dfrac{40 \times 6}{8}\right) + (240 \times 2)$ kN m $= 510$ kN m.

$Z = \dfrac{I}{y} = \dfrac{83\,385 \times 10^4}{250} = 333 \cdot 54 \times 10^4$ mm^4.

$\therefore$ $f = \dfrac{510 \times 1\,000 \times 1\,000}{3\,335\,400} = \mathbf{153}$ **N/mm^2.** [*Page* 270.]

(43)

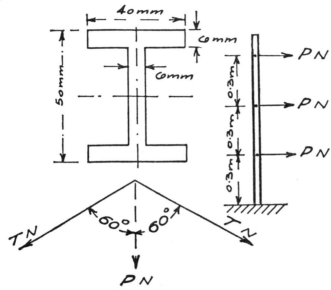

FIG. 405.

I_{XX} for section $= \dfrac{40 \times 50^3}{12} - \dfrac{34 \times 38^3}{12} = 261\ 196$ mm^4.

$Z_{XX} = \dfrac{I_{XX}}{y} = \dfrac{261\ 196}{25} = 10\ 448$ mm^3.

$(P \times 900) + (P \times 600) + (P \times 300) =$ max. B.M. $= fZ = 120 \times 10\ 448$.
$P = 697$ N.
$2T \cos 60° = 697$.

∴ $T =$ max. permissible tension $= $ **697 N.** [*Page* 252.]

(44) B.M.$_{max}$ will occur in portion of beam carrying the additional load.

Let 'w' kN per metre run $=$ additional load.
$R_A \times 3 = (3 \times 1\cdot5) + (1\cdot5w \times 2\cdot25)$.   ∴ $R_A = (1\cdot5 + 1\cdot125w)$ kN.

Position of B.M.$_{max}$ (from left end) $= \dfrac{1\cdot5 + 1\cdot125w}{w + 1}$ m.

S.M.(M)—13*

$$\text{B.M.}_{\cdot\text{max}} = \left[(1\cdot5 + 1\cdot125w) \times \left(\frac{1\cdot5 + 1\cdot125w}{w + 1} \right) - (1\cdot5 + 1\cdot125w) \right.$$

$$\left. \times \left(\frac{1\cdot5 + 1\cdot125w}{2(w + 1)} \right) \right] = (1\cdot5 + 1\cdot125w) \times \left(\frac{1\cdot5 + 1\cdot125w}{2(w + 1)} \right) = 3\cdot6.$$

$$\therefore \quad w = \mathbf{4 \ kN/m.}$$

Max. B.M. due to 1 kN/m load $= \dfrac{3 \times 3}{8} = 1\cdot125$ kN m.

Treating 4 kN/m load as concentrated at centre (for construction purposes) B.M. at 0·75 m from left end $= \dfrac{6 \times 0\cdot75 \times 2\cdot25}{3} = 3\cdot375$ kN m.

$R_L \times 3 = (6 \times 2\cdot25) + (3 \times 1\cdot5).$ $\therefore$ $R_L = 6$ kN.
$R_R \times 3 = (6 \times 0\cdot75) + (3 \times 1\cdot5).$ $\therefore$ $R_R = 3$ kN.
B.M. and S.F. diagrams are shown in Fig. 406. [*Page* 229.]

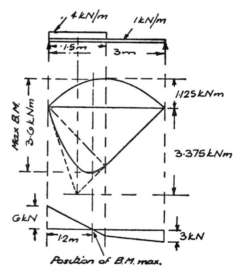

FIG. 406.

(45) If 'f' N/mm^2 be max. stress at given section,

$$\frac{f}{75} = \frac{4\cdot2}{50}. \qquad \therefore \quad f = 6\cdot3 \text{ N/mm}^2.$$

$$\left(\frac{W}{2} \times 600\right) - \left(\frac{W}{2 \cdot 4} \times 0 \cdot 6 \times 0 \cdot 3 \times 1\,000\right) = \frac{fbd^2}{6}$$

$$= \frac{6 \cdot 3 \times 50 \times 150 \times 150}{6}$$

W = **5 250** N. [*Page* 269.]

(46) W = each load. Central-span moment = $\frac{1}{2}$ the support moment. If '*l*' = central span.

$$\left[\left(\frac{3W}{2} \times \frac{l}{2}\right)\right] - \left[W \times \left(\frac{l}{2} + 2\right)\right] = \frac{W \times 2}{2}$$

$\therefore$ l = **12 m**. [*Page* 221.]

(47) Total weight of wall = $\dfrac{225}{1\,000} \times \dfrac{5 \times 0 \cdot 866 \times 5}{2} \times 16$

$$= 38 \cdot 98 \text{ kN}.$$

B.M. at **mid-span** = $(19 \cdot 49 \times 1 \cdot 5) - \left(19 \cdot 49 \times \dfrac{2 \cdot 5}{3}\right)$ = **12·99 kN m.**

Triangular load acts through its C.G.

B.M. at support. Height of all at support = 1 tan 60° = 1·732 m.

Weight of overhanging wall = $\left(\frac{1}{2} \times 1 \times 1 \cdot 732 \times \dfrac{225}{1\,000} \times 16\right)$

$$= 3 \cdot 12 \text{ kN}.$$

B.M. at **support** = $3 \cdot 12 \times \frac{1}{3}$ = **1·04 kN m.**

(48) Least radius = $\dfrac{d}{4} = \dfrac{140 \text{ mm}}{4}$ = 35 mm.

$$\frac{-}{r} = \frac{4 \cdot 2 \times 1\,000}{35} = 120. \qquad p_c = 60 \text{ N/mm}^2.$$

Safe concentric load = $\dfrac{60}{1\,000} \times \dfrac{\pi \times 140^2}{4}$ = **923·6 kN.** [*Page* 296.]

(49) Try 152 mm × 152 mm × 37 kg section.

r_{yy} = 3·87 cm. $\dfrac{l}{r} = \dfrac{375}{3 \cdot 87}$ = 96·9.

p_c by interpolation = 82·7 N/mm².

Safe axial load $= \dfrac{82 \cdot 7 \times 100 \times 47 \cdot 4}{1\,000} =$ **392 kN.**

∴   The section is not suitable.
Try **203 mm × 203 mm × 46 kg section.**
$r_{yy} = 5 \cdot 11$ cm.

$$\frac{l}{r} = \frac{375}{5 \cdot 11} = 73 \cdot 4.$$

p_c by interpolation $= 111$ N/mm^2.

Safe axial load $= \dfrac{111 \times 100 \times 58 \cdot 8}{1\,000} =$ **653 kN.**

Hence this **section is suitable.**   [*Page* 297.]

(50) $\dfrac{W}{50\,625} + \dfrac{W \times 37 \cdot 5}{Z} = \dfrac{4 \cdot 2}{1\,000}$

$$Z = \frac{bd^2}{6} = \frac{225 \times 225 \times 225}{6} = 1\,898\,437 \cdot 5 \text{ mm}^3.$$

$\dfrac{W}{50\,625} + \dfrac{W \times 37 \cdot 5}{1\,898\,437 \cdot 5} = \dfrac{4 \cdot 2}{1\,000}.$ ∴ $W =$ **106·3 kN.**

[*Page* 303.]
Alternatively, as the load is at the edge of the middle-third, the

average stress $= \dfrac{4 \cdot 2}{2} = 2 \cdot 1$ N/mm^2.

∴ $W = \dfrac{2 \cdot 1 \times 50\,625}{1\,000} =$ **106·3 kN.** [*Page* 347.]

(51) Normal component of resultant thrust $= 120 \cos 30° = 103 \cdot 92$ kN.

Pressure at toe $= \dfrac{V}{b}\left(1 + \dfrac{6e}{b}\right) = \dfrac{103 \cdot 92}{2 \cdot 8}\left(1 + \dfrac{6 \times 0 \cdot 3}{2 \cdot 8}\right)$

$\qquad\qquad\quad =$ **60·97 kN/m^2.**

Pressure at heel $= \dfrac{V}{b}\left(1 - \dfrac{6e}{b}\right) = \dfrac{103 \cdot 92}{2 \cdot 8}\left(1 - \dfrac{6 \times 0 \cdot 3}{2 \cdot 8}\right)$

$\qquad\qquad\quad =$ **13·26 kN/m^2.** [*Page* 355.]

(52) Wind thrust $= (1\cdot5 \times 4\cdot5 \times 1) = \textbf{6·75 kN}$ acting at **2·25 m** from the bottom of the wall.

Water thrust $= \frac{1}{2}wh^2 = \frac{1}{2} \times 10 \times 4\cdot5^2 = \textbf{101·25 kN}$ acting at **1·5 m** from bottom of wall.

Earth thrust $= \frac{1}{2}wh^2\dfrac{1 - \sin\theta}{1 + \sin\theta} = \frac{1}{2} \times 14\cdot4 \times 4\cdot5^2 \times \frac{1}{3} = \textbf{48·6 kN}$.

acting at **1·5 m** from bottom of wall. [*Pages* 332, 333, 342.]

(53) Earth thrust $= 34\cdot56$ kN (retained earth only).
Overturning moment $= 34\cdot56$ kN $\times 1\cdot2$ m $= 41\cdot47$ kN m.
Thrust against wall due to superimposed load $= (10 \times \frac{1}{3} \times 3\cdot6) = 12$ kN.
Overturning moment $= 12$ kN $\times 1\cdot8$ m $= 21\cdot6$ kN m.
Total overturning moment $= \textbf{63·07 kN m}$. [*Page* 360.]

(54) x m $=$ thickness of wall.
$$W = 2 \times x \times 1 \times 20 = 40x \text{ kN}.$$

$$P = 2 \times 1 \times \frac{750}{1\,000} = 1\cdot5 \text{ kN}.$$

$40x \times x/6 = 1\cdot5 \times 1$.
$$x = \textbf{0·473 m}. \; [\textit{Page } 352.]$$

(55) $R_L \times 12 = (72 \times 9) + (48 \times 6) + (24 \times 3)$
$R_L = 84$ kN (net reaction).
$R_R \times 12 = (24 \times 9) + (48 \times 6) + (72 \times 3)$
$R_R = 60$ kN (net reaction).

$(F_A \times 3) + (72 \times 3) = (84 \times 6)$. [Moments about opposite joint.]
$\therefore$ $F_A = \textbf{96 kN}$. Member 'A' is a strut.
$F_B = $ S.F. $\times \sqrt{2} = (60 - 24)\sqrt{2} = 36\sqrt{2}$ kN.
    $= \textbf{50·9 kN}$. Member 'B' is a tie. [*Page* 315.]

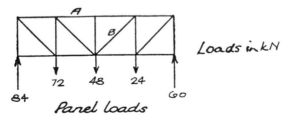

Loads in kN

Panel loads

FIG. 407.

(56) Direct stress $= \dfrac{\text{Total load}}{\text{Area of section}} = \dfrac{660 \times 1\,000}{114 \times 100}$ N/mm^2

$\qquad\qquad\qquad = 57{\cdot}9$ N/mm^2.

Bending stresses:

XX axis. $f = \dfrac{M}{Z} = \dfrac{100 \times 1\,000 \times 200}{1\,099 \times 1\,000}$ N/mm$^2 = 18{\cdot}2$ N/mm^2.

YY axis. $f = \dfrac{M}{Z} = \dfrac{80 \times 1\,000 \times 75}{379 \times 1\,000}$ N/mm$^2 = 15{\cdot}8$ N/mm^2.

Max. compressive stress in steel

$\qquad = (57{\cdot}9 + 18{\cdot}2 + 15{\cdot}8)$ N/mm^2
$\qquad = \mathbf{91{\cdot}9\ N/mm^2}.\ [\mathit{Page}\ 304.]$

(57) Direct stress $= \dfrac{\text{Load}}{\text{Area}} = \dfrac{200 \times 1\,000}{114 \times 100}$ N/mm$^2 = 17{\cdot}54$ N/mm^2.

Bending stresses:

XX axis. $f = \dfrac{M}{Z} = \dfrac{200 \times 1\,000 \times 150}{1\,099 \times 1\,000}$ N/mm$^2 = 27{\cdot}3$ N/mm^2

YY axis. $f = \dfrac{M}{Z} = \dfrac{200 \times 1\,000 \times 50}{379 \times 1\,000}$ N/mm$^2 = 26{\cdot}4$ N/mm^2.

Max. compressive stress $= (17{\cdot}54 + 27{\cdot}3 + 26{\cdot}4)$ N/mm^2
$\qquad\qquad\qquad\qquad\quad = \mathbf{71{\cdot}2\ N/mm^2}.$

CHAPTER XVII

TEST PAPERS WITH WORKED SOLUTIONS

AT the end of the chapter the numerical answers to the questions set in the test papers are first given. The answers are then fully worked out but the calculations are given in concise form in order to keep the chapter within reasonable limits. When points of theory are raised in a question, page references are given in the solutions.

Test Paper No. 1

(1) Define the term 'centre of gravity.'

Calculate the position of the centre of gravity of the plated T-section given in Fig. 1. Check the calculated position by drawing a link polygon.

(2) A floor is to be composed of timber joists placed at 300 mm centres. The total inclusive floor load is 3·8 kN/m² of floor. The effective span of the joists is 3 m. Assuming a working stress of 7 N/mm² for the timber and a joist breadth of 50 mm, calculate the necessary depth for the joists. Draw the bending moment and shear force diagrams for one of the joists.

(3) Briefly indicate the principal differences in properties between liquid pressure and earth pressure. Calculate the thickness of the wall (Fig. 2) so that it may be just stable from the point of view of tension at the base 'AB.' Verify the calculated thickness by a graphical method.

(4) Fig. 3 gives a sketch graph of the results of a tensile test on a mild steel specimen.

Calculate (i) Young's modulus, (ii) ultimate stress, (iii) the total extension expressed as a percentage of the gauge length.

Using a factor of safety of 4, obtain the safe axial load for a tie-bar of this quality steel, 100 mm wide × 20 mm thick.

(5) Draw the stress diagram for the roof truss given in Fig. 4. Distinguish between 'struts' and 'ties' by writing, alongside the members, a plus sign for a strut and a minus sign for a tie.

Write down the force in the member indicated by a cross.

(6) Determine the position and magnitude of the maximum bending moment in the beam shown in Fig. 5.

Draw the shear force diagram for the beam.

(7) Show, graphically and by calculation method, that the concurrent system of forces given in Fig. 6 has no resultant vertical effect. What force, introduced into the system, would reduce it to equilibrium?

(8) Obtain, by means of a link polygon, the support reactions for the simply supported beam given in Fig. 7. Check the graphical results by calculation method.

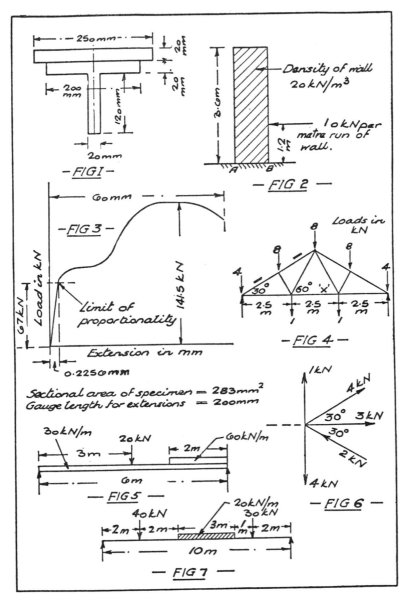

Test Paper No. 2

(1) Give the important assumptions upon which the theory of bending is based.

Draw the bending moment and shear force diagrams for the beam shown in Fig. 1.

Calculate the necessary section modulus for the beam. Maximum permissible stress = 6 N/mm².

(2) Draw the stress diagram for the cantilever truss given in Fig. 2.

Tabulate the forces in the respective members of the truss. Distinguish between the members which are in tension and those which are in compression.

(3) Explain briefly the meaning of the following: (i) compressive strain, (ii) shear stress, (iii) working stress, (iv) Young's modulus.

An extensometer, fitted for use with either 200 mm or 250 mm gauge length, was used to determine 'E' for a mild steel specimen. The following test results were recorded:

$$\text{Sectional area of specimen} = 380 \text{ mm}^2.$$
$$\text{Axial load applied} = 19\cdot95 \text{ kN.}$$
$$\text{Corresponding extension} = 0\cdot05 \text{ mm.}$$

Assuming 'E' to be 210 kN/mm², find the gauge length actually used.

(4) Express, by calculation method, the conditions of equilibrium for a non-concurrent co-planar system of forces.

Calculate the force in each bar of the loaded frame given in Fig. 3.

(5) The two steel beams given in Fig. 4 have the same sectional area.

Demonstrate the economy of the rolled steel joist type of section by comparing the total uniformly distributed loads the respective beams will carry for the same effective span.

(6) Explain the meaning of the term 'bearing' as applied to rivets in a riveted joint.

Calculate the value of one rivet in the following circumstances: rivet dia. = 20 mm in 22 mm hole, plate thickness = 16 mm, rivet in double shear. f_s = 100 N/mm² and f = 290 N/mm².

Obtain the maximum safe value for 'P' in the joint given in Fig. 5. Assume the rivet stresses given above and 155 N/mm² as the working stress in tension in the tie-bars.

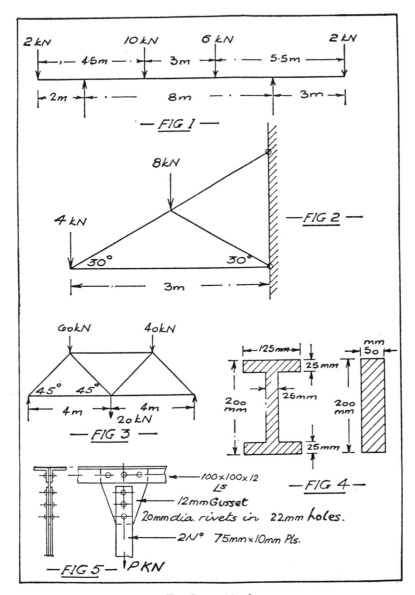

— FIG 1 —

— FIG 2 —

— FIG 3 —

— FIG 4 —

— FIG 5 —

TEST PAPER NO. 2.

Test Paper No. 3

(1) Fig. 1 shows a king-post truss carrying positive wind and dead loads. Draw the stress diagram for the truss.

Obtain the force in the member indicated by a cross. Is this member a strut or a tie? Assuming the trusses to be spaced at 4 m centres, calculate the wind pressure (normal to roof slope) and the dead load respectively, in N/m^2 of roof surface, corresponding to the joint loads given.

(2) A beam 'AB' (Fig. 2) rests on two supports. The left and right end supports are composed, respectively, of steel and concrete.

If 'A' and 'B' are still at the same (new) level when load 'W' is applied calculate 'W' and 'x', given that the load induces a stress of 0.7 N/mm^2 in the concrete.

$E_{steel} = 210 \text{ kN/mm}^2$, $E_{concrete} = 14 \text{ kN/mm}^2$.

(3) Find the value of 'w,' in the timber beam example given in Fig. 3, in order that the maximum fibre stress in the timber at a beam section, 0·6 m from the left end, shall be 6.19 N/mm^2. The beam is 75 mm wide × 150 mm deep.

Draw the bending moment and shear force diagrams for the beam, assuming the calculated load.

(4) Test the given retaining wall (Fig. 4) for tension at point 'B,' Obtain the compressive stress at 'A.'

Draw a diagram showing the variation of pressure on the subsoil under the wall.

(5) Fig. 5 shows a composite beam section.

Neglecting rivet holes, find the position of the centre of gravity of the section (i) by calculation, (ii) by any graphical method.

Properties of section:

$$\text{U.C. Area} = 58.8 \text{ cm}^2.$$
$$\text{Angles. Area of one angle} = 20.3 \text{ cm}^2.$$
$$\text{C.G. from back of angle} = 2.66 \text{ cm}.$$

(6) Prove the formula for the maximum bending moment for a simply supported beam with uniformly distributed load.

A 203 mm × 133 mm × 30 kg U.B. is used as a simply supported beam for an effective span of 2·4 m. Inclusive of its own weight, it carries a uniform load of 60 kN. What additional point load could it carry if the load were placed (i) at the centre of span, (ii) at 0·6 m from the left end? The maximum stress must not exceed 165 N/mm^2.

Z_{XX} for the U.B. = 279 cm^3.

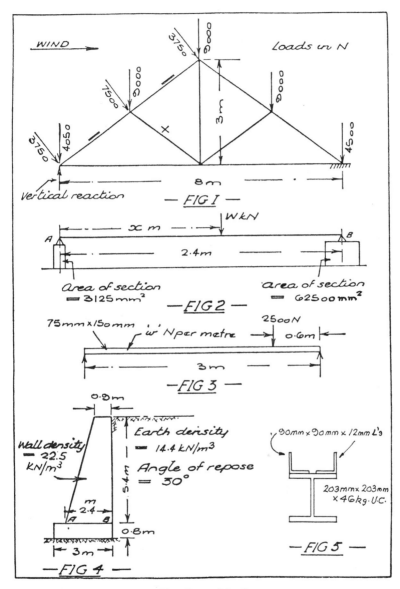

WIND →

Loads in N

— FIG 1 —

Vertical reaction

8 m

3 m

3750 6000 4500

7500 9000 6000

3750 4050

— FIG 2 —

A WkN B

x m

2.4m

Area of section = 3125 mm²

Area of section = 62500 mm²

— FIG 3 —

75mm × 150mm 'w' N per metre 2500N

0.6m

3 m

— FIG 4 —

0.9m

Wall density = 22.5 kN/m³

Earth density = 14.4 kN/m³

Angle of repose = 30°

5.4m

m 2.4

A B

0.8m

3 m

— FIG 5 —

90mm × 90mm × 12mm L's

203mm × 203mm × 46 kg U.C.

TEST PAPER NO. 3.

Test Paper No. 4

(1) Design suitable timber beams for the floor shown in Fig. 1. Find the necessary section modulus for each of the steel beams 'AB' and 'CD' respectively. Select suitable U.B.s from the section tables given on pages 258–65.

(2) A stone wall and lintol (constructed in reinforced concrete) are of 300 mm uniform thickness. Taking the particulars given in Fig. 2, construct the B.M. and S.F. diagrams for the lintol. The density of both wall and lintol may be assumed to be 22 kN/m³.

(3) Draw the stress diagram for the braced girder (Fig. 3). Obtain the force in the member indicated by a cross.

(4) Using the method of sections, find the force in each of the members 'A,' 'B,' and 'C' in the truss given in Fig. 4. Distinguish between struts and ties.

(5) Employing the method of the link polygon, construct the B.M. and S.F. diagrams for the simply supported beam (Fig. 5).

Write down the scales clearly, showing how the bending moment scale was derived.

(6) Calculate the necessary density for a vertical wall of rectangular section, 2 m high, if no tendency for tension is to be developed at the base joint, assuming it to be 0·5 m thick, and to be subjected to a horizontal wind pressure of 700 N/m².

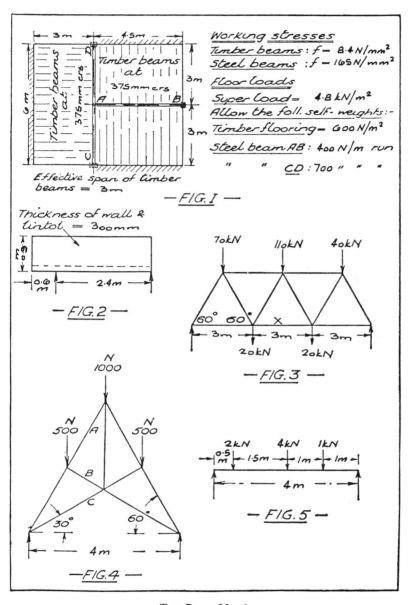

Working stresses
Timber beams : f — 8·4 N/mm²
Steel beams : f — 165 N/mm²
Floor loads
Super load = 4·8 kN/m²
Allow the foll. self-weights:-
Timber flooring — 600 N/m²
Steel beam AB : 400 N/m run
"   "   CD : 700 " " "

Effective span of timber beams = 3 m

— FIG. 1 —

Thickness of wall & lintol = 300 mm

— FIG. 2 —

— FIG. 3 —

— FIG. 4 —

— FIG. 5 —

TEST PAPER No. 4.

Test Paper No. 5

(1) For the given overhanging beam (Fig. 1) calculate (*a*) the support reactions, (*b*) the position and value of maximum bending moment.

Draw the shear force diagram for the beam.

(2) *Calculate* the support reactions for the truss shown in Fig. 2. The reaction at the left end is assumed to be vertical.

(3) A bolt, in double shear, is used with a plate thickness of 12 mm. Assuming working stresses of 95 N/mm² and 300 N/mm² for shear and bearing respectively, calculate the bolt dia. in order that it shall be equally strong in 'bearing' and 'shear.'

Calculate the safe uniform load for the beam 'AB' (Fig. 3) from the point of view of the *bolts* in its end connections.

(4) Draw the stress diagram for the roof truss given in Fig. 4. Write, alongside the members, 'S' for compression and 'T' for tension. Find the force in the member indicated by a cross.

(5) A specimen of timber, 20 mm × 20 mm in section, was tested as a simple beam with central-point loading. The effective span was 600 mm. Assuming the modulus of rupture of the timber to be 48 N/mm², calculate the load at fracture.

What factor of safety is represented by the beam shown in Fig. 5, if the same quality of timber is used? The beam is 75 mm wide × 250 mm deep.

(6) Given the expression $\dfrac{5}{384} \dfrac{Wl^3}{EI}$, find the maximum deflection of

a timber beam with simply supported ends, of width = 50 mm and depth = 150 mm, and of 3 m effective span. The beam carries a uniform load of 4·5 kN total value. 'E' for the timber = 11 000 N/mm².

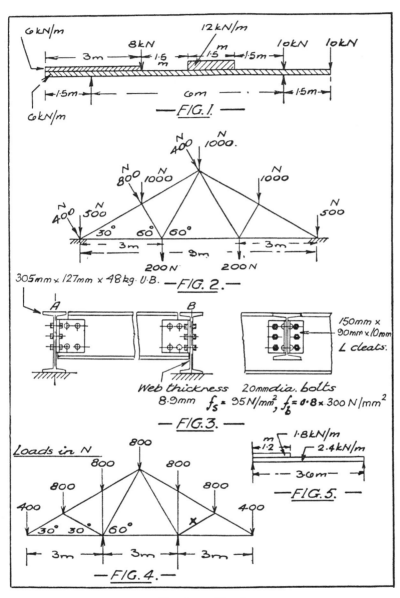

6 kN/m

8 kN

12 kN/m

10 kN 10 kN

3m

1.5 m

1.5 m

1.5 m

6 kN/m

1.5 m

6 m

1.5 m

— FIG. 1. —

N 400

N 1000.

N 800

N 1000

N 1000

N 400

N 500

30°

60°

60°

N 500

3m

9m

3m

200 N

200 N

305 mm × 127 mm × 48 kg. U.B. — FIG. 2. —

A

B

150 mm ×
90 mm × 10 mm
L cleats.

Web thickness 20 mm dia. bolts
8.9 mm $f_s = 95 N/mm^2$, $f_b = 0.8 \times 300 N/mm^2$

— FIG. 3. —

Loads in N

800

800

800

800

800

400

400

30° 30°

60°

x

3m

3m

3m

— FIG. 4. —

m 1.2

1.8 kN/m
2.4 kN/m

3.6 m

— FIG. 5. —

TEST PAPER NO. 5.

Test Paper No. 6

(1) Calculate the maximum permissible depth of water behind the given retaining wall (Fig. 1) if there is to be no tension at the base 'AB.'

(2) Obtain the force in each of the members marked 'A,' 'B,' 'C,' and 'D' respectively in Fig. 2 (i) by calculation, (ii) by drawing a stress diagram.

(3) Calculate the support reactions and draw the stress diagram for the roof truss given in Fig. 3.

(4) Draw the bending moment and shear force diagrams for the cantilever shown in Fig. 4.

(5) Define the terms 'effective column length' and 'slenderness ratio.'

Calculate the safe concentric load for a mild steel solid circular column of 100 mm dia. and 2·7 m effective height, given:

l/r	p_c (N/mm^2)
100	79
110	69

(6) Show that the floor, shown diagrammatically in Fig. 5, can safely support a total inclusive floor load of 7·6 kN/m².

Working stresses: Timber = 8·4 N/mm².
$\qquad\qquad\qquad$ Steel $\;\;$ = 165 N/mm².

Z_{XX} for 254 mm × 146 mm × 43 kg U.B. = 504 cm³.

(7) A short concrete pillar (Fig. 6), in addition to its own weight, carries the reactions of two steel beams.

Draw the distribution of pressure diagram for the base 'ZZ.'

Density of concrete = 23 kN/m³.

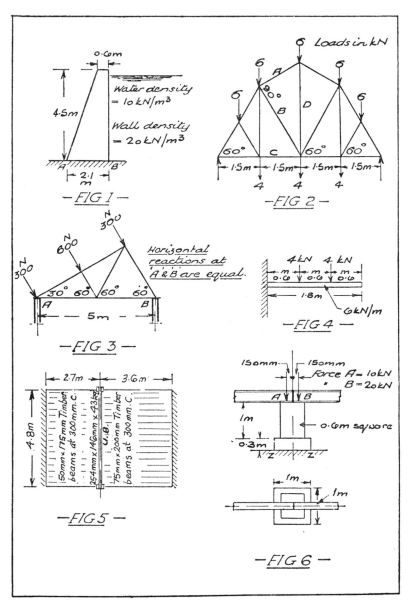

- FIG 1 -

Water density = 10 kN/m³

Wall density = 20 kN/m³

4·5m

0·6m

2·1 m

A B

Loads in kN

6

6 6

6 6

A

90°

B D

60° C 60° 60°

1·5m 1·5m 1·5m 1·5m

4 4 4

- FIG 2 -

30° N

60° N

30° N

Horizontal reactions at 'A' & 'B' are equal.

30° 60° 60° 60°

A 5m B

- FIG 3 -

4kN 4kN

0·6 m 0·6 m 0·6 m

1·8m

6kN/m

- FIG 4 -

2·7m 3·6 m

4·8 m

150mm × 175mm Timber beams at 300mm.C.

254mm × 146mm × 43kg G.B.

75mm × 200mm Timber beams at 300mm.C.

- FIG 5 -

150mm 150mm

Force A = 10kN
" B = 20kN

A B

1m

0·3m

Z Z

0·6m square

1m

1m

- FIG 6 -

TEST PAPER NO. 6.

Answers to Problems in Test Papers

Test Paper No. 1 (*page* 390)

Numerical Answers

(1) C.G. lies on the vertical axis of symmetry of the section, 124 mm above the bottom.

(2) Necessary depth for joists = 150 mm.

$\text{B.M.}_{\text{max}} = 1\cdot2825$ kN m

$\text{S.F.}_{\text{max}} = \pm1\cdot71$ kN

(3) Width of wall = 1 m

(4) (i) $E = 210$ kN/mm^2

(ii) Ultimate stress = $0\cdot5$ kN/mm^2

(iii) 30%

Safe axial load = 250 kN.

(5) Force in given member = 15 kN.

(6) B.M._{max} occurs at $3\cdot33$ m from left end.

$\text{B.M.}_{\text{max}} = 226\cdot7$ kN m

S.F. values: Left end = 120 kN

Right end = 200 kN.

(7) Equilibrant = $4\cdot732$ kN acting horizontally towards the left.

(8) Left-end reaction = 65 kN.

Right-end reaction = 65 kN.

Detailed Solutions

(1) Definition of C.G. (page 90).

Area of plate   =   5 000 mm^2

Area of flange   =   4 000 mm^2

Area of vertical leg =   2 400 mm^2

Total area = 11 400 mm^2.

Let $\bar{y}$ be height of C.G. above bottom of section.

$$11\,400\bar{y} = (2\,400 \times 60) + (4\,000 \times 130) + (5\,000 \times 150)$$

$$\bar{y} = \frac{1\,414\,000}{11\,400} = 124 \text{ mm.}$$

For link polygon construction, treat the areas as 'acting' towards the right (see page 120).

(2) Load per joist $= \left(3 \text{ m} \times \dfrac{300}{1\,000} \text{ m}\right) \times 3 \cdot 8 \text{ kN/m}^2 = 3 \cdot 42 \text{ kN}.$

$$\text{B.M.}_{\text{max}} = \frac{Wl}{8} = \frac{3 \cdot 42 \times 3}{8} \text{ kN m} = 1 \cdot 2825 \text{ kN m}.$$

$$M = \frac{fbd^2}{6}. \qquad \therefore \quad 1 \cdot 2825 \times 10^6 = \frac{7 \times 50 \times d^2}{6}$$

$$d^2 = \frac{1 \cdot 2825 \times 10^6 \times 6}{7 \times 50} = 22\,000.$$

$$\therefore \quad d = 150 \text{ mm}.$$

(Forms of B.M. and S.F. diagrams as in Fig. 276).

$$\text{B.M.}_{\text{max}} = 1 \cdot 2825 \text{ kN m}.$$

$$\text{S.F.}_{\text{max}} = \pm \frac{W}{2} = \pm 1 \cdot 71 \text{ kN}.$$

(3) Differences in liquid and earth pressures (Chapter XV).
Let 'b' m = necessary thickness of wall.
Weight of wall per metre run $= (3 \cdot 6 \times b \times 20) = 72b \text{ kN}.$
Taking moments about outer middle-third point:

$$(10 \times 1 \cdot 2) = 72b \times b/6 = 12b^2.$$

$$\therefore \quad b^2 = \frac{10 \times 1 \cdot 2}{12} = 1. \qquad \therefore \quad b = 1 \text{ m}.$$

Graphical method of solution (page 344).
(4) (i) Taking the stress and corresponding strain at the limit of proportionality:

$$\text{Stress} = \frac{\text{Load}}{\text{Area}} = \frac{67}{283} = 0 \cdot 237 \text{ kN}.$$

$$\text{Strain} = \frac{\text{Extension}}{\text{Original length}} = \frac{0 \cdot 2256 \text{ mm}}{200 \text{ mm}} = 0 \cdot 001\,128.$$

$$E = \frac{\text{Stress}}{\text{Strain}} = \frac{0 \cdot 237}{0 \cdot 001\,128} \text{ kN/mm}^2 = 210 \text{ kN/mm}^2.$$

(ii)   Ultimate stress $= \dfrac{\text{Max. load in test}}{\text{Original section area}} = \dfrac{141 \cdot 5}{283}$

$= 0 \cdot 5 \text{ kN/mm}^2.$

(iii) Percentage elongation $= \dfrac{60 \text{ mm}}{200 \text{ mm}} \times 100 = 30.$

Working stress $= \dfrac{\text{Ultimate stress}}{\text{Factor of safety}} = \dfrac{0 \cdot 5 \text{ kN/mm}^2}{4}.$

$= 0 \cdot 125 \text{ kN/mm}^2$

Sectional area of bar $= 2\ 000 \text{ mm}^2.$

$\therefore$ Safe axial load $= (2\ 000 \times 0 \cdot 125) = 250 \text{ kN}.$

(5) (See Fig. 408).

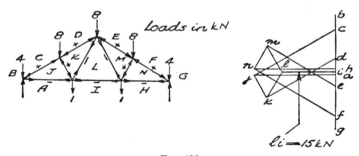

FIG. 408.

(6) $R_L \times 6 = (180 \times 3) + (20 \times 3) + (120 \times 1)$
   $R_L = 120 \text{ kN}.$
   $R_R \times 6 = (180 \times 3) + (20 \times 3) + (120 \times 5)$
   $R_R = 200 \text{ kN}.$

Load on beam from left end up to a point 3·33 m from left end $=$ $(100 + 20) = 120 \text{ kN}$, hence B.M.$_{max}$ occurs at 3·33 m from left end (see page 230).

B.M.$_{max} = (120 \times 3 \cdot 33) - (100 \times 1 \cdot 67) - (20 \times 0 \cdot 33)$
   $= 226 \cdot 7 \text{ kN m}.$

Shear diagram.—S.F. at left end $= 120 \text{ kN}.$
           S.F. at right end $= 200 \text{ kN}.$

[Construction of S.F. diagrams, page 230.]

(7) Total up force = $(1 + 4 \cos 60° + 2 \cos 60°)$ kN = 4 kN.
Total down force = 4 kN, hence there is vertical equilibrium.

Resultant force = $(4 \cos 30° + 3 - 2 \cos 30°)$ kN.
  = $(2 \times 0.866 + 3)$ = 4.732 kN, acting towards the
right, horizontally. Equilibrant of system is 'equal and opposite' to the
resultant.

(8) $R_L \times 10 = (40 \times 8) + (60 \times 4.5) + (30 \times 2)$
            = 320 + 270 + 60 = 650
      R_L = 65 kN.
  $R_R \times 10 = (30 \times 8) + (60 \times 5.5) + (40 \times 2)$
            = 240 + 330 + 80 = 650
      R_R = 65 kN.

Link-polygon construction (page 124).
Treat uniform load as '65 kN' concentrated at its centre of length for
graphical solution.

Test Paper No. 2 (*page* 392)

Numerical Answers

(1) Maximum bending moment = 17.25 kN m.
  Max. positive shear force   = 8.5 kN.
  Max. negative shear force   = −7.5 kN.
  Necessary section modulus = 2 875 × 10³ mm³.

(2) (See Fig. 409.)

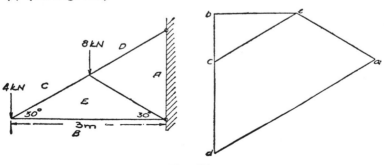

Fig. 409.

Member.	Force.	Type.
	kN	
CE	8·00	Tie
DA	16·00	Tie
AE	8·00	Strut
BE	6·93	Strut

(3) 200 mm.

(4) (See detailed solution below.)

(5) Safe uniform load for rolled steel type $= 1·656$ times that for rectangular type.

(6) Strength or value of one rivet $= 76$ kN.

Safe value of 'P' $= 164·3$ kN.

Detailed Solutions

(1) Assumptions in theory of bending (page 240).

Form of B.M. and S.F. diagrams as on page 221.

If 'A' and 'B' are reaction points and 'C,' 'D,' 'E,' and 'F' the load points in order from left:

$$(R_A \times 8) + (2 \times 3) = (6 \times 2·5) + (10 \times 5·5) + (2 \times 10)$$
$$8R_A + 6 = 15 + 55 + 20 = 90$$
$$8R_A = 90 - 6 = 84. \quad \therefore \quad R_A = 10·5 \text{ kN.}$$
$$(R_B \times 8) + (2 \times 2) = (10 \times 2·5) + (6 \times 5·5) + (2 \times 11)$$
$$8R_B + 4 = 25 + 33 + 22 = 80$$
$$8R_B = 80 - 4 = 76. \quad \therefore \quad R_B = 9·5 \text{ kN.}$$

B.M.$_C$ = 0. B.M.$_A$ = (2×2) = 4 kN m (negative).

B.M.$_B$ = (2×3) = 6 kN m (negative).

B.M.$_D$ = $(10·5 \times 2·5) - (2 \times 4·5)$ kN m = $(26·25 - 9)$ kN m
$= 17·25$ kN m.

B.M.$_E$ = $(9·5 \times 2·5) - (2 \times 5·5)$ kN m = $(23·75 - 11)$ kN m
$= 12·75$ kN m.

B.M.$_F$ = 0.

S.F. *values:* Between 'C' and 'A' $= -2·0$ kN.

Between 'A' and 'D' $= +8·5$ kN.

Between 'D' and 'E' $= -1·5$ kN.

Between 'E' and 'B' $= -7·5$ kN.

Between 'B' and 'F' $= +2·0$ kN.

Max. B.M. $= 17{\cdot}25$ kN m.

$$Z = \frac{M}{f} = \frac{17{\cdot}25 \times 1\,000 \times 1\,000}{6} = 2\,875 \times 10^3 \text{mm}^3$$

(2) (See diagram in numerical answers.)
(3) Definitions (Chapter IX).
Let x mm $=$ gauge length.

$$\text{Stress} = \frac{\text{Load}}{\text{Area}} = \frac{19{\cdot}95}{380} = 0{\cdot}0525 \text{ kN/mm}^2$$

$$\text{Strain} = \frac{0{\cdot}05 \text{ mm}}{x \text{ mm}} = \frac{0{\cdot}05}{x}$$

$$\therefore \quad 210 = \frac{0{\cdot}0525}{0{\cdot}05/x}. \qquad \therefore \quad x = \frac{21 \times 0{\cdot}05}{0{\cdot}0525} \text{ mm} = 200 \text{ mm}.$$

(4) Conditions of equilibrium (page 78).

Notation for braced girder: *Inclined members*, left to right, 1, 2, 3, and 4. *Flange members:* bottom left 5, bottom right 7, top flange member 6.

$$R_L \times 8 = (60 \times 6) + (20 \times 4) + (40 \times 2) = 360 + 80 + 80$$
$$= 520$$
$$\therefore \quad R_L = 65 \text{ kN}.$$
$$R_R \times 8 = (40 \times 6) + (20 \times 4) + (60 \times 2) = 240 + 80 + 120$$
$$= 440$$
$$\therefore \quad R_R = 55 \text{ kN}.$$

Reduction coefficient for inclined members $= \dfrac{\text{Length of diagonal}}{\text{Height of truss}}$

Length of diagonal $= \sqrt{2^2 + 2^2} = \sqrt{8}$.

$$\therefore \quad \text{Reduction coefficient} = \frac{\sqrt{8}}{2} = \sqrt{2}$$

$F_1 = 65\sqrt{2} = 91{\cdot}9$ kN (strut).

$F_2 = (65 - 60)\sqrt{2} = 5\sqrt{2}$ kN $= 7{\cdot}07$ kN (tie).
$F_3 = (55 - 40)\sqrt{2} = 15\sqrt{2}$ kN $= 21{\cdot}2$ kN (tie).
$F_4 = 55\sqrt{2} = 77{\cdot}8$ kN (strut).

$F_5 \times 2 = 65 \times 2$. $\quad \therefore \quad F_5 = 65$ kN (tie, moments about opposite joint).

$F_7 \times 2 = 55 \times 2$. $\therefore$ $F_7 = 55$ kN (tie, moments about opposite joint).

$(F_6 \times 2) + (60 \times 2) = 65 \times 4$. $\therefore$ $F_6 \times 2 = 260 - 120 = 140$.
$\therefore$ $F_6 = 70$ kN (strut).

(5) Rectangular beam:

$$Z = \frac{bd^2}{6} = \frac{50 \times 200^2}{6} = \frac{2\,000\,000}{6} \text{ mm}^3$$

$$I_{XX} \text{ for U.B. type} = \frac{BD^3}{12} - \frac{bd^3}{12} = \left(\frac{125 \times 200^3}{12} - \frac{100 \times 150^3}{12} \right) \text{ mm}^4$$

$$= \frac{6\,625 \times 10^5}{12}$$

$$Z_{XX} = \frac{6\,625 \times 10^5}{12 \times 100} = \frac{3\,312\,500}{6} \text{ mm}^4$$

The rolled steel type will carry $\dfrac{3\,312\,500}{2\,000\,000} = 1{\cdot}656 \times$ the load that the rectangular type will carry.

(6) Bearing in rivets (page 182).

$$\text{D.S. value of rivet} = 2 \times \frac{\pi d^2}{4} \times f_s \text{N} = \frac{2 \times \pi \times (22^2)}{4} \times \frac{100}{1\,000} \text{ kN}$$

$$= 76 \text{ kN.}$$

$$\text{Bearing value} = d \times t \times f_b = 22 \times 16 \times \frac{290}{1\,000} \text{ kN} = 102 \text{ kN.}$$

$\therefore$ Value of one rivet = 76 kN.

Riveted joint:

(i) *Connection between plates and gusset plate.*
D.S. value of one rivet = 76 kN.

$$\text{Bearing value} = dtf_b = \frac{22 \times 12 \times 290}{1\,000} = 76{\cdot}6 \text{ kN.}$$

$\therefore$ Actual value = 76 kN.
$\therefore$ Rivet strength of connection = 3×76 kN = 228 kN.

The weakest section for the plates is through the bottom rivet.

$$\text{Tensile strength} = 2(75 - 22) \times 10 \times \frac{155}{1\,000}\,\text{kN}$$

$$= 2 \times 53 \times 10 \times 0 \cdot 155 = 164 \cdot 3\,\text{kN}.$$

(ii) *Connection between gusset plate and angles.*
Rivets in double shear and bearing in 12 mm plate.
Rivet strength = 76 kN.
∴ Strength of connection = 3 × 76 = 228 kN.
Safe value of 'P' = smallest of calculated strengths = 164·3 kN.

Test Paper No. 3 (*page* 394)

Numerical Answers

(1) (See Fig. 410.)
Member 'X' is a strut. Force = 15 400 N.
Wind pressure = 750 N/m² of roof surface.
Dead load = 900 N/m² of roof surface.

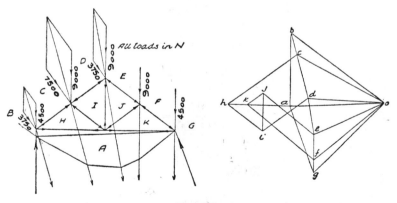

FIG. 410.

(2) W = 76·5625 kN.
   x = 1·37 m.
(3) w = 2 000 N per metre.
Calculations for B.M. and S.F. diagrams given below.
(4) No tension at point 'B.' (Eccentricity = 0·31 m.)
Compressive stress at 'A' = 148 3 kN/m².

Foundation pressures:

At toe of wall = 129·3 kN/m².
At heel of wall = 40·4 kN/m².

(5) C.G. is 15·4 cm above bottom of section.
(6) (i) 46·7 kN.
     (ii) 72·2 kN.

Detailed Solutions

(1) *Wind pressure.*

Area of roof surface corresponding to a full joint $= \left(\dfrac{5 \text{ m}}{2} \times 4 \text{ m} \right)$

$$= 10 \text{ m}^2.$$

$\therefore$ Wind pressure $= \dfrac{7\,500 \text{ N}}{10 \text{ m}^2} = 750 \text{ N/m}^2.$

Dead load:

Area of roof slope per joint = 10 m².

$\therefore$ Dead load $= \dfrac{9\,000 \text{ N}}{10 \text{ m}^2} = 900 \text{ N/m}^2.$

(2) Load carried by concrete = stress × area $= \dfrac{0·7 \times 62\,500}{1\,000} \text{ kN}$

$$= 43·75 \text{ kN}.$$

Strain in concrete $= \dfrac{\text{Stress}}{\text{E}} = \dfrac{0·7}{14\,000} = \dfrac{1}{20\,000}.$

$\therefore$ Strain in steel $= \dfrac{1}{20\,000}.$

Stress in steel = E × strain $= \left(210\,000 \times \dfrac{1}{20\,000} \right) \text{N/mm}^2$

$$= 10·5 \text{ N/mm}^2.$$

Load carried by steel $= \left(\dfrac{10·5 \times 3\,125}{1\,000} \right) \text{kN} = 32·8125 \text{ kN}.$

$\text{W} = \text{R}_\text{A} + \text{R}_\text{B} = (32·8125 + 43·75) \text{ kN} = 76·5625 \text{ kN}.$

Moments about A:

$$R_B \times 2 \cdot 4 = W \times x$$

$$\therefore \quad x = \frac{43 \cdot 75 \times 2 \cdot 4}{76 \cdot 5625} \text{ m} = 1 \cdot 37 \text{ m.}$$

(3) Left-end reaction $= \left(\dfrac{w \times 3}{2} + \dfrac{2\,500 \times 0 \cdot 6}{3} \right) = (1 \cdot 5w + 500) \text{ N}$

B.M. at 0·6 m from left end $= [(1 \cdot 5w + 500) \times 0 \cdot 6] - (0 \cdot 6w \times 0 \cdot 3)$
$= (0 \cdot 9w + 300 - 0 \cdot 18w) \text{ N m} = (0 \cdot 72w + 300) \text{ N m.}$

$$\therefore \quad 0 \cdot 72w + 300 = \frac{fbd^2}{6} = \frac{6 \cdot 19 \times 75 \times 150 \times 150}{6 \times 1\,000}$$

$$0 \cdot 72w + 300 = 1\,740$$
$$w = 2\,000 \text{ N per metre.}$$

Left-end reaction $= 3\,500$ N.
Right-end reaction $= 5\,000$ N.
B.M. diagram may be drawn thus: Construct a parabola of height 2 250 N m above base and a triangle of height (at load point) 1 200 N m below base. The total depth of the diagram will represent bending amount values.

The S.F. diagram is directly constructed by usual rules (page 220).
(4) *Joint AB*—deal with portion of wall above AB only.

Earth thrust $= \frac{1}{2} \times 14 \cdot 4 \times 5 \cdot 4^2 \times \dfrac{1 - 0 \cdot 5}{1 + 0 \cdot 5} = 69 \cdot 98$ kN.

Weight of wall per metre $= \left(\dfrac{2 \cdot 4 + 0 \cdot 9}{2} \times 5 \cdot 4 \times 22 \cdot 5 \right)$ kN

$$= 200 \cdot 48 \text{ kN.}$$

The solution may be carried through graphically (as on page 334), or by calculation method.

Calculation method:

$$\bar{x} = \text{distance of C.G. from back of wall}$$
$$8 \cdot 91\bar{x} = (4 \cdot 86 \times 0 \cdot 45) + (4 \cdot 05 \times 1 \cdot 4) = 7 \cdot 857$$
$$\bar{x} = 0 \cdot 882 \text{ m}$$

Shift of resultant $= \dfrac{69 \cdot 98 \times 1 \cdot 8}{200 \cdot 48} \text{ m} = 0 \cdot 628 \text{ m.}$

Resultant cuts base at $(0.882 + 0.628)$ m $= 1.51$ m from back of wall.

Outer middle-third point $= (\frac{2}{3} \times 2.4)$ m $= 1.6$ m from back of wall.

∴   Resultant cuts base just inside middle-third.   ∴   There will be no tension at 'B.'

Eccentricity $= (1.51 - 1.2) = 0.31$ m.

Compressive stress at A $= \dfrac{V}{b}\left(1 + \dfrac{6e}{b}\right) = \dfrac{200.48}{2.4}\left(1 + \dfrac{6 \times 0.31}{2.4}\right)$

$$= 148.3 \text{ kN/m}^2.$$

Total wall section:

Earth thrust $= \frac{1}{2} \times 14.4 \times 6.2^2 \times \frac{1}{3} = 92.26$ kN.

Weight of wall $= (8.91 + 2.4) \times 22.5 = 254.48$ kN.

$(4.86 + 4.05 + 2.4) \bar{x} = (4.86 \times 0.45) + (4.05 \times 1.4) + (2.4 \times 1.5)$

$$\bar{x} = 1.013 \text{ m}.$$

Shift $= \left(92.26 \times \dfrac{6.2}{3}\right) \div 254.48 = 0.749$ m.

Resultant cuts base at $(1.013 + 0.749)$ m $= 1.762$ m from back of wall.

∴   Eccentricity $= 0.262$ m.

Foundation pressures:

At toe: $\dfrac{V}{b}\left(1 + \dfrac{6e}{b}\right) = \dfrac{254.48}{3}\left(1 + \dfrac{6 \times 0.262}{3}\right) = 129.3 \text{ kN/m}^2.$

At heel: $\dfrac{V}{b}\left(1 - \dfrac{6e}{b}\right) = \dfrac{254.48}{3}\left(1 - \dfrac{6 \times 0.262}{3}\right) = 40.4 \text{ kN/m}^2.$

The pressure variation diagram is linear, as in Fig. 388. There is no uplift at the base.

(5) The C.G. is on the vertical axis of symmetry.

Let $\bar{y} =$ height above base.

Total area of section $= (58.8 + 20.3 + 20.3)$ cm^2.

$$= 99.4 \text{ cm}^2.$$

∴ $99.4\bar{y} = (58.8 \times 10.16) + (40.6 \times 22.98) = 1\,530$

∴ $\bar{y} = 15.4$ cm.

Graphical solution may be effected by the method given on page 105, or by link-polygon construction.

(6) Formula for given beam example is proved on page 212.

(i) Moment of resistance of section $= fZ = \left(\dfrac{165 \times 279}{100}\right)$ kN m

$$= 46 \text{ kN m.}$$

B.M. due to U.D. load $= \dfrac{60 \times 2 \cdot 4}{8} = 18$ kN m.

$\therefore$ Available B.M. for central load $= (46 - 18)$ kN m $= 28$ kN m.
Let W kN $=$ load

$$\frac{W \times 2 \cdot 4}{4} = 28$$

$$W = \frac{28 \times 4}{2 \cdot 4} = 46 \cdot 7 \text{ kN.}$$

(ii) Note that the B.M.$_{max}$ values due to the two component load systems must not be added in this case, as the max. values do not occur at the same beam section.

Let 'W' kN be the safe additional load. Assume, as a first trial, that the B.M.$_{max}$ will occur at the concentrated-load point.

$$R_L \times 2 \cdot 4 = (W \times 1 \cdot 8) + (60 \times 1 \cdot 2)$$
$$2 \cdot 4 R_L = 1 \cdot 8W + 72$$
$$R_L = (0 \cdot 75W + 30) \text{ kN}$$

B.M. at load point $= [(0 \cdot 75W + 30) \times 0 \cdot 6 - (15 \times 0 \cdot 3)]$ kN m
$\therefore$ $(0 \cdot 75W + 30) \times 0 \cdot 6 - 4 \cdot 5 = 46$
$$0 \cdot 45W + 18 = 46 + 4 \cdot 5 = 50 \cdot 5$$
$$0 \cdot 45W = 32 \cdot 5$$
$$W = 72 \cdot 2 \text{ kN.}$$

Using the zero shear rule for position of max. B.M., we find that our assumption for B.M.$_{max}$ position was correct.

For the B.M.$_{max}$ to occur in the position of span between the load point and right end, the right-end reaction must be less than (1·8 m ×

$2 \cdot 5$ kN/m) $= 45$ kN, i.e. $\left(\dfrac{W \times 0 \cdot 6}{2 \cdot 4} + 30\right)$ less than 45.

$\therefore$ 0·25W less than 15
$\therefore$ 'W' less than 60 kN.

The beam can safely carry an additional 72·2 kN at 0·6 m from the left end.

Test Paper No. 4 (*page* 396)

Numerical Answers

(1) Suitable dimensions for timber beams, 75 mm wide × 150 mm deep.

Steel beam AB.—Necessary $Z = 255$ cm³, say, a 254 mm × 102 mm × 25 kg U.B. ($Z = 265$ cm³).

Steel Beam C.D.—Necessary $Z = 580$ cm³, say 406 mm × 140 mm × 39 kg U.B. ($Z = 624·7$ cm³).

(2) B.M. at left support = 1·07 kN m (negative).

Max. B.M. (for right span) occurs at 1·125 m from the right-end reaction and = 3·77 kN m.

S.F. at left support = $-3·56$ kN (to left) and 7·58 kN (to right).

S.F. at right support = $-6·68$ kN.

(3) Force in given bar = 150 kN (tension).

(See Fig. 411.)

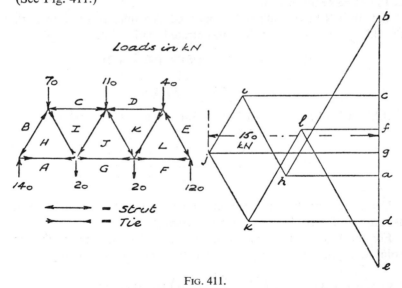

FIG. 411.

(4) Member A. Strut. Force = 1 299 N.
    Member B. Strut. Force =   250 N.
    Member C. Tie.   Force = 1 000 N.

(5) B.M. at 2 kN load = 2 kN m.
  B.M. at 2 kN load = 5 kN m.
  B.M. at 1 kN load = 3 kN m.
Method of construction given on page 233.
(6) Necessary density of wall = 16 800 N/m^3.

Detailed Solutions

(1) *Timber beams:*
Area supported by one beam = 3 m × 0·375 m = 1·125 m^2.
U.D. load carried = (1·125 × 5·4) kN = 6·075 kN.

$$\text{Max. B.M.} = \frac{Wl}{8} = \frac{6\cdot075 \times 3}{8} = 2\cdot278 \text{ kN m.}$$

$$M = \frac{fbd^2}{6}. \quad \therefore \quad 2\cdot278 \times 1\,000 \times 1\,000 = 8\cdot4 \times \frac{bd^2}{6}.$$

$$\therefore \quad bd^2 = 1\cdot627 \times 10^6.$$

If 'b' = 75 mm, d^2 = 21 700. $\therefore$ d = say 150 mm.
Steel beam 'AB':
Loads carried: U.D. load + self-weight

$$= (13\cdot5 \text{ m}^2 \times 5\cdot4 \text{ kN/m}^2) + (4\cdot5 \times 0\cdot4)$$
$$= (72\cdot9 + 1\cdot8) \text{ kN} = 74\cdot7 \text{ kN.}$$

$$\text{Max. B.M.} = \frac{W}{8} = \frac{74\cdot7 \times 4\cdot5}{8} \text{ kN m} = 42 \text{ kN m.}$$

$$M = fZ. \quad \therefore \quad 42 \times 1\,000 = 165 \times Z. \quad \therefore \quad Z = 255 \text{ cm}^3.$$

Steel beam CD:
Loads carried: U.D. load from left floor bay + self-weight + central-point load (= R_A).
Load from left bay = (6 × 1·5 × 5·4) kN = 48·6 kN.
Self-weight = 0·7 × 6 = 4·2 kN.
$\therefore$ Total U.D. load = 52·8 kN.

$$\text{Max. B.M. due to U.D. load} = \frac{Wl}{8} = \left(\frac{52\cdot8 \times 6}{8}\right) \text{ kN m}$$

$$= 39\cdot6 \text{ kN m.}$$

B.M.$_{max}$ due to central load $= \left(\dfrac{37 \cdot 35 \times 6}{4}\right)$ kN m $= 56 \cdot 03$ kN m.

$$\left(R_A = \frac{W}{2} = \frac{74 \cdot 7 \text{ kN}}{8} = 37 \cdot 35 \text{ kN}\right).$$

∴ Total max. B.M. $= (39 \cdot 6 + 56 \cdot 03)$ kN m $= 95 \cdot 63$ kN m.

$$Z = \frac{M}{f} = \frac{95 \cdot 63 \times 1\,000}{165} = 580 \text{ cm}^3.$$

(2) Total weight of wall plus lintol $= (3 \times 0 \cdot 9 \times 0 \cdot 3 \times 22)$ kN $= 17 \cdot 82$ kN.

Weight per metre run $= 5 \cdot 94$ kN.

$R_L \times 2 \cdot 4 = 17 \cdot 82 \times 1 \cdot 5.$ ∴ $R_L = 11 \cdot 14$ kN.
$R_R \times 2 \cdot 4 = 17 \cdot 82 \times 0 \cdot 9.$ ∴ $R_R = 6 \cdot 68$ kN.

B.M. at left support $= [(5 \cdot 94 \times 0 \cdot 6) \times 0 \cdot 3] = 1 \cdot 07$ kN m.

B.M.$_{max}$ in right span occurs at x m from right support, where $5 \cdot 94 \times x = R_R = 6 \cdot 68$.

$$\therefore \quad x = \frac{6 \cdot 68}{5 \cdot 94} = 1 \cdot 125 \text{ m.}$$

B.M.$_{max} = \left[(6 \cdot 68 \times 1 \cdot 125) - \left(6 \cdot 68 \times \dfrac{1 \cdot 125}{2}\right)\right]$ kN m $= 3 \cdot 77$ kN m.

$\dfrac{Wl}{8}$ in right span for construction purposes $= \dfrac{5 \cdot 94 \times 2 \cdot 4 \times 2 \cdot 4}{8}$ kN m

$$= 4 \cdot 28 \text{ kN m.}$$

(3) *By calculation* (moments about opposite joint)
$R_L \times 9 = (70 \times 7 \cdot 5) + (20 \times 6) + (110 \times 4 \cdot 5) + (20 \times 3)$
$$\qquad\qquad\qquad\qquad\qquad\qquad\qquad + (40 \times 1 \cdot 5)$$
$\qquad R_L = 140$ kN.
$(F_X \times 2 \cdot 598) + (20 \times 1 \cdot 5) + (70 \times 3) = (140 \times 4 \cdot 5).$
$F_X = 150$ kN.

(4) *Member 'A'.*—Take moments about the intersection point of members 'B' and 'C.'

$\qquad (F_A \times \text{arm}) + (500 \times 1) = R_L \times 2$
$\qquad (F_A \times 2 \tan 30°) + 500 = 1\,000 \times 2 = 2\,000$

$$\therefore \quad F_A = \frac{1\,500}{1 \cdot 155} \text{ N} = 1\,299 \text{ N (strut).}$$

Member 'B.'—Take moments about the intersection point of members 'A' and 'C.'

$$(F_B \times 2) = 500 \times 1$$
$$\therefore \quad F_B = 250 \text{ N (strut)}.$$

Member 'C.'—Take moments about the interesection point of members 'A' and 'B.'

$$(F_C \times 2 \sin 30°) = R_L \times 1$$
$$\therefore \quad F_C \times 1 = 1\,000 \times 1$$
$$\therefore \quad F_C = 1\,000 \text{ N (tie)}.$$

Struts and ties are determined as explained on page 310.

(5) See page 233 for method by link polygon.

$$R_L \times 4 = (2 \times 3{\cdot}5) + (4 \times 2) + (1 \times 1) = 7 + 8 + 1 = 16$$
$$R_L = 4 \text{ kN}$$
$$R_R \times 4 = (1 \times 3) + (4 \times 2) + (2 \times 0{\cdot}5) = 3 + 8 + 1 = 12$$
$$R_R = 3 \text{ kN}$$

B.M. at 2 kN load = $4 \times 0{\cdot}5 = 2$ kN m
B.M. at 4 kN load = $(4 \times 2) - (2 \times 1{\cdot}5) = 8 - 3 = 5$ kN m
B.M. at 1 kN load = $(3 \times 1) = 3$ kN m

(6) Let 'x' N/m^3 = density.
Weight of wall per metre run = $(2 \times 0{\cdot}5 \times x) = 1x$ N
Total wind thrust per metre run = $(2 \times 1 \times 700) = 1\,400$ N
Taking moments about outer middle-third point:

$$1x \times \frac{0{\cdot}5}{6} = 1\,400 \times 1$$

$$\frac{x}{12} = 1\,400$$

Density = 16 800 N/m^3.

Test Paper No. 5 (*page* 398)

Numerical Answers

(1) (See Fig. 412.)
   Left-end reaction = 55·25 kN.
Right-end reaction = 62·75 kN.
Max. B.M. occurs at 3·125 m from left support.
B.M.$_{max}$ = 39·14 kN m.

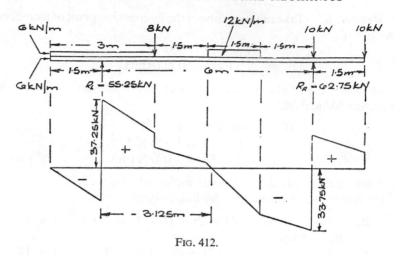

FIG. 412.

(2) $R_L = 3\,123\cdot7$ N.

$R_R = 2\,779$ N making an angle of $73\frac{1}{2}°$ (nearly) with the horizontal.

(3) Bolt diameter $= 24\cdot4$ mm, say 24 mm (this is the nearest bolt size which satisfies the conditions).

Safe load for beam $= 376\cdot8$ kN.

(4) Force in member $= 800$ N.   Member is a strut.

(See diagram in detailed solutions later.)

(5) Central load at fracture $= 427$ N.

Factor of safety $= 8\cdot2$.

(6) Max. deflection $= 10\cdot2$ mm.

Detailed Solutions

(1) $(R_L \times 6) + [(6 \times 1\cdot5) \times 0\cdot75] + (10 \times 1\cdot5) = [(6 \times 7\cdot5) \times 3\cdot75] + [(6 \times 3) \times 6] + (8 \times 4\cdot5) + [(12 \times 1\cdot5) \times 2\cdot25]$

$6R_L + 6\cdot75 + 15 = 168\cdot75 + 108 + 36 + 40\cdot5$

$\qquad 6R_L = 331\cdot5. \qquad \therefore \quad R_L = 55\cdot25$ kN.

$(R_R \times 6) + [(6 \times 1\cdot5) \times 0\cdot75] = [(6 \times 7\cdot5) \times 3\cdot75]$
$\qquad\qquad + (10 \times 7\cdot5) + (10 \times 6) + [(12 \times 1\cdot5) \times 3\cdot75] + (8 \times 1\cdot5)$

$\qquad\qquad 6R_R = 376\cdot5. \qquad \therefore \quad R_R = 62\cdot75$ kN.

$R_L + R_R = 118$ kN which is correct.

B.M. at left support $= (12 \times 1\cdot5 \times 0\cdot75) = 13\cdot5$ kN m $(-)$

B.M. at right support $= (10 \times 1\cdot5) + (9 \times 0\cdot75) = 21\cdot75$ kN (—).

$B.M._{max}$ *in central span* (using zero shear rule).

Up to beginning of 12 kN/m load, load on span

$$= [(6 \times 4\cdot5) + (6 \times 3) + 8] = 53 \text{ kN}$$

$(55\cdot25 - 53) = 2\cdot25$ kN more required.

At 18 kN/m run, this means 0·125 m.

∴ $B.M._{max}$ occurs at 3·125 m from the left-end support.

$$B.M._{max} = (55\cdot25 \times 3\cdot125) - \left(6 \times 4\cdot625 \times \frac{4\cdot625}{2}\right) - (18 \times 3\cdot125)$$

$$- (8 \times 1\cdot625) - (0\cdot125 \times 12 \times 0\cdot125/2)$$
$$= 172\cdot65 - 133\cdot51 = 39\cdot14 \text{ kN m.}$$

Hence absolute maximum bending moment = 39·14 kN m.

Construct S.F. diagram by rules given on page 220.

(2) *Reactions due to vertical loads:*

Total load on truss = 4 400 N.

∴ $R_L = R_R = 2\ 200$ N (acting vertically).

Reactions due to inclined loads:

The resultant inclined load = 1 600 N. We may assume this resultant force to act at any point in its line of action. A convenient point is where it cuts the lower tie, i.e. at the point where the left '200 N' acts.

Resolving vertically, $V = 1\ 600 \cos 30° = 1\ 385\cdot6$ N.

$$V_L = \frac{1\ 385\cdot6 \times 6}{9} = 923\cdot7 \text{ N}$$

$$V_R = \frac{1\ 385\cdot6 \times 3}{9} = 461\cdot9 \text{ N}$$

$$H_L = 0 \quad H_R = 1\ 600 \cos 60° = 800 \text{ N.}$$

Total reactions:

$R_L = (2\ 200 + 923\cdot7) = 3\ 123\cdot7$ N.

Total $V_R = (2\ 200 + 461\cdot9) = 2\ 661\cdot9$ N.

Total $H_R = 800$ N.

∴ $R_R = \sqrt{2\ 661\cdot9^2 + 800^2} = 2\ 779$ N.

If reaction make 'θ' with horizontal.

$$\tan \theta = \frac{V}{H} = \frac{2\,661 \cdot 9}{800} = 3 \cdot 3274$$

$$\theta = 73\tfrac{1}{2}° \text{ nearly.}$$

(3) For equal strengths $\dfrac{2\pi d^2}{4} f_s = dt f_b$

$$\therefore \quad \frac{2 \times \pi \times d^2 \times 95}{4} = d \times 12 \times 300$$

$$\therefore \quad d = \frac{12 \times 300}{47 \cdot 5\pi} = 24 \cdot 1 \text{ mm, say 24.}$$

In the given connection the bolts are in single shear, or bearing in either 10 mm angle thickness or 8·9 mm web thickness, i.e. a lesser thickness of 8·9 mm.

Single-shear value of one bolt $= \dfrac{\pi d^2 f_s}{4} = \dfrac{\pi \times 20^2 \times 95}{4} = 29\,845$ N.

Bearing value in 8·9 mm plate thickness $= dt f_b$

$$= 20 \times 8 \cdot 9 \times 240 \text{ N} = 42\,720 \text{ N.}$$

∴   Value of one bolt = 29·85 kN.
∴   Strength of 6 bolts = 179·1 kN.
∴   Safe load for beam = 2 × 179·1 = 358·2 kN.
(4) (See Fig. 413.)
(5) Let 'W' N = central-point load at fracture.

$$\frac{Wl}{4} = \frac{fbd^2}{6}. \quad \therefore \quad \frac{W \times 600}{4} = \frac{48 \times 20 \times 20^2}{6}$$

∴   W = 427 N.
$R_L \times 3 \cdot 6 = (2 \cdot 16 \times 3) + (8 \cdot 64 \times 1 \cdot 8) = (6 \cdot 48 + 15 \cdot 552) = 22 \cdot 032$
$\qquad R_L = 6 \cdot 12$ kN.
$R_R \times 3 \cdot 6 = (2 \cdot 16 \times 0 \cdot 6) + (8 \cdot 64 \times 1 \cdot 8) = (1 \cdot 296 + 15 \cdot 552) = 16 \cdot 848$
$\qquad R_R = 4 \cdot 68$ kN.

B.M.$_\text{max}$ will occur at $\dfrac{4 \cdot 68}{2 \cdot 4} = 1 \cdot 95$ m from right end.

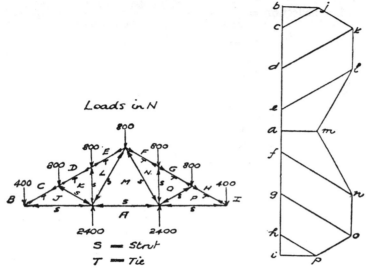

FIG. 413.

$$\text{B.M.}_{\text{max}} = (4.68 \times 1.95) - \left(4.68 \times \frac{1.95}{2}\right) \text{kN m}$$

$$= (4.68 \times 0.975) \text{ kN m} = 4.57 \text{ kN m}$$

$$4.57 \times 1\,000 \times 1\,000 = \frac{f \times 75 \times 250 \times 250}{6}$$

$$f = \frac{4.57 \times 6 \times 10^6}{75 \times 250 \times 250} \text{ N/mm}^2$$

$$= 5.86 \text{ N/mm}^2$$

$$\text{Factor of safety} = \frac{48 \text{ N/mm}^2}{5.86 \text{ N/mm}^2} = 8.2$$

(6) Max. deflection $= \dfrac{5}{384} \dfrac{Wl^3}{EI}$.

$$W = 4.5 \text{ kN}, \qquad l = 3\,000 \text{ mm}, \qquad E = 11\,000 \text{ N/mm}^2.$$

$$I = \frac{bd^3}{12} = \frac{50 \times 150 \times 150 \times 150}{12} \text{ mm}^4 = 14\,062\,500 \text{ mm}^4$$

$$\text{Max. deflection} = \frac{5}{384} \times \frac{4 \cdot 5 \times 1\,000 \times 3\,000 \times 3\,000 \times 3\,000}{11\,000 \times 14\,062\,500}$$

$$= 10 \cdot 2 \text{ mm}$$

Test Paper No. 6 (*page* 400)

Numerical Answers

(1) Depth of water permissible = 3·626 m.

(2) 'A' = 11·5 kN. Strut.
    'B' = 0·87 kN. Strut.
    'C' = 10·4 kN. Tie.
    'D' = 5·5 kN. Tie. (See Fig. 414.)

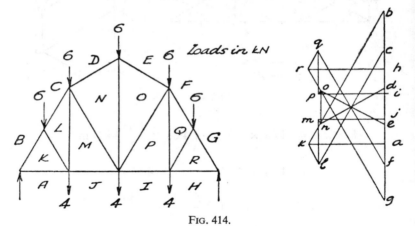

FIG. 414.

(3) $R_A = R_B$ = 600 N. (See Fig. 415.)

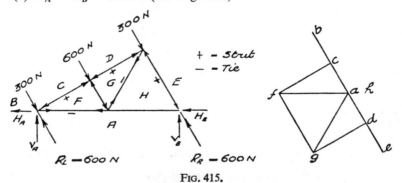

FIG. 415.

Each reaction makes 60° with horizontal.

(4) B.M. at support = 16·92 kN m.

S.F. at support = 18·8 kN.

(See Fig. 416.)

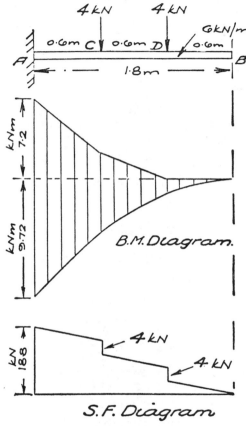

FIG. 416.

Details of calculation given in solutions later.

(5) 558 kN. Definitions (see pages 289 and 290.)

(6) Max. stress in 50 mm × 175 mm beams = 8·15 N/mm².

Max. stress in 75 mm × 200 mm beams = 7·39 N/mm².

Max. stress in steel beam = 137 N/mm².

(7) Max. pressures. Left edge of base = 36·18 kN/m².

Right edge of base = 54·18 kN/m².

(See Fig. 417).

S.M.(M)—15

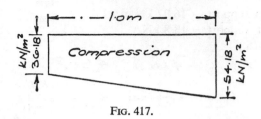

FIG. 417.

Detailed Solutions

(1) Weight of wall per metre $= \frac{1}{2}(2 \cdot 1 + 0 \cdot 6) \times 4 \cdot 5 \times 20 = 121 \cdot 5$ kN
If 'h' m $=$ depth of water, water thrust per metre run of wall $= \frac{1}{2}wh^2$
$= \frac{1}{2} \times 10 \times h^2$ kN.
Let $\bar{x} =$ distance of C.G. of wall from back

$$6 \cdot 075 \bar{x} = (2 \cdot 7 \times 0 \cdot 3) + (3 \cdot 375 \times 1 \cdot 1) = 4 \cdot 5225$$
$$\bar{x} = 0 \cdot 746 \text{ m.}$$

Taking moments about outer middle-third point:

$$\frac{1}{2} \times 10 \times h^2 \times h/3 = 121 \cdot 5 \times \text{'shift'}$$
$$= 121 \cdot 5 \times [(\tfrac{2}{3} \times 2 \cdot 1) - 0 \cdot 746]$$

$$\therefore \quad h^3 = \frac{2 \times 3 \times 121 \cdot 5 \times 0 \cdot 654}{10} = 47 \cdot 67$$

$$\therefore \quad h = 3 \cdot 626 \text{ m.}$$

(2) $$R_L = R_R = \frac{42}{2} = 21 \text{ kN.}$$

Imagine a section plane cutting members 'A,' 'B,' and 'C.'
Member A.—Moments about mid-point of bottom tie.
$(F_A \times 3) + (6 \times 1 \cdot 5) + (4 \times 1 \cdot 5) + (6 \times 2 \cdot 25) = 21 \times 3.$
$3F_A + 9 + 6 + 13 \cdot 5 = 63$
$3F_A = 63 - 28 \cdot 5 = 34 \cdot 5. \qquad \therefore \quad F_A = 11 \cdot 5 \text{ kN.}$
Member 'A' is a strut.

Member B.—Moments about intersection point of 'A' and 'C,' i.e.
a point level with the bottom tie and 3 m to the left of the left reaction
point.
Arm of moment for $F_B = 6 \cos 30° = 5 \cdot 196$ m.

Assuming member 'B' to be pulling at the section plane:
$(F_B \times 5{\cdot}196) + (6 \times 4{\cdot}5) + (4 \times 4{\cdot}5) + (6 \times 3{\cdot}75) = 21 \times 3.$
$5{\cdot}196F_B + 67{\cdot}5 = 63.$ $\therefore$ $5{\cdot}196F_B = -4{\cdot}5.$
$F_B = -0{\cdot}87$ kN.
The negative sign indicates that F_B is not pulling at the section plane, hence member 'B' is a strut.

Member C.—Moments about intersection point of members 'A' and 'B.'
Arm of moment $= 1{\cdot}5 \tan 60° = 1{\cdot}5 \times \sqrt{3} = 2{\cdot}598$ m.

$$\therefore \quad F_C \times 2{\cdot}598 + (6 \times 0{\cdot}75) = 21 \times 1{\cdot}5$$
$$2{\cdot}598F_C = 31{\cdot}5 - 4{\cdot}5 = 27$$
$$F_C = 10{\cdot}4 \text{ kN.}$$

Member 'C' is a tie.

Member D.—Consider vertical equilibrium at the top end of the member.
$$2 \times F_A \cos 60° = 6 + F_D$$
$$\therefore \quad F_D = (2 \times 11{\cdot}5 \times \tfrac{1}{2}) - 6 = 5{\cdot}5 \text{ kN.}$$

As member 'D' pulls downwards it is a tie.

(3) Resultant wind load $= 1\,200$ N.

Resolving the resultant, assuming it to act at the middle point of bottom tie:
$$V = 1\,200 \cos 30° = 1\,200 \times 0{\cdot}866 = 1\,039{\cdot}2 \text{ N}$$
$$\therefore \quad V_A = V_B = 519{\cdot}6 \text{ N.}$$

$$H_A = H_B = \frac{1\,200 \cos 60°}{2} = 300 \text{ N.}$$

$$R_A = R_B = \sqrt{519{\cdot}6^2 + 300^2} = 600 \text{ N.}$$

If 'θ' be inclination of each reaction to horizontal,

$$\tan \theta = \frac{V}{H} = \frac{519{\cdot}6}{300} = 1{\cdot}732$$

$$\therefore \quad \theta = 60°.$$

(4) B.M. at support due to U.D. load $= 10{\cdot}8 \times 0{\cdot}9 = 9{\cdot}72$ kN m.
    B.M. at D   due to point loads $= 0$.
    B.M. at C   due to point loads $= 2{\cdot}4$ kN m.
    B.M. at A   due to point loads $= (4 \times 1{\cdot}2) + (4 \times 0{\cdot}6)$
                                        $= 7{\cdot}2$ kN m.

The B.M. diagrams for the load systems are shown in Fig. 416, one being drawn above the base line and the other below. The net B.M. diagram is therefore the addition of the two diagrams, i.e. B.M. values are obtained by scaling the total depth of the combined diagrams. If desired a new diagram may be obtained by plotting ordinates afresh from a horizontal base line, the ordinate at any given point being the sum of the corresponding ordinates in the component diagrams. The S.F. diagram is plotted directly.

$$\text{S.F.}_{max} = [(6 \times 1 \cdot 8) + 4 + 4]\text{kN} = 18 \cdot 8 \text{ kN}.$$

(5) Least radius of gyration $= \dfrac{\text{Diameter}}{4} = \dfrac{100 \text{ mm}}{4} = 25 \text{ mm}$

$$\text{Slenderness ratio} = \dfrac{2\,700 \text{ mm}}{25 \text{ mm}} = 108$$

l/r	$p_c \, (N/mm^2)$
100	79
110	69
Diff. = $\overline{10}$	Diff. = $\overline{10}$
Diff. = 8	Diff. = $\dfrac{10}{10} \times 8 = 8$

$\therefore$ For $l/r = 108$, $p_c = (79 - 8) = 71 \text{ N/mm}^2$.

$$\text{Sectional area of column} = \frac{\pi \times D^2}{4} = \frac{\pi \times 100^2}{4} = 7\,854 \text{ mm}^2$$

$$\therefore \quad \text{Safe axial load} = \left(\frac{7\,854 \times 71}{1\,000}\right) = 558 \text{ kN}.$$

(6) *Timber beams, 2·7 m bay:*
Assuming 7·6 kN/m² total load:
Load carried by one beam $= (7 \cdot 6 \times 2 \cdot 7 \times 0 \cdot 3) = 6 \cdot 156 \text{ kN}.$

$$\therefore \quad \frac{6 \cdot 156 \times 1\,000 \times 2 \cdot 7 \times 1\,000}{8} = \frac{f \times 50 \times 175 \times 175}{6}$$

$$= \frac{6 \cdot 156 \times 2 \cdot 7 \times 10^6 \times 6}{8 \times 50 \times 175 \times 175} \text{ N/mm}^2 = 8 \cdot 14 \text{ N/mm}^2.$$

Timber beams. 3·6 m *bay:*

Load carried by one beam = (7·6 × 3·6 × 0·3) = 8·208 kN.

$$\therefore \quad \frac{8·208 \times 1\,000 \times 3·6 \times 1\,000}{8} = \frac{f \times 75 \times 200 \times 200}{6}$$

$$f = \frac{8·208 \times 3·6 \times 10^6 \times 6}{8 \times 75 \times 200 \times 200} \, \text{N/mm}^2 = 7·39 \, \text{N/mm}^2.$$

Steel beam:

$$\text{Load carried} = \left(\frac{6·3}{2} \times 4·8 \times 7·6\right) \text{kN} = 115 \, \text{kN}.$$

$$M = fZ. \quad \therefore \quad f = \frac{M}{Z}$$

$$M = \frac{115 \times 4·8}{8} \, \text{kN m} = 69 \, \text{kN m}$$

$$\therefore \quad f = \frac{69 \times 1\,000}{504} = 137 \, \text{N/mm}^2$$

The given floor load is therefore safe.

(7) Volume of concrete in pier = (1 × 0·6 × 0·6) + (1 × 1 × 0·3)

$$= 0·66 \, \text{m}^3.$$

Weight = (0·66 × 23) kN = 15·18 kN.

$$\text{Direct stress on base 'ZZ'} = \frac{\text{Total load}}{\text{Area}}$$

$$= \frac{10 + 20 + 15·18}{1} \, \text{kN/m}^2 = 45·18 \, \text{kN/m}^2$$

$$\text{Bending stress} = \frac{M}{Z}$$

B.M. = [(20 × 0·15) − (10 × 0·15)] kN m = 1·5 kNm

$$Z = \frac{bd^2}{6} = \frac{1 \times 1^2}{6} \, \text{m}^3 = \frac{1}{6} \, \text{m}^3$$

∴   Bending stress = 1·5 × 6 kN/m² = 9 kN/m².

Max. compressive stress = (45·18 + 9) = 54·18 kN/m².

Min. compressive stress = (45·18 − 9) = 36·18 kN/m².

BRITISH STANDARDS

THE following list of B.S. has been extracted, by kind permission of the British Standards Institution, from the sectional lists of British Standards dealing with 'Building,' etc. Copies of the B.S. may be obtained from the British Standards Institution, Sales Branch, 2 Park Street, London W1A 2BS.

Only the basic B.S. or Code number has been listed. From time to time, amendments and additions to these documents have been published.

B.S.   **Materials for Bridges and Buildings, etc.**

4: —— Structural steel sections.
Part 1: 1972. Hot-rolled sections. Metric and Imperial.
Part 2: 1969. Hot-rolled hollow sections. Metric.

153: —— Steel girder bridges.
Parts 1 and 2: 1958.
Part 1. Materials and workmanship.
Part 2. Weighing, shipping and erection.
Part 3A: 1954. Loads.
Parts 3B and 4: 1958.
Part 3B. Stresses.
Part 4. Design and construction.

405: 1945 Expanded metal (steel) for general purposes.

449: —— The use of structural steel in buildings.
Part 1: 1970. Imperial.
Part 2: 1969. Metric.

648: 1964 Schedule of weights of building materials. Metric and Imperial.

4360: 1968 Weldable structural steels.
Part 2: 1969. Metric.

4449: 1969 Hot rolled steel bars for the reinforcement of concrete. Metric.

4482: 1969 Hard drawn mild steel wire for the reinforcement of concrete. Metric.

B.S.   **Cement, Lime and Plasters**

12: —— Portland cement (ordinary and rapid hardening).
Part 2: 1971. Metric.

146: 1958 Portland-blastfurnace cement.

890: 1966 Building limes.

915: 1947 High alumina cement.

1014: 1961 Pigments for cement, magnesium oxychloride and concrete.

1191: —— Gypsum building plasters.
Part 1: 1967. Excluding premixed lightweight plasters.
Part 2: 1967. Premixed lightweight plasters.

1370: 1958 Low heat Portland cement.

1881: —— Methods of testing concrete.
Parts 1–5: 1970. Metric.
Part 6: 1971. Metric.

B.S.   **Aggregates**

812: 1967 Methods for sampling and testing of mineral aggregates sands and fillers.

877: 1967 Foamed or expanded blastfurnace slag lightweight aggregate for concrete.

882, 1201: 1965 Aggregates from natural sources for concrete (including granolithic).

1047: 1952 Air-cooled blastfurnace slag coarse aggregate for concrete.

1165: 1966 Clinker aggregate for concrete.

1198, 1199, 1200: 1955 Building sands from natural sources.

B.S.   **Timber**

373: 1957 Testing small clear specimens of timber.

565: 1963 Glossary of terms relating to timber and woodwork.

881, 589: 1955 Nomenclature of commercial timbers, including sources of supply.

1186: —— Quality of timber and workmanship in joinery.
Part 1: 1952. Quality of timber.
Part 2: 1955. Quality of workmanship.

1860: —— Structural timber. Measurement of characteristics affecting strength.
Part 1: 1959. Softwood.

C.P.   **Codes of Practice**

3: —— Chapter V, Part 1: 1967. Dead and imposed loads. Metric.
Chapter V. Part 2: 1970. Wind loads. Metric.

101: 1972 Foundations and substructures for non-industrial buildings of not more than four storeys.

111: —— Structural recommendations for loadbearing walls.
Part 2: 1970. Metric.

112: —— Structural use of timber.
Part 2: 1971. Metric.

114: —— Structural use of reinforced concrete in buildings.
Part 2: 1969. Metric.

FABRICATION OF STEELWORK

IN various parts of the book we have referred to steel frames, built-up steel girders, etc. The reader may be interested to learn how such steelwork is fabricated.

Fig. 418 shows the use of a torque-controlled spanner to tighten the nut on a high strength friction grip bolt.

Fig. 419 illustrates the joining of units together by the metal-arc welding process. The metal required for deposition at the weld forms part of an electrical circuit. It is in the form of rods known as 'electrodes'. In the photograph the welding operator is holding the electrode a small distance away from the joint being welded. An electric arc completes the circuit and the electrode metal is gradually being deposited to form the necessary connection.

Fig. 420 shows an automatic welding machine in action.

Fig. 418.

FIG. 419.

FIG. 420.

Figs. 418 to 420 are reproduced by courtesy of the British Constructional
Steelwork Association Ltd.

Answers to Chapter Exercises

Exercises (1) (*page* 14)

(1) 52 N 32° (nearly) to vertical.
(2) 100 N 53° (nearly).
(3) Vertical force = 500 N.
(4) 7·39 kN bisecting angle between rope directions.
(5)   (i) 20 kN 37° (nearly) to horizontal.
    (ii) 174N 13° (nearly) to horizontal.
    (iii) 908 N 82½° (nearly) to horizontal.
    All the above resultants act downwards towards the right.
(6) 2 126 N 13½° (nearly) to vertical.
(8) 2½ kN vertically downwards.
(9) 2 000 N (nearly).
(10) 2·05 kN.
(11) 88·65 kN at 24° (nearly) to the vertical. Resultant cuts base at 1·09 m from back of wall.

Exercises (2) (*page* 28)

(1)   (i) 448 N acting upwards towards the right.
       366 N acting downwards towards the right.
    (ii) 8·66 kN vertically downwards.
       5·00 kN horizontally towards right.
    (iii) 2 kN upwards left
       2·83 kN vertically downwards.
(2) Forward pull = 103·9 N.
(3) 154·5 N.
(4) 44·64 N.
(5)   (i) H = 4 kN (to right).   V = 6·93 kN (upwards).
    (ii) H = 500 N (to right).   V = 866 N (downwards).
    (iii) H = 1·532 kN (to left).   V = 1·286 kN (downwards).
(6) H = 2·134 kN (to right).
    V = 1·77 kN (downwards).
(7) H = 646·4 N (to right).
    V = 1 519·6 N (downwards).
    Resultant = 1 651 N making an angle of 67° (nearly) with the horizontal.

(8) Horizontal sliding tendency = 3 420 N.
∴ frictional resistance should be 5 130 N.

(9) 60°.

(10) 3 000 N.

Exercises (3) (*page* 41)

(1) Left-hand rafter. 500 N Strut.
Right-hand rafter. 866 N Strut.

(2) 433 N.

(3) Pull in tie = 4 kN.
Thrust in jib = 7 kN.
Jib is in compression.

(4) Pull in tie = 4 kN.
Thrust in jib = 12 kN.

(5) (i) 63½° with horizontal.
(ii) Reaction at upper hinge = 20 N.
Reaction at lower hinge = 44·72 N.

(6) Left reaction = 462 N acting downwards towards the left at 30° to the horizontal.
Right reaction = 462 N acting vertically downwards.

(7) Tension in rope = 60 N.
Reaction at hinge = 60 N acting upwards towards the left at 30° to the horizontal.

(8) Force polygon closes. By calculation method $\sum H = 0$ and $\sum V = 0$.

(9) X = 256 N.   Y = 1 079 N.
Both members are struts.

(10) Member DE is a strut. Force = 9 kN.
Member EA is a tie   Force = 1·414 kN.

Exercises (4) (*page* 65)

(1) 60 N m A.C.W.
5·196 kN m C.W.
9·6 kN m A.C.W.

(2) (*a*) 70·7 N.
(*b*) 50 N.

(3) 68 N m C.W.

(4) 25 mm.

(5) 30 N.

(6) Reaction = 1 212·4 N.

(7) (i) Left end = 8 kN.   Right end = 10 kN.

　　(ii) Left end = 26·8 kN.　　Right end = 25·2 kN.
　　(iii) Left end = 17·2 kN.　　Right end = 18·8 kN.
(8) Left end = 55·8 kN.　　Right end = 41·4 kN.
(9) Left end = 4 050 N.　　Right end = 4 250 N.
(10) 1·25 m.
(11) R_L = 12 kN.　　R_R = 15 kN.
　　Moment = 27 kN m. Moments agree in magnitude, one is C.W. and the other A.C.W.
(12) The magnitude of the resultant = 2·88 N.
(13) A = $26\frac{2}{3}$ kN.
　　B = $13\frac{1}{3}$ kN.
　　C = $13\frac{1}{3}$ kN.
　　D = $6\frac{2}{3}$ kN.

Total = 60   kN.

(14) P_{max} = 76·25 N.

Exercises (5) (*page* 85)

(1) (i) 9 kN vertically downwards at 120 mm to right of 3 kN force.
　　(ii) 17 kN vertically downwards at $12\frac{4}{17}$ m to right of 2 kN force.
(2) (i) 2 kN horizontally towards the right at 12 m vertically above the 2 kN force.
　　(ii) 90 N acting, parallel to system, towards the right in line with the 80 N force.
(3) Resultant = 12 kN in line with the 2 kN load.
　　Reactions: Left end = 7·2 kN.　　Right end = 4·8 kN.
(4) 2 200 N, acting parallel to system towards the left at 5·5 m from apex of truss.
(5) Resultant load acts at 2·4 m from 600 kN load. Overall length = 6 m.
(6) 10·2 m from right end.
(7) Force in couple　　= 12 N.
　　Arm of couple　　= $93\frac{1}{3}$ mm.
　　Moment of couple = 1 120 N mm.
(8) C = T = 30 kN.
(10) Pull in string = 7·5 N.
　　Hinge reaction: Vertical component　　= 3·25 N.
　　　　　　　　　　Horizontal component = 6·495 N.
　　Magnitude = 7·26 N acting upwards towards the right at $26\frac{1}{2}°$ (nearly) to the horizontal.

(11) W = 12 N.

    $x = 46\frac{2}{3}$ mm.

(12) R_A = 2 000 N.

    R_B = 10 440 N acting at 73° 18′ to the horizontal, upwards towards the left.

    [Vertical component   = 10 000 N

    Horizontal component =   3 000 N].

Exercises (6) (*page* 112)

(1) 90 mm from '4 N' mass.

(2) (*a*) 8·17 m.   (*b*) 2 m.

(3) 29·4 mm from back of each leg.

(4) (i) 41·9 mm below top of section on vertical axis of symmetry.

   (ii) 20·5 mm from left edge of angle section.

      33·0 mm from bottom edge of angle section.

   (iii) 24·2 mm from left edge of section, on horizontal axis of symmetry.

   (iv) 58·9 mm from top of section, on vertical axis of symmetry.

(5) 124·4 mm above bottom of section on vertical axis of symmetry.

(6) 3·25 m from back of wall.

(7) 0·446 m from left edge of section.

    0·5 m from left bottom edge of section.

(8) (i) R_L = 108 kN.

       R_R = 216 kN.

   (ii) R_L = 264 kN.

       R_R = 168 kN.

(9) 2 m from line joining centres of the two left-hand columns.

    $6\frac{2}{3}$ m above horizontal line through the '10 kN' column.

(10) 171·2 mm above base, on vertical axis of symmetry.

(11) 12 mm from left edge of section.

    11·8 mm from bottom edge of section.

    The C.G. is at mid-thickness of the casting.

(12) 1·104 m from left edge of base.

Exercises (7) (*page* 127)

(1) Resultant = 185 N acting at 2·03 m from the top of mast at an angle of 22° (downwards towards the left) with the horizontal.

(2) 2·8 m.

(3) 54·2 mm from left edge of section and 29·2 mm from bottom edge.

(5) 2·11 m from back of wall.

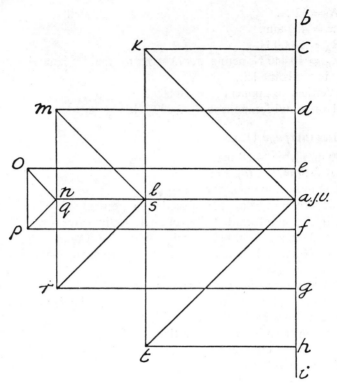

FIG. 421.

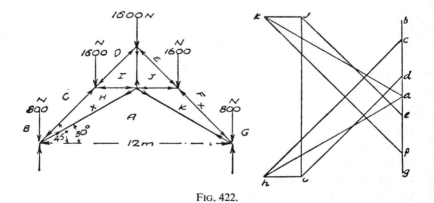

FIG. 422.

(7) (i) $R_L = 6.5$ kN. $R_R = 7.5$ kN.

    (ii) $R_L = 7$ kN. $R_R = 6$ kN.

(8) $R_L = 11.5$ kN. $R_R = 12.5$ kN.

(9) Free-end reaction (vertical) = 1 559 N.

    Fixed-end reaction = 2 381 N acting at 41° to the horizontal.

(11) $R_L = 26.25$ kN.

    $R_R = 41.25$ kN.

Exercises (8) (*page* 153)

(1) See Fig. 421.

Member.	Force.	Type.
JK & TU	28·28 kN	Tie
LM & RS	16·97 kN	Tie
NO & QP	5·66 kN	Tie
AJ & AU	0·00 kN	—
CK & HT	20·00 kN	Strut
AL & AS	20·00 kN	Tie
DM & GR	32·00 kN	Strut
AN & AQ	32·00 kN	Tie
EO & FP	36·00 kN	Strut
BJ & IU	24·00 kN	Strut
KL & TS	20·00 kN	Strut
MN & RQ	12·00 kN	Strut
OP	8·00 kN	Strut

(2) See Fig. 422.

    6 558 N (tie) in left member (AH).

    8 030 N (strut) in right member (FK).

(3) See Fig. 423.

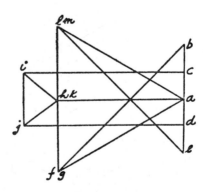

Fig. 423.

    AF = 10·93 kN (tie).

    IJ = 4·00 kN (strut).

    JK = 2·83 kN (tie).

(4) See Figs. 424 and 425.
(5) $R_L = 17\frac{1}{3}$ kN.
   $R_R = 20\frac{2}{3}$ kN.
   Force in member $= 1.9$ kN (tie).
   See Fig. 426

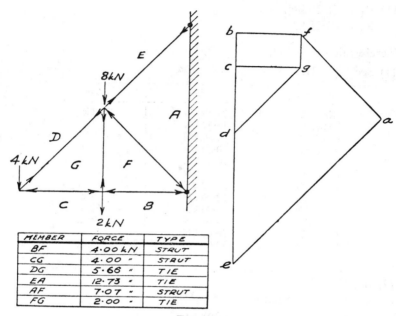

MEMBER	FORCE	TYPE
BF	4·00 kN	STRUT
CG	4·00 "	STRUT
DG	5·66 "	TIE
EA	12·73 "	TIE
AF	7·07 "	STRUT
FG	2·00 "	TIE

FIG. 424.

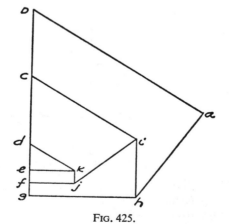

FIG. 425.

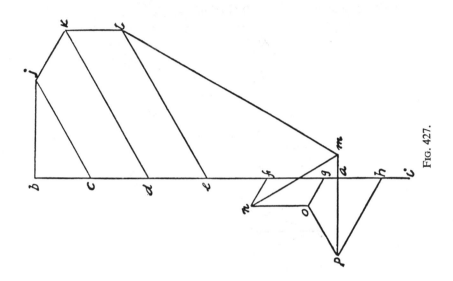

FIG. 427.

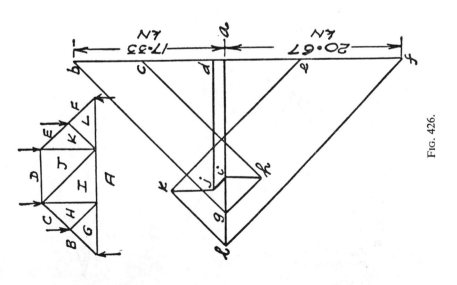

FIG. 426.

(6) See Fig. 427.
(7) Force in member = 1 200 N (tie).
    See Fig. 428.

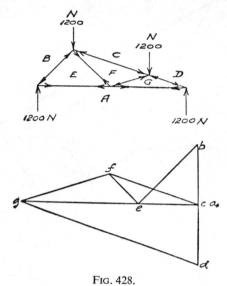

FIG. 428.

(8) See Fig. 429.

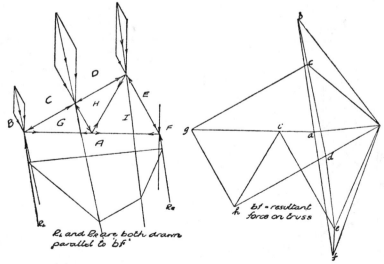

FIG. 429.

(9) Dead loads: Full joint load = 7·5 kN; ½ joint load (at eaves) = 3·75 kN. Wind loads at apex and eaves = 1·5 kN. Central joint = 3·0 kN.

(10) Forces in inclined members from left:
13·86 kN (strut), 5·775 kN (tie), 5·775 kN (strut), 6·93 kN (strut), 6·93 kN (tie), 11·55 kN (strut). See Fig. 430.

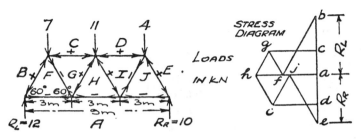

FIG. 430.

Exercises (9) (*page* 190)

(1) 150 N/mm².
    24 mm thickness.
(2) 22 mm.
(3) 2·5 m square.
(4) 330 mm square.
(5) 10.
(6) 8·4 kN/mm².
(7) 1·5 mm.
(8) (i) 90 N/mm².
    (ii) 300 kN.
(9) (i) 210 kN/mm².
    (ii) 500 N/mm².
    (iii) 12·5 mm.
(10) 16 mm dia.
(11) Concrete load = 10 kN.
    Timber load   = 13·5 kN.
(12) 207·3 kN.
(13) 234.
    [Rivet strength = 249 kN].
(14) Value of one bolt = 36·1 kN.
    Max. safe reaction load = 6 × 36·1 = 216·6 kN.

Exercises (10) (*page* 215)

(1) B.M.$_{max}$ = 2·4 kN m (−).
    S.F.$_{max}$ = 2 kN.

(2) (i) B.M.$_{max}$ = 25 kN m (−).
    (ii) S.F.$_{max}$ = 6 kN.
    (iii) 9 kN m.
    (iv) 6 kN.

(3) 10 000 N/m².
    Max. S.F. = 9 000 N.

(4) B.M.$_{max}$ = 75 kN m.
    S.F.$_{max}$ = ±50 kN.

(5) B.M.s at load points from left:
    8·25 kN m, 13·5 kN m, 9·75 kN m.
    R$_L$ = 5·5 kN.   R$_R$ = 6·5 kN.

(6) R$_L$ = 7·4 kN.   R$_R$ = 7·6 kN.
    B.M.s at load points from left:
    8·88 kN m, 15·36 kN m, 17·04 kN m, 9·12 kN m.
    B.M. at given section = 16·62 kN m.
    S.F. at given section = 1·4 kN.

(7) B.M.$_{max}$ = 36 kN m.
    S.F.$_{max}$ = 12 kN.
    B.M. at given section = 32 kN m.
    S.F. at given section = 4 kN.

(8) Load = 2 kN per metre run.
    B.M.$_{max}$ = 25 kN m.

(9) B.M.$_{max}$ = 1·08 kN m in timber beam.
    S.F.$_{max}$ = ±1·8 kN in timber beam.
    B.M.$_{max}$ = 34·56 kN m in steel beam.
    S.F.$_{max}$ = ±28·8 kN in steel beam.

(10) B.M.$_{max}$ = 39·6 kN m.

(11) (i) 8 kN.
     (ii) 4·5 kN.

(12) B.M. = 7 kN m.
     S.F. = 1·944 kN.
     B.M.$_{max}$ = 7·29 kN m.
     S.F.$_{max}$ = ±9·72 kN.

(13) B.M. (mid-height) = 45 kN m.
     S.F. (mid-height) = 9 kN.
     B.M. (base) = 180 kN m.
     S.F. (base) = 18 kN.

(14) 1·2 m and 1·8 m.
(15) 4 kN.
　(i) $5\frac{7}{9}$ kN.
　(ii) 4 kN.

Exercises (11) (*page* 234)

(1) B.M. at 1·8 m = 3·3 kN m.
　S.F. at 1·8 m = −0·5 kN.
　B.M.'s at load points from left:
　0, −3·0 kN m (at reaction), 3·6 kN m, 3·0 kN m, −2·4 kN m (at reaction), 0.
　Reactions: Left end = 7·5 kN, right end = 6·5 kN.
(2) Reactions: Left end = 1 575 N, right end = 2 625 N.
　B.M.'s: Left support = 450 N m. Right support = −1 800 N m.
　B.M.$_{max}$ occurs at 2·4375 m from left support (for central span).
　B.M.$_{max}$ 738·3 N m. Absolute B.M.$_{max}$ = 1 800 N m.
(3) Reactions: Left end = 13·8 kN, right end = 22·2 kN.
　B.M.'s: Left support = −6kN m, right support = −12 kN m.
　B.M. at point load = 13·5 kN m.
　B.M. diagrams for overhangs are parabolae. The diagram for central span consists of straight lines.
(4) Reactions: Left end = 8 kN, right end = 4 kN.
　B.M.$_{max}$ occurs at $2\frac{1}{3}$ m from left end.
　B.M.$_{max}$ $13\frac{1}{3}$ kN m.
(5) Reactions: Left end = 14 kN, right end = 10 kN.
　B.M.$_{max}$ = 24·5 kN m at 3·5 m from left end.
(6) Reactions: Left end = 10 kN, right end = 8 kN.
　(6 + 8) kN ecxceeds left-end reaction value.
(7) Reactions: Left end = 21 kN, right end = 15 kN.
　B.M.$_{max}$ = 76·5 kN m at 9 m from left end.
(8) Reactions: Left end = 34·5 kN, right end = 15·5 kN.
　A B.M.$_{max}$ occurs at '5 kN' load and equals 32·5 kN m.
　Absolute B.M.$_{max}$ = 37·5 kN m at left support.
(9) Reactions: Left end = 17 kN, right end = 18 kN.
　B.M.$_{max}$ = 30·25 kN m at 5·5 m from left support.
(10) Reactions: Left end = 27·87 kN, right end = 22·13 kN.
　B.M. at left support = −50 kN m.
　B.M. at right support = −32 kN m.
　B.M. at 6 kN load = −10·67 kN m.

B.M. at 8 kN load $= -1{\cdot}33$ kN m.

[B.M. diagram is wholly beneath base line.]

(11) Reactions: Left end = 9 kN, right end = 6 kN.

B.M.$_{max}$ occurs at 4 m from left end = 18 kN m.

(12) $50{\cdot}625$ kN m. Each reaction = $67{\cdot}5$ kN.

B.M. at each support $= -16{\cdot}875$ kN m.

(13) Total wind load = 1 000 N.

B.M. at base = 2 000 N m.

S.F.$_{max}$ = 1 000 N.

[Treat as cantilever with partial uniform load.]

(14) B.M.$_{max}$ occurs at $2{\cdot}31$ m from left end.

B.M.$_{max}$ = $11{\cdot}09$ kN m.

(15) B.M. value at top of mast = 0.

B.M. value at 300 N   = 400 N m.

B.M. value at 400 N   = 1 400 N m.

B.M. value at base   = 800 N m.

Max. B.M. = 1 400 N m.

(See Fig. 431).

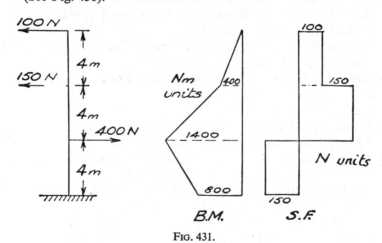

Fig. 431.

Exercises (12) (*page* 278)

(1)   (i) Max. fibre stress = 8 N/mm^2.

(ii) $22{\cdot}5$ kN.

(iii) 100 mm.

(2) 1 575 N.

(3) 1 772 N.

(4) 50 mm × 175 mm.
(5) Reactions: Left end = 1 940 N, right end = 1 460 N.
B.M.$_{max}$ = 1 320 N m.
$f = 7.04$ N/mm^2.
(6) 7.91 kN/m^2.
(7) B.M. available for end load = (M.R. -7.524×10^6 N mm).
= $(12.6 - 7.524) \times 10^6 = 5.076 \times 10^6$ N mm.
Load = 3.384 kN.
(8) 186 kN.
93 kN central load.
(9) (i) Z_{XX} = 231 cm^3.
(ii) I_{XX} = 33 330 cm^4. The answer of 33 330 is at variance with
the value given on page 259, which is 33 324. The difference
is due to an approximation in rounding off the value of
Z_{XX}.
(iii) 406.4 mm.
(10) Reactions: Left end = 76 kN, right end = 64 kN.
B.M.$_{max}$ = 72.2 kN m at 1.9 m from left end.
$Z = 516$ cm^3.
(11) Reactions: Left end = 60 kN, right end = 60 kN.
B.M.$_{max}$ occurs at '60 kN' load = 91.8 kN m.
$Z = 557$ cm^3.
(12) 406 mm × 140 mm × 39 kg U.B. (Z required = 591 cm^3).
(13) $f = 106$ N/mm^2.
(14) B.M.$_{max}$ is at '400 kN' load = 120 kN m.
Necessary 'Z' for each beam = 182 cm^3.
(15) B.M.$_{max}$ = 2 353 N m.
$b = 75$ mm, $d = 150$ mm.
(16) For bay with 75 mm × 175 mm beams: 11.1 kN/m^2.
For bay with 75 mm × 225 mm beams: 11.2 kN/m^2.
For steel beam strength: 12.69 kN/m^2.
∴ safe floor load = 11.1 kN/m^2.
(17) Max. B.M. = 1 500 N m.
'Z' of section = 41 430 mm^3.
$f = 36.2$ N/mm^2.
(18) 'Z' of section = 21 510 mm^3.
$W = 2 500$ N.
(19) 'Z' of section = 700 cm^3.
108 kN.
(20) 300 mm.

Exercises (13) (*page* 305)

(1) (i) 25 mm.
    (ii) 21·65 mm.
    (iii) 6·79 cm.
    (iv) 3·64 cm.

(2) (i) 80
    (ii) 39·1.

(3) $p_c = 112·5 \text{ N/mm}^2$.
    Load = 853 kN.

(4) 3·25 m.

(5) 254 mm × 254 mm × 167 kg will carry 2 871 kN.

(6) Direct stress   = $2·24 \text{ N/mm}^2$.
    Bending stress = $1·34 \text{ N/mm}^2$.
    Max. comp. stress = $3·58 \text{ N/mm}^2$.
    Min. comp. stress = $0·9 \text{ N/mm}^2$.

(7) Direct stress   = 30 N/mm^2.
    Bending stress = 30 N/mm^2.
    ∴ total stress varies from 60 N/mm^2 to zero.

(8) Direct stress   = $108·2 \text{ kN/m}^2$.
    Bending stress = $30·0 \text{ kN/m}^2$.
    At 'A' total   = $138·2 \text{ kN/m}^2$.
    At 'B' total   = $78·2 \text{ kN/m}^2$.

(9) 77·7 mm.

(10) 250 mm.
    (i) 0·09 mm.
    (ii) $Z = 1·534 × 10^6 \text{ mm}^3$.
       $W = 114·6$ kN.

(11) Component at right angles to joint = 1 879·4 N.
    Compressive stress at 'A' = $0·0183 \text{ N/mm}^2$.

Exercises **(14)** (*page* 328)

(1) A = 1 600 N strut.
    B =   800 N strut.
    C = 2 078 N tie.

(2) Reactions: Left end = 1 450 N, right end = 1 150 N.
    A = 1 700 N strut.
    B = 1 270 N tie.
    C =   837 N tie.

(3) See Fig. 432.

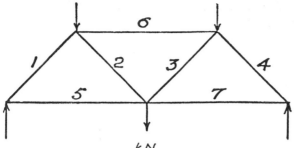

kN

MEMBER	FORCE	TYPE
1	16·97	STRUT
2	2·83	TIE
3	5·66	TIE
4	14·14	STRUT
5	12·00	TIE
6	14·00	STRUT
7	10·00	TIE

FIG. 432.

(4) A = 1 000 N strut.
(5) See Fig. 433.

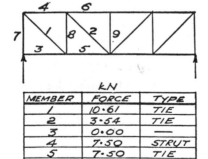

kN

MEMBER	FORCE	TYPE
1	10·61	TIE
2	3·54	TIE
3	0·00	—
4	7·50	STRUT
5	7·50	TIE
6	10·00	STRUT
7	9·50	STRUT
8	6·50	STRUT
9	4·00	STRUT

FIG. 433.

(6) A = 2 600 N tie.
   B = 1 300 N tie.
     = 2 800 N strut.

(7) Left rafter = 200 N.
Right rafter = 346·4 N.
(8) A = 3 464 N strut.
B = 2 000 N tie.

Exercises (15) (*page* 363)

(1) Wind load = 2 kN.
Overturning moment = 2 kN m.
(2) 39·2 kN.
(3) (i) 32·4 kN.
(ii) 38·88 kN m.
Earth pressure at base = 18 kN/m².
(4) Water thrust = 45 kN.
x = 2·12 m.
(5) Earth thrust = 12·8 kN.
Weight of wall = 56·7 kN.
Resultant = 58·13 kN at 16° (nearly) to vertical.
Resultant cuts base at 0·678 m from back of wall.
Compression everywhere along base.
(6) (i) W = 10·8 kN.
Wind pressure = 720 N/m².
(ii) 2·3 m.
(7) Earth thrust = 77·76 kN.
Weight of wall = 176·25 kN.
Resultant cuts base at 0·383 m from centre.
Max. ground pressure = 121 kN/m².
Resistance to sliding = 88·12 kN.
Wall is stable, but resistance to sliding does not exceed the sliding
tendency by a sufficient margin.
(8) (i) 112·5 kN/m².
(ii) 150 kN/m².
(iii) 200 kN/m².
(9) Weight of wall = 226·8 kN.
C.G. from back of wall = 0·933 m.
Eccentricity of resultant = 0·178 m.
V = 271·8 kN.
Max. compressive stress = 163·6 kN/m².

INDEX